Biophysical labeling methods in molecular biology

Physical labels such as stable nitroxide radicals, luminescent and photochromic chromophores, so-called Mössbauer atoms, and electron-dense assemblies of heavy atoms have proved to be effective tools in solving many problems at the molecular level in biological systems. These physical labels are used as "molecular rulers" to measure the distances between chosen groups and to measure the size, form, and microrelief of objects. By providing information about these factors, the label provides information that can help the scientist to understand the structures of membranes, nucleic acids, enzymes, and proteins and how they function.

This volume covers all aspects of this field: the theoretical bases, the experimental techniques, and how to interpret the resulting data. It also critically discusses some recent results obtained with these techniques and gives an analysis of likely developments in the future.

Biophysical labeling methods in molecular biology

GERTZ I. LIKHTENSHTEIN

Department of Chemistry,
Ben-Gurion University

Institute of Chemical Physics,
Russian Academy of Science

CAMBRIDGE UNIVERSITY PRESS
Cambridge, New York, Melbourne, Madrid, Cape Town, Singapore, São Paulo

Cambridge University Press
The Edinburgh Building, Cambridge CB2 2RU, UK

Published in the United States of America by Cambridge University Press, New York

www.cambridge.org
Information on this title: www.cambridge.org/9780521431323

First published 1993
This digitally printed first paperback version 2005

A catalogue record for this publication is available from the British Library

Library of Congress Cataloguing in Publication data
Likhtenshteĭn, G. I. (Gerts Il'ich)
Biophysical labeling methods in molecular biology / Gertz
I. Likhtenshtein.
p.   cm.
Includes bibliographical references and index.
ISBN 0-521-43132-8
1. Biophysical labeling.   2. Molecular biology – Technique.
I. Title.
QH324.9.B49L55   1993
574.8′8 – dc20   92-19487

ISBN-13 978-0-521-43132-3 hardback
ISBN-10 0-521-43132-8 hardback

ISBN-13 978-0-521-01787-9 paperback
ISBN-10 0-521-01787-4 paperback

Contents

Preface   *page* ix
Abbreviations   xi

1. The method of spin labeling   1
    1.1. Introduction   1
    1.2. Structure of nitroxide labels and probes   3
    1.3. ESR signals of NRs: magnetic parameters   6
    1.4. Methods of measurement of the ESR signal parameters   10
        1.4.1. Stationary methods   10
        1.4.2. Pulse methods   17
    1.5. Rotational diffusion of nitroxides   21
        1.5.1. General   21
        1.5.2. Elements of the theory of the ESR spectra of
               rotating nitroxides   22
        1.5.3. Very slow rotation   24
        1.5.4. Slow-motion regions   29
        1.5.5. Fast-rotation regions   31
        1.5.6. Rotations in different regions   34
        1.5.7. High-frequncy low-amplitude dynamics   36
        1.5.8. Superslow motion   37
    1.6. Nitroxides as dielectric, pH, and redox probes   38
    1.7. Nitroxides in ESR tomography   41
    1.8. Spin traps   43

2. Double-labeling techniques   46
    2.1. General   46
    2.2. Effects of spin–spin interactions on the parameters of
         ESR spectra   48
        2.2.1. Principal effects   48
        2.2.2. On the parameters of ESR signals of paramagnetics   50
        2.2.3. Spin–spin interactions in biradicals and polyradicals
               and paramagnetic complexes of metals with
               nitroxide ligands   53

v

2.3. Determination of the distance between spins 57
2.4. The spin label–spin probe method 62
   2.4.1. General 62
   2.4.2. Selection of spin probes 66
   2.4.3. Investigation of steric, electrostatic, and exchange
       effects 67
   2.4.4. Determination of the immersion depth of a radical
       center 70
   2.4.5. NRs in oxymetry 71
2.5. Nuclear magnetic resonance of paramagnetic systems 74

3. Fluorescent labeling methods 80
3.1. General 80
   3.1.1. Absorption spectra 80
   3.1.2. Fluorescence and phosphorescence 83
3.2. Chemical properties of fluorescent labels and probes 85
3.3. Rotational diffusion of fluorescent chromophores 93
   3.3.1. Depolarization of fluorescence 93
3.4. Fluorescence and molecular dynamics of the medium 96
3.5. Study of local acidity and electrostatic and polar
   properties of biological objects 99
   3.5.1. Measurement of pH 99
   3.5.2. Measurements of electric charge density,
       transmembrane potential, and ion concentration 100
   3.5.3. Measurement of polarity: on the dynamic polarity
       scale 102
3.6. Inductive resonance energy transfer as a method of
   investigating structures and dynamics of biological
   objects 104
   3.6.1. Mechanism of inductive resonance energy transfer 104
   3.6.2. Estimation of the distance between donor and
       acceptor groups 106
   3.6.3. Orientation factor 107
3.7. Dynamic quenching of fluorescence as an approach to
   the study of molecular dynamics 111
3.8. Charge transfer complexes, excimers, and exciplexes as
   luminescent probes 112
3.9. Study of slow translational diffusion: photobleaching
   and fluctuation techniques 114

4. Triplet labeling methods 116
4.1. Peculiarities of triplet excited states 116
4.2. Structures and chemical properties of triplet probes 118

4.3. Exchange interactions with participation of excited
     triplet states: elements of theory 120
4.4. Static exchange: experimental data 124
4.5. Dynamic exchange processes 126
     4.5.1. Elements of theory 126
     4.5.2. Experimental data 127
4.6. Photochrome probes 130
4.7. The triplet probe–photochrome labeling method 133

5. Mössbauer spectroscopy, electron scattering, and other
   labeling methods 136
   5.1. Mössbauer labels 136
        5.1.1. Physical principles 136
        5.1.2. Dynamic effects in Mössbauer spectroscopy 139
   5.2. NMR probes 141
   5.3. Total tritium labeling technique 143
   5.4. Electron-scattering labels 144
        5.4.1. General 144
        5.4.2. Physical grounds 146
        5.4.3. Modification of biological objects by electron-
               scattering labels 148
        5.4.4. Electron microscopy determination of shape and
               size of electron-scattering particles 154

6. Studies of proteins and enzymes: structure, dynamics, and
   mechanism of action 158
   6.1. Active centers of enzymes 158
        6.1.1. Serine proteases 158
        6.1.2. Nitrogenase 161
        6.1.3. Dehydrogenases 166
        6.1.4. Cytochrome P-450 168
        6.1.5. Myosin and actin 170
        6.1.6. Other enzymes and proteins 171
   6.2. Conformational changes in proteins and enzymes 175
        6.2.1. Large-scale and allosteric conformational changes 175
        6.2.2. Transglobular conformational transition 177
   6.3. Molecular dynamic properties of proteins and enzymes 180
        6.3.1. General 180
        6.3.2. Experimental data 182
        6.3.3. Dynamics and functional activities of proteins 188
   6.4. Physical labeling as a tool for studying the electron
        transfer mechanism 195
        6.4.1. General 195

6.4.2. Delocalization of spin density and local polarity
     in proteins   196
6.4.3. Collisions between molecules: steric factor   197
6.4.4. Mechanisms of dynamic adaptation at electron
     transfer   198

7. Structure and dynamics of membranes   201
  7.1. Model membranes   203
    7.1.1. Structure of model membranes: localization of
       labels and probes   203
    7.1.2. Molecular dynamic properties and conformational
       transitions in model membranes   209
    7.1.3. Mixed and protein-lipid model membranes   217
  7.2. Biological membranes   221
    7.2.1. Erythrocyte membranes   221
    7.2.2. Sarcoplasmic reticulum   223
    7.2.3. Rhodopsin membranes   225
    7.2.4. Microsomes   226
    7.2.5. Acetylcholine receptor   228
    7.2.6. Membranes of chromatophores of photosynthetic
       bacteria   229
    7.2.7. Other membranes   231

8. Nucleic acids and other biological systems: biological assays   233
  8.1. Nucleic acids   233
    8.1.1. Modification of nucleic acids with physical labels   233
    8.1.2. Investigation of microstructure and
       conformational changes in nucleic acids   236
  8.2. Polysaccharides   239
    8.2.1. Glycoproteins   240
    8.2.2. Cotton fibers and cellulose   241
  8.3. Spin-labeled, physiologically active compounds   243
  8.4. Cells, tissues, organisms   248
    8.4.1. Distribution of labels: microcomponent
       localization of cells   248
    8.4.2. Redox properties of cells   250
  8.5. Biological assays   252
  8.6. Biological analyses   255
    8.6.1. Biologically active ions and compounds   255
    8.6.2. Immunological assays   259

Conclusion   262
References   266
Index   303

Preface

About 200 years ago the German poet and philosopher J. W. Goethe noted that Nature is not only a great artist but also a skillful master. The contemporary generation of scientists who work in the field of molecular biology can appreciate the external beauty of nature, the internal perfection of biological structures and physicochemical processes taking place in nature, and the enormous difficulty of studying them.

Modern molecular biology faces extremely complicated experimental problems. Proteins, biological membranes, nucleic acids, polysaccharides, and other ingredients of a biological cell interact, form sophisticated structures, and accomplish numerous catalytic, regulatory, and other functions. Many of the specific problems to be solved arise in the study of these systems.

In investigations of biological systems, one uses a broad arsenal of physical and chemical methods. Of particular importance in this arsenal is the approach of selective modification of biological objects with various labels capable of providing information on their structure, molecular dynamics, and mechanisms of actions. The necessity of such an approach is caused by the specificity of biological systems. Rather than seek complete information, a researcher usually aims to learn the main structural and dynamic properties important in the functional activity of a system.

These days the method of physical labeling is used to solve many structural problems in biophysical and biochemical laboratories all over the world. The most popular methods use spin and fluorescent labels and probes. The achievements in this field in the 1970s have been summarized in a number of monographs and reviews. However, since then considerable progress has been made; in addition to the scientific areas that seem relatively conventional by now, new ones have appeared. In particular, methods of triplet, photochromic, electron scattering, and Mössbauer labeling have been put forward and developed in the author's laboratory at the Institute of Chemical Physics (Chernogolovka). A method of total labeling of protein and other object surfaces originated at the same institute.

It has proven to be particularly effective to combine various types of

ix

labels. Consistent use of this approach has led to the solution of a number of complicated problems such as deciphering the active centers of enzymes (nitrogenase, cytochrome P-450, photosynthetic reaction centers) and elucidating the structure and molecular dynamics of proteins, membranes, and nucleic acids.

In the author's opinion, methods developed in various laboratories to solve specific problems are of general importance and may prove to be useful for studying a wide range of biological objects.

In the first five chapters of the present monograph, the general experimental and theoretical grounds are expounded for various methods using physical labels and probes (spin, fluorescent, triplet, photochromic, Mössbauer, and electron scattering), as well as the technique of total labeling. Also elucidated therein are the main principles of chemically modifying the objects under study to accommodate or contain the labels. The concluding chapters consider the principal results in the field of enzymatic catalysis, molecular biology, and biophysics obtained by means of the method of physical labeling.

The author believes that the effort involved in writing this monograph will have been worthwhile if it arouses interest in biophysical labeling and makes it easier for chemists, biochemists, and biophysicists to understand the principles of this method to such an extent that they will use it creatively to solve their own scientific problems.

The author has used a three-level approach to describe some of the topics, taking into account the variations in the educational training of his readers:

1. a qualitative consideration of the phenomenon on which the method is based,
2. presentation of formulas, figures, and schemes in their simplest and most suitable form for practical application,
3. a more rigorous physical and mathematical substantiation.

Greater attention has been paid to the first two levels.

The author is very grateful to his colleagues Drs. A. V. Kulikov, A. I. Kotelnikov, L. A. Levchenko, L. A. Syrtsova, O. V. Belonogova, A. P. Sadkov, V. I. Fogel, V. M. Mekler, S. A. Marakushev, V. R. Bogatyrenko, S. I. Druzhinin, and E. S. Cherepanova, who have shared the bad and good times over many years in investigations of biological objects by the method of physical labeling. Finally, the author is deeply indebted to T. V. Kamenskaya and L. V. Vorobiev for their constant help in the preparation of the manuscript.

Abbreviations

ANS	1,2,4-triaminonaphtholsulfonic acid
ATP	adenine triphosphate
BSA	bovine serum albumin
DAF	delayed annihilation fluorescence
DHSPC	dihydrostearylphosphatidylcholine
2D-ELDOR	two-dimensional ELDOR
DMPE	dimyristoylphosphatidylethanolamine
DMPG	dimyristoylphosphatidylglycerol
DMPC	dimyristoylphosphatidylcholine
DPPC	dipalmitoylphosphatidylcholine
ELDOR	electron–electron double resonance
ENDOR	electron–nuclear double resonance
ESEM	electron spin-echo method
ESE MT	electron spin echo with magnetization transfer
EXAFS	extended X-ray absorption fine structure
FL	fluorescent label
FP	fluorescent probe
FRAP	fluorescence recovery after photobleaching
FRAT	free-radical assay technique
HF	high frequency
HFI	hyperfine interaction
HSA	human serum albumin
Ig	immunoglobulin
IR	ion-relaxator
MESL	mercarbide electron-scattering label
MW	microwave
NMR	nuclear magnetic resonance
NMME	nuclear magnetic modulation effect
NQI	nuclear quadrupole interaction
NR	nitroxide radical
PCMB	*p*-chloromercuribenzoate
PE	phosphatidylethanolamine
PI	phosphatidylinositol

PM	paramagnetic metal
PS	phosphatidylserine
RC	reaction center
SL	spin label
SP	spin probe
SR	sarcoplasmic reticulum
SR ESR	saturation recovery ESR
ST ESR	saturation transfer ESR
TCTA	2,4,6-trichloro-1,3,5-triazine
TEMPO	2,2,6,6-tetramethylpiperidine-1-oxyl
TEMPOLE	4-hydroxy-2,2,6,6-tetramethylpiperidine-1-oxyl
TEMPONE	4-oxo-2,2,6,6-tetramethylpiperidine-1-oxyl
TL	triplet label
TP	triplet probe
TTLT	total tritium labeling technique
A_{iso}	isotropic HFI constant
$A_{x,y,z}$	anisotropic HFI constant
a_B	Bohr radius
α_e	electrostatic factor
$\alpha_{e,g}$	polarizability
β_e	electron Bohr magneton
β_n	nuclear Bohr magneton
D	Debye unit
$D_{R,\perp,\parallel}$	rotational diffusion coefficient
D_{tr}	translational diffusion coefficient
Δ	line width of ESR spectrum
ΔF_0	standard free energy
$\Delta F^{\neq}$	free energy of activation
ΔF_r	reorganization energy
ΔH_{pp}	peak-to-peak line width
$\Delta H_{\frac{1}{2}}$	line width between the points of maximum slope
ΔH_L	Lorentzian line width
ΔH_G	Gaussian line width
Δv_{max}	spectral shift
$\Delta \varphi_m$	transmembrane potential
δ	chemical shift
E	electric field strength
ε_0	dielectric constant
ε	extinction coefficient
f_g	geometric steric factor
g	g-factor
g_e	g-factor of a free electron
γ_e	electron gyromagnetic ratio

γ_n	nuclear gyromagnetic ratio
H	Hamiltonian
H_m	modulation amplitude
H_1	MW amplitude
H_z	strength of a magnetic field
h	Planck constant
η	viscosity
I	spectral line intensity
I_f	fluorescence intensity
I_n	magnetic quantum number of nucleus
I_{ph}	phosphorescence intensity
J	exchange integral
$j(\omega)$	correlation function
K_Q	Stern-Volmer constant
K_q	quenching rate constant
k_{cl}	collision rate constant
k_e	spin exchange rate constant
k_{tr}	electron transfer rate constant
$\ae$	nonadiabaticity coefficient
$\ae_{D,A}$	spin wave function
λ	line width
λ'	spin-orbital coupling constant
$\bar{M}$	transition dipole moment
M_r	molecular mass
m	magnetic quantum number
m_i	quantum number
μ	ionic strength
μ_e	magnetic moment of an electron
μ_{el}	electron dipole moment
N_A	Avogadro number
n	refraction index
ν_c	correlation frequency
ν_L	Larmor frequency
ν_m	modulation frequency
ν_r	resonance frequency
P	degree of polarization
p	p-orbital
Q	quadrupole moment
ρ	spin density
S	order parameter
S_i	excited singlet state
S_{jj}	overlap integral
s	s-orbital

σ	charge density
T_c	transition temperature
T_i	excited triplet state
T_{1e}	electron spin–lattice relaxation time
T_{2e}	electron transverse relaxation time
T_{1n}	nuclear spin–lattice relaxation time
T_{2n}	nuclear transverse relaxation time
T_p	passage time
τ_D	Debye relaxation time
τ_c	correlation time
τ_{ch}	characteristic time
τ_{col}	collision time
τ_f^*	fluorescence lifetime
τ_M	chemical exchange time
τ_{ph}^*	phosphorescence lifetime
τ_r	electric dipole relaxation time
τ_R	rotation diffusion time
τ_{rs}	residence time
V	voltage
V'	resonance integral
V_2'	absorption ESR spectrum with a 90° phase shift
φ	electric potential
φ_0	quantum yield
$\varphi_{D,A}$	wave function
ω_L	Larmor frequency

1

The method of spin labeling

1.1. Introduction

At present there probably is no biophysical method that has spread so fast and so widely as spin labeling. This method is based on a combination of the unique physical and chemical properties of nitroxide radicals (NRs).

Among the myriad compounds that are products of modern organic chemistry there is hardly a class of substances that could have so diverse and unexpected applications as do NRs: they are used as "molecular rulers" to measure the distances between paramagnetic centers, the sizes and shapes of grooves and slits in the enzyme active centers, and the depth of the center immersion into the biological matrix. NRs can serve as specific microscopic seismic stations to monitor the finest changes in the microrelief of biological objects and conformational waves at their functioning. Radicals may be used to probe the rigidity and mobility of certain portions of enzymes, membranes, and other objects and to study the conformational dynamics of the macromolecular shape, i.e., the macromolecule's "breath." They help in the study of the spatial distributions of certain substances, e.g., water, nitroxides, dioxygen, etc. (ESR tomography).

In many other cases, when a local microscopic property (e.g., dielectric constant, donor–acceptor capacity, strength of the electric or magnetic field, redox ability, hydrogen ion concentration, delocalization of spin density and electron density) should be measured in a given portion of a biological object, NRs may serve as efficient tools.

The majority of spin labels and probes are stable and compact compounds that contain a nitroxide fragment (Figure 1.1), producing distinct and intensive signals of electron spin resonance (ESR). These compounds attracted the attention of many scientists immediately after their discovery in the 1960s (Neiman et al. 1962; Rozantsev 1970; Buchachenko & Wasserman 1973). Skillful researchers invented a cascade of fine and sophisticated methods for measuring the parameters of the ESR signals. A whole cohort of talented theoreticians developed procedures to analyze ESR signals to obtain information on the local properties of the microenvironment.

a.

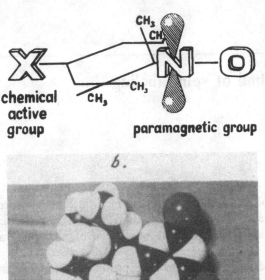

Figure 1.1. Chemical formula for a nitroxide spin label (a) and the Stuart-Briegleb model of radical SLXI (b).

It was extremely fortunate that the problem of NR chemistry fascinated many outstanding chemists; their dazzling syntheses brought to life a variety of labels and probes that could be used to solve a number of specific problems of molecular biology.

The main physical principles of spin labeling will be considered in further sections of this chapter. An explorer needs to know the following to use the method at the utmost efficiency:

1. The ways the ESR signals of NRs appear and are registered, the parameters that characterize the ESR spectra, the way they are determined, and the physical properties of the system they reflect.

2. The effects that arise from the electron–nucleus spin interaction and their dependence on the rotational and translational diffusion of radicals and on their mutual disposition.

3. The ESR signal parameters that need to be measured to obtain the data on dielectric, donor–aceptor, redox, and base–acidic properties of the medium under study.

4. The chemical structures and properties of various types of NRs used as covalently bound spin labels and probes to be adsorbed.

1.2. Structure of nitroxide labels and probes

At present the following derivatives of NRs are those used most extensively as spin labels and probes (Rozantsev 1970, 1990; Gaffney 1976; Volodarsky 1988; Keana 1984; Keana *et al.* 1990; Hideg & Hankowszky 1989; Hideg 1990; Aurich 1990):

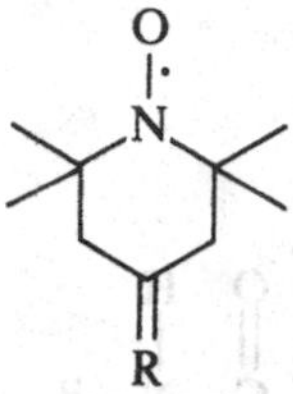

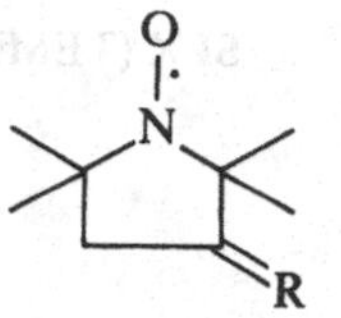

These are the piperidine (I), pyrrolidine (II), doxyl (III), proxyl (IV), aceto-xyl (V), imidazole (VI), and imidazolineoxide (VII) derivatives of NRs.

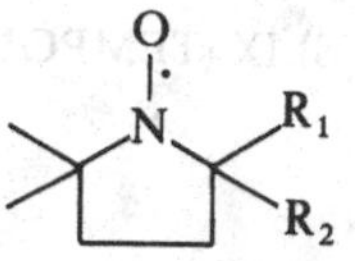

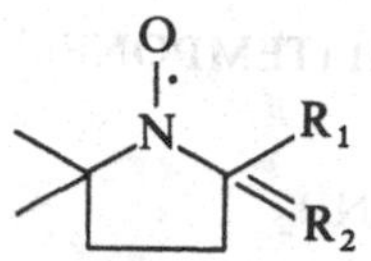

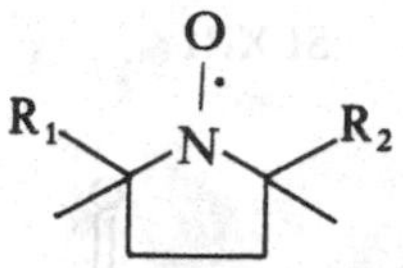

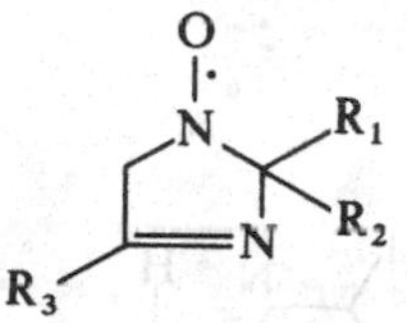

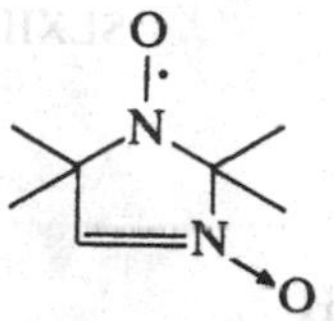

The actual efficiency of the spin-labeling method depends, to a great extent, on the capacity of the NR to add specifically while the free valence remains intact (Neiman *et al.* 1962). In many cases, the active groups of the labels are fragments of known reagents whose selective actions have been extensively tested by experimentation.

Today there is a wide variety of spin labels that can react selectively with the functional groups of proteins. Each label contains three main components (Figure 1.1): (1) a paramagnetic fragment producing an intense ESR signal, (2) a chemically active residue that promotes covalent or co-ordinational addition to a certain protein group, and (3) a bridging group of atoms (a "leg") whose structure (length, flexibility, hydrophobicity, polarity) markedly affects the label's behavior after its addition to the protein macromolecule. Formulas of some spin labels and probes are given below.

According to the existing terminology, a label is a compound that binds covalently to an object under study; a probe either adds on noncovalently or diffuses freely in the medium (R⋅ is a pyrrolidine or piperidine nitroxide residue, respectively).

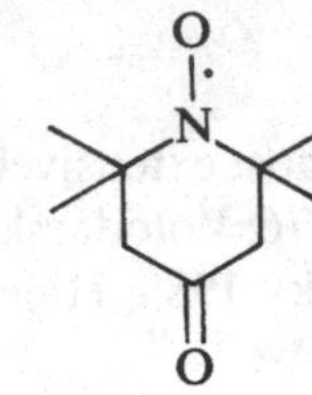

SLVIII (TEMPONE) 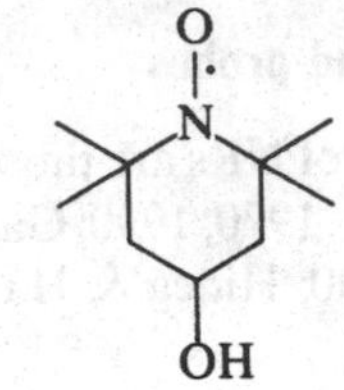   SLIX (TEMPOLE) 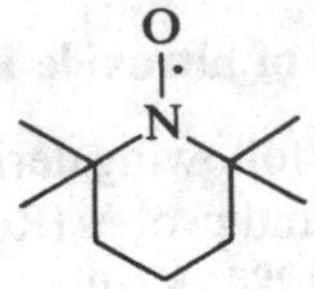   SLX (TEMPO)

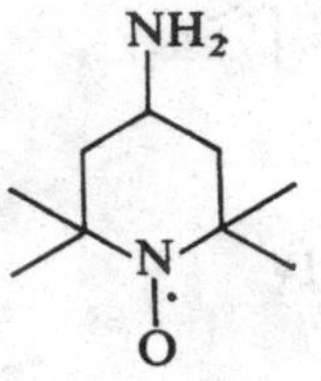

SLXI 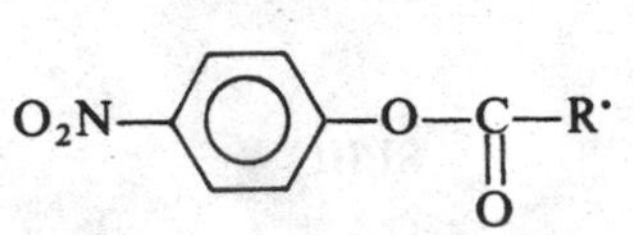   SLXII 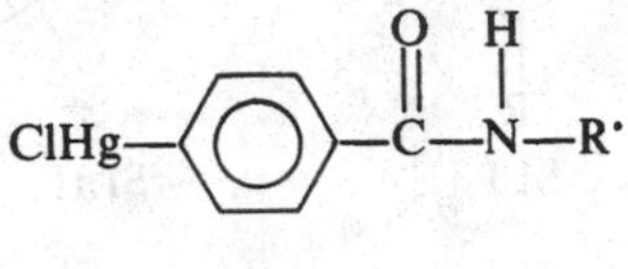   SLXIII

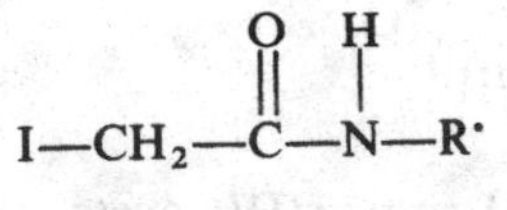

SLXIV 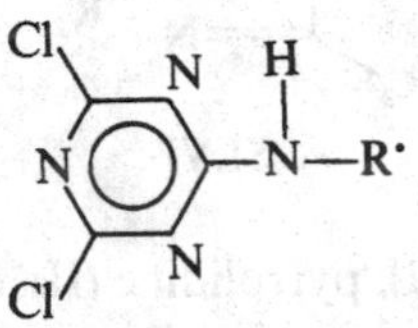   SLXV 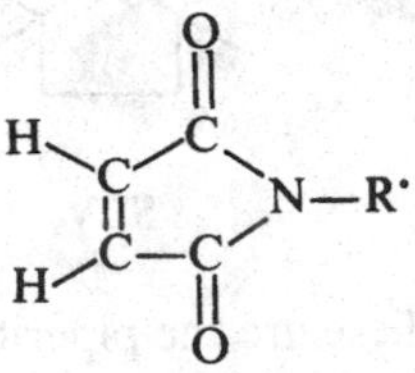   SLXVI

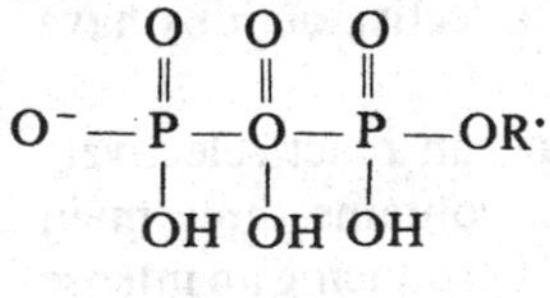

SLXVII 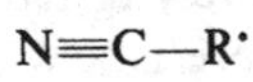   SLXVIII

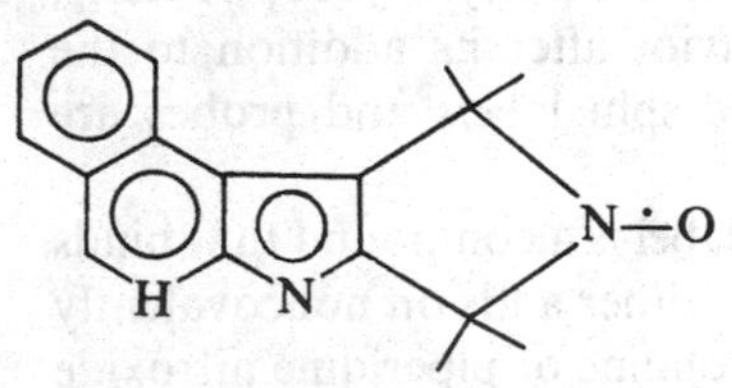

SLXIX 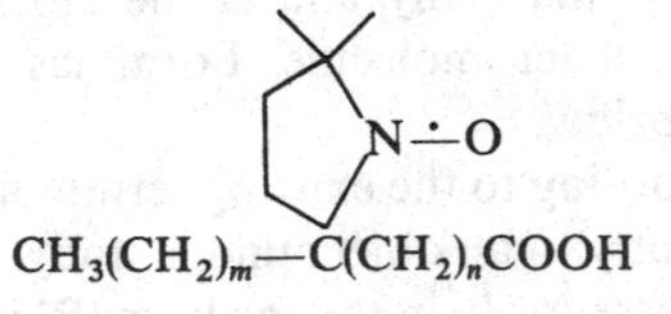   SLXX

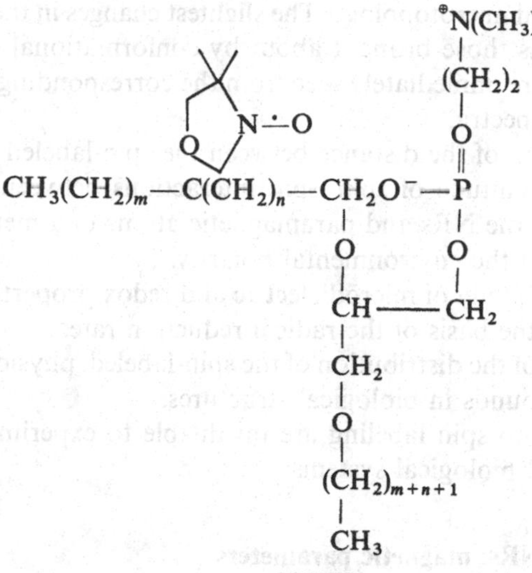

$$CH_3(CH_2)_m-C(CH_2)_n-CH_2O^--P$$

with side structures showing the nitroxide ring labeled Q, N—O, and the phosphate chain bearing $^\oplus N(CH_3)_3$, $(CH_2)_2$, O, P, O, O, CH—CH$_2$, CH$_2$, O, $(CH_2)_{m+n+1}$, CH$_3$

SLXXI

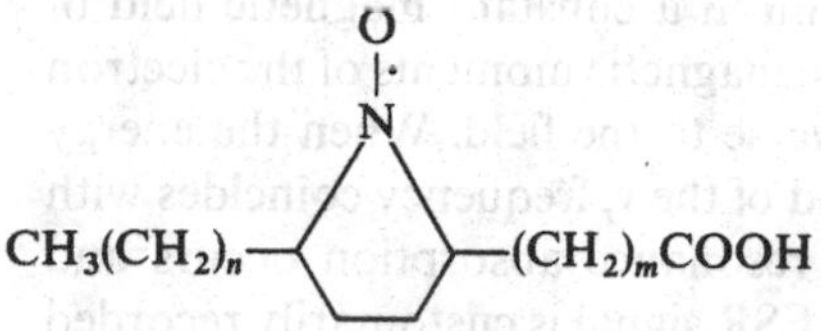

$$CH_3(CH_2)_n--(CH_2)_mCOOH$$

SLXXII

$$O_2N--N-(CH_2)_n-\overset{O}{\overset{\|}{C}}-R^\bullet$$

with NH$_2$ and H substituents

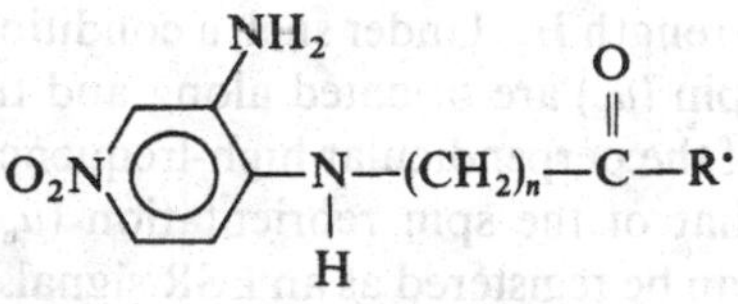

SLXXIII

Compounds IX–XVIII contain fragments of reagents conventionally used in biochemistry for functional groups of proteins, whereas probes XIX–XXII are used in the study of membranes. More detailed descriptions of NR synthesis and the properties of NRs may be found in the references cited earlier.

Spin labels and probes may be selectively added to the functional groups of proteins, lipids, nucleic acids, saccharides, etc., introduced to hydrophobic zones of membranes, or localized on the electrostatically charged portions of molecular systems.

The following kinds of data on the local properties of the spin-labeled systems may be obtained from analyses of typical ESR spectra of NRs (McConnell & McFarland 1970; Freed 1976; Berliner 1976, 1979, 1990; Likhtenshtein 1976a, 1979a, 1988a,b, 1990a,b; Watts 1985; Rassat 1990):

    1. Calculation of the parameters of rotational diffusion of the NR, permitting the determination of microviscosity and study of the

environmental microtopology. The slightest changes in these properties, such as those brought about by conformational or phase
transitions, are immediately seen from the corresponding changes
in the ESR spectra.

2. Determination of the distance between the spin-labeled portions
   from the magnitude of spin–spin interactions between radicals
   and between the NRs and paramagnetic atoms of a metal (PM).
3. Evaluation of the environmental polarity.
4. Quantitative study of microdielectric and redox properties of the
   medium on the basis of the radical reduction rate.
5. Elucidation of the distribution of the spin-labeled, physiologically
   active compounds in biological structures.

The possibilities due to spin labeling are invaluable to experimenters in
studying complicated biological systems.

1.3. ESR signals of NRs: magnetic parameters

The phenomenon of ESR involves the absorption or dispersion of a high-
frequency electromagnetic field (v_r) by a system of particles with the intrinsic spin moment of an unpaired electron in a constant magnetic field of
strength H_z. Under such a condition, the magnetic moments of the electron
spin (μ_e) are oriented along and transverse to the field. When the energy
of the perpendicular high-frequency field of the v_r frequency coincides with
that of the spin reorientation $(\mu_e H_z)$, resonance absorption occurs and
can be registered as an ESR signal. The ESR signal is customarily recorded
as a plot of the first derivative of the absorption versus the H_z value at
constant (τ_r) (Figure 1.2). The resonance condition is described by the
following equation:

$$hv_r = g\beta \cdot H_z,$$ (1.1)

where β is the Bohr magneton, g is the g-factor (for a free electron, $g_0 =$
2.0023), and h is the Planck constant.

The value of the g-factor is a fundamental characteristic that depends on
the structure of a paramagnetic particle and its interaction with the environment. In the case of asymmetric particles, three values of the g-factor,
g_x, g_y, and g_z (Figure 1.3), correspond to various orientations of the particle
relative to the constant magnetic field direction and are used to completely
describe resonance phenomena.

According to the Bloch model (which is convenient for the study of a
number of important properties of a spin system), the spin precession in
a constant magnetic field proceeds at the Larmor frequency (v_L). When
v_r and v_L coincide, the spins turn over 180° and under thermal chaotic

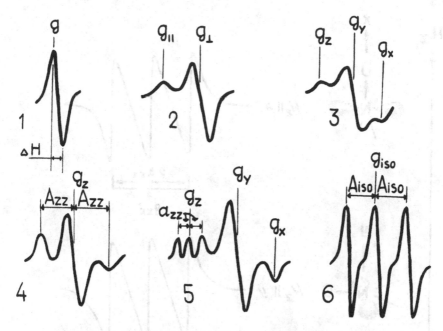

Figure 1.2. Examples of the ESR spectra of paramagnetic centers differing in the set of parameters, the g-factor, and hyperfine interaction (A): 1, 2, and 3, centers with isotropic, axial, and rhombic symmetry of the g-tensor, respectively; 4 and 5, spectra of the magnetically diluted matrix of the NR taken in 3-cm and 2-mm bands, respectively; 6 spectrum of the NR in solution.

action return to their initial position. The latter process (relaxation) is characterized by two relaxation times.

The change in the longitudinal component of magnetization proceeds with the characteristic time T_{1e}, which is called the time of longitudinal or spin–lattice relaxation. When the magnetization changes because of precessing spin dephasing relative to the direction perpendicular to the magnetic field, the process is characterized by the time of transverse relaxation, T_{2e}. Both T_{1e} and T_{2e} greatly affect the properties of the ESR signal. Thus, the width of an individual component of the spectra (the spin packet width, Rd) is related to the T_{2e} value:

$$\Delta H_{\frac{1}{2}} = \sqrt{3}/4\pi T_{2e}. \tag{1.2}$$

The T_{1e} value is rather important for the ability of the ESR signal to be saturated, because at sufficient power of microwave (MW) radiation and long times of spin–lattice relaxation, spins in the upper excited state

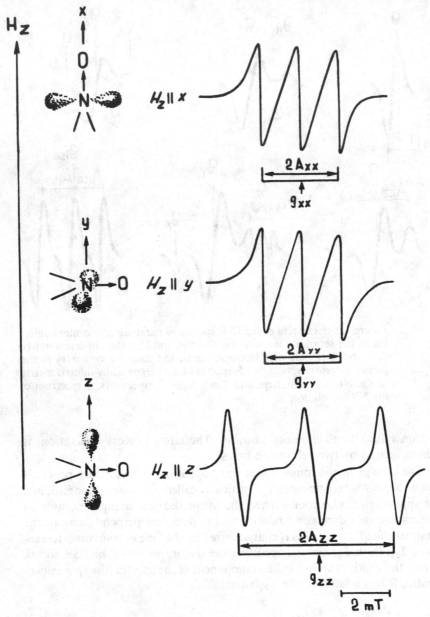

Figure 1.3. ESR spectra of an NR with different orientations relative to the direction of the magnetic field (z), taken in 3-cm band.

fail to relax. Then the populations of the upper and lower levels equalize, and the ESR signal amplitude decreases.

The T_{1e} and T_{2e} values depend on the efficiency of transition of magnetic energy from the initial spins to other spins or vibrations of the lattice electronic dipoles.

Radicals whose nuclei possess a magnetic moment (of quantum number I_n) will also be characterized by hyperfine interactions (HFI). One of the manifestations of the HFI is the isotopic Fermi or contact interaction, caused by the presence of electron density at the nucleus (i.e., in the s-orbital). When electron density is found at the nucleus, the nuclear magnetic moment induces an additional magnetic field that depends on the orientation of the nuclear moment relative to the constant magnetic field. This dependence on orientation results in the splitting of the ESR signal into components. For systems with $I_n = \frac{1}{2}$ (e.g., ^1H or ^{15}N nuclei), two components appear: a low-field component arising from $I_n = +\frac{1}{2}$ and a high-field component arising from $I_n = -\frac{1}{2}$. The ESR spectra of the radicals with the ^{14}N nuclei ($I_n = 1$) are characterized by three components, $m_1, m_0,$ and m_{-1}, with corresponding quantum members of $+1$, 0, and -1. The value of the isotopic hyperfine splitting (A_{iso}), also depends on the structure and properties of the environment.

Another kind of HFI is connected with the dipole–dipole interaction of the magnetic moments of the electron and the nucleus. For asymmetric molecules, the interaction depends on the orientation direction, which links interacting spins relative to the direction of the constant magnetic field (Figure 1.3) and is characterized in its general case by the parameters A_x, A_y, and A_z corresponding to the orientations of the main axes of the radical relative to the magnetic field.

The interactions considered above give rise to the appearance of a system of energy levels in the magnetic field that can be described by the following Hamiltonian:

$$H = \beta(g_x H_x \hat{S}_x + g_y H_y \hat{S}_y + g_z H_z \hat{S}_z) + h(A_x \hat{S}_x \hat{I}_x + A_y \hat{S}_y \hat{I}_y + A_z \hat{S}_z \hat{I}_z), \quad (1.3)$$

where $\hat{S}_x, \hat{S}_y, \hat{S}_z$ and $\hat{I}_x, \hat{I}_y, \hat{I}_z$ are operands of the electronic and nuclear spins corresponding to different axes, $g_x, g_y,$ and g_z are values of the g-factor tensors, and $A_x, A_y,$ and A_z are values of the HFI.

Practically all the approaches in spin labeling are based on measurements of the parameters described and analyses of the effects produced on them by various structural and dynamic factors. However, usually nitroxide fragments in natural biological and model samples are not arranged regularly. Therefore the ESR spectrum is, as a rule, composed of a great number of microscopic signals corresponding to various orientations of the nitroxide fragment relative to the magnetic field. The positions and intensities of these microsignals depend on the intensity of rotational diffusion of the fragments, the microheterogeneity of the environment, exchange and dipole interactions between spins, values of electron and nuclear relaxation times, etc.

The considerable complexity and resulting high information content of ESR spectroscopy of NR are the two salient features that both cause difficulties and promote achievements in the spin-labeling method.

1.4. Methods of measurement of the ESR signal parameters

In measuring the ESR signal, one commonly uses a system consisting of an electromagnet, a cavity, a microwave (MW) source (a klystron), and recorder. The sample to be studied is placed into the cavity and irradiated by MW power of a fixed frequency v_r from the klystron through a waveguide. The resonance conditions (1.1) are attained by gradually increasing the magnetic field strength H_z increasing the electric current in the coils around the constant magnet. At the resonance, one observes absorption and accompanying dispersion of the MW radiation, which results in an imbalance of the cavity counter.

One of the main parameters of the ESR radiospectrometer is the MW amplitude (H_1). This amplitude is obtained in most commercial radiospectrometers from standing electromagnetic waves in a rectangular or cylindrical cavity. The optimal regime consists of standing waves with a loop in the center of the cavity, and it is achieved when the cavity size is approximately equal to the wavelength of the MW radiation. In such a device, a major limitation is the possibility of parasitic absorption or dispersion of the MW radiation due to dielectric loss in the medium, e.g., in aqueous solutions. An alternative without this limitation is the loop gap or Fabry-Pérot cavity, in which the electric and magnetic loops appear in different parts of the unit (Hyde & Feix 1989).

Two versions of ESR are commonly used: stationary and pulsed. Each is described in the following sections.

1.4.1. Stationary methods

In stationary methods, MW radiation is commonly modulated at a frequency v_m and amplitude H_m by means of extra modulation coils. The value of the first harmonic of the modulation signal is proportional to the first derivative of the absorption signal (A). Therefore the routine measurements as a rule are presented as the $A(H_z)$ plot. In order to obtain an undistorted ESR signal, the H_m value should not exceed about one-third of the width of the measured line.

The conditions for recording the ESR spectra may vary depending on the ratio between the parameters of the ESR spectrometer and those of the spin relaxation of the system under study (T_{1e}, T_{2e}). Two situations of particular interest are fast passage and adiabatic passage. Under the condition of fast passage the time of the magnetic-field changes is much less than that of spin relaxation: $T_p = H_1/H_m v_m \ll (T_{1e} T_{2e})^{\frac{1}{2}}$. An adiabatic passage of the ESR signal is monitored when the passage time is more than that of the Larmor precession, T_L, relative to the direction of the MW field, i.e., $T_p \gg T_L = I/\gamma_e H_1$, where $\gamma_e = 1.76 \cdot 10^7 \, \text{Rd s}^{-1}$ is the gyromagnetic ratio for the electrons.

The conditions of fast passage for NRs with $(T_{1e} T_{2e})^{\frac{1}{2}} \approx 10^{-5} \, \text{s}$ at $T = 77 \, \text{K}$ are attained at $v_m \geqslant 10^5 \, \text{Hz}$, $H_1 \approx H_m$. At room temperature $(T_{1e} T_{2e})^{\frac{1}{2}} \approx 10^{-7} \, \text{s}$, and these conditions are not fulfilled.

The H_1 and H_2 values in commercial spectrometers may vary within the range of several oersted to operate in various regimes at a fixed value of the modulation frequency, $v_m = 10^5 \, \text{s}^{-1}$, for example.

The ability of ESR signals to become saturated in the MW field is determined by the value of the MW amplitude at H_1 and the spin–spin relaxation parameters (T_{1e}, T_{2e}). An analysis of the experimental dependence of the ESR signal amplitude on H_1 allows a determination of the relaxation parameters of an individual line from the formula

$$A = a_2 T_{2e} H_1/(1 + b_2 T_{2e}^2 + c_2 H_1^2 T_{1e} T_{2e}), \qquad (1.4)$$

where a_2, b_2, and c_2 are the included magnetic parameters of NR. The dependence of A on the amplitude of the MW field is described by a saturation curve (Figure 1.4).

Special algorithms have been developed to analyze the saturation curves of complex spectra (Safronov *et al.* 1969; Lebedev & Muromtsev 1972; Kulikov & Likhtenshtein, 1977). The algorithms may be used to determine T_{1e} and T_{2e} from the saturation curves recorded under fast passage. The parameters of the saturation curves, H_1^* and H_1^{**}, being equal to the respective MW field values, at which the A amplitude reaches the half-height (Figure 1.4e), may be used for the T_{1e} and T_{2e} calculations in the following sequence:

1. The individual line width $\Delta H_{\frac{1}{2}}$ is calculated (in oersteds) from the data on the dependence of this value on the parameter $q = H_1^{**}/H_1^*$:

q	$\Delta H_{\frac{1}{2}}$	q	$\Delta H_{\frac{1}{2}}$	q	$\Delta H_{\frac{1}{2}}$	q	$\Delta H_{\frac{1}{2}}$
18	4.4	28	1.05	36	0.64	50	0.34
20	2.9	30	0.90	38	0.57	60	0.24
22	1.8	32	0.80	40	0.51	65	0.20
24	1.5	34	0.70	45	0.42	100	0.10

2. The $\Delta H_{\frac{1}{2}}$ value is used for calculation of the parameter β from

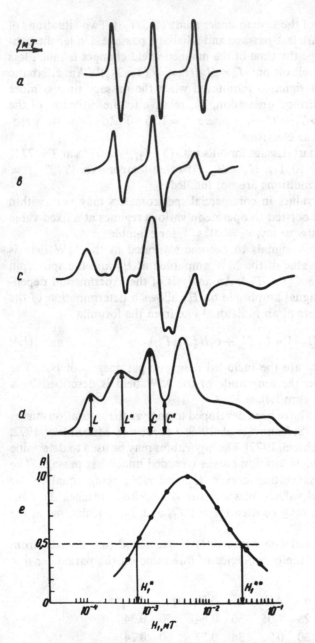

Figure 1.4. ESR spectra and saturation curves of NRs: (a) radical SLXIV in aqueous solution; (b) radical SLXIV on myoglobule in solution; (c) radical SLXIX on serum albumin in solution admixed with the unattached radical; (d) an absorption spectrum with a phase shift in radical SLXIV; (e) saturation curves for the central component of radical SLXIV in alcohol at 77K (Likhtenshtein 1988b).

the following data:

$\Delta H_{\frac{1}{2}}$	β	$\Delta H_{\frac{1}{2}}$	β	$\Delta H_{\frac{1}{2}}$	β
1.0	24.8	1.6	32.0	2.5	40.0
1.2	27.0	1.8	33.6	3.0	45.0
1.4	29.0	2.0	35.0	3.5	50.0
				4.5	60.0

3. T_{1e} is calculated from $T_{1e} = \beta 10^{-9}/(H_1^*)^2$ s.

Though not all the requirements for correct use of the method described are met in the case of NRs, the data thus obtained agree well with direct measurements by pulse techniques. Thus, values of $\Delta H_{\frac{1}{2}} = 0.1$ Oe and $T_{1e}^0 = 3.3 \times 10^{-4}$ s were obtained for NRs in a water-glycerin mixture at 77K, whereas the methods of pulse saturation and spin echo gave $T_{1e}^0 = 3.8 \times 10^{-4}$ s and $\Delta H_{\frac{1}{2}} = 0.05$ Oe, respectively.

For radicals and paramagnetic ions with very fast spin relaxation rates the T_{1e} value can be indirectly determined by the following procedure according to equation (2.18). The dependence of $1/T_{1e}$ on the value of the electron spin relaxation of the second paramagnetic, τ_{1e}, is expected to have an extremum (maximum). Under the condition of the maximum dipole contribution attained, for example, by the temperature increase $\Delta(1/T_{1e}) \approx \mu^2 \gamma_e^2/15r^6 \Delta\omega$, $\tau_{1e} = \Delta\omega$ can be calculated for the maximum temperature (T_{max}). Then equation (2.18) can be used to estimate τ_{1e} for each temperature above and below T_{max}.

The method just described can be efficient for studying the structures of multicenter systems (see Chapters 2 and 6).

The T_{1e} and T_{2e} values for NRs may be determined from the Lorentzian width (ΔH_L) and its capacity to be saturated (Lebedev & Muromtsev, 1972). The ΔH_L value is derived from

$$\Delta H_L/\Delta H_{\frac{1}{2}} + (\Delta H_G/\Delta H_{\frac{1}{2}})^2 = 1, \tag{1.5}$$

where $\Delta H_{\frac{1}{2}}$ is the width of the experimental line between the points of the maximum slope of the first derivative of the spectrum without saturation, and ΔH_G is the width of the Gaussian line, which is calculated from $\Delta H_G = \Sigma a_i^H$ (where a_i^H is the proton hyperfine splitting constant). The T_{1e} value can be estimated from the ΔH_L^2-versus-H_1^2 plot.

The values of the magnetic parameters, the g factors, and the hyperfine splitting constants (A_i) are evaluated by analyzing the positions and shapes of the ESR signal lines, which also depend on the processes of radical rotational diffusion and electron spin–spin interactions. In the absence of the latter effects, the magnetic parameters may be evaluated directly either from the spectrum or by comparison of calculated and experimental data.

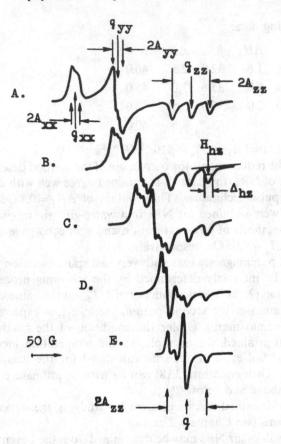

Figure 1.5. Computer-simulated spectra of NR at six MW ESR frequencies (Antsiferova & Valova 1991): (A) 250 GHz ($H_0 = 8.25 \cdot 10^4$ G, $\lambda \simeq 1.2$ mm); (B) 150 GHz ($H_0 = 4.95 \cdot 10^4$ G, $\lambda \simeq 2$ mm); (C) 70 GHz ($H_0 = 2.5 \cdot 10^4$ G, $\lambda \simeq 4$ mm); (D) 35 GHz ($H_0 = 1.25 \cdot 10^4$ G, $\lambda \simeq 8$ mm, Q-band); (E) 9.5 GHz ($H_0 = 3.3 \cdot 10^3$ G, $\lambda \simeq 3$ cm, X-band); (F) 1 GHz ($H_0 = 3.5 \cdot 10^2$ G, $\lambda \simeq 3$ dm, L-band). The isotropic Brownian diffusion is used in simulations. Input parameters are $\tau = 1.1 \cdot 10^{-7}$ s, $\Delta H_0 = 3$ G, $g_x = 2.0084$, $g_y = 2.0060$, $g_z = 2.0022$, $A_x = A_y = 6.73$ G, $A_z = 37.27$ G.

Several types of radiospectrometers with different resonance frequency bands can be used in the laboratory (Figure 1.5): 0.1 GHz or 30 cm (S-band), 0.4 GNz or 7 cm (L-band), 9.5 GHz or 3 cm (X-band), 35 GHz or 8 mm (Q-band), as well as radiospectrometers of high resolution of 142.5 GHz or 2 mm.

According to the resonance conditions, equation (1.1), the position of the ESR spectrum line at any fixed frequency v_r and g-factor is determined by the strength of the magnetic field (H_z). It is obvious that the difference

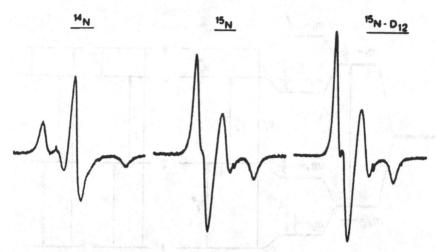

Figure 1.6. ESR spectra of ^{14}N-, ^{15}N-, and (^{15}N, D$_{12}$)-5-NS bound to bovine serum albumin (BSA) at molar ratios of 0.3–5 NS:1.0 BSA. The spectra are 100-G displays recorded at 23°C with identical instrument settings of 10-mW microwave power and 100-kHz field modulation of 0.6-G amplitude (Park & Trommer 1989).

in the signal position (ΔH_z) at different g-factors will be proportional to the v_r value. Thus, radiospectrometers in the 2-mm band have much higher resolution with respect to the g-factor.

In the conventional X-band, higher resolving power may be attained by substituting ^{15}N and ^{2}H for the natural isotopes in the NR used (Figure 1.6). High-frequency 2-mm radiospectroscopy is much more informative than the conventional 3-cm band (Figure 1.2) and affords the possibility of direct determination of the A and g parameters from the spectra (Grinberg *et al.* 1983; Krinichnyi 1990). Under such conditions, superconducting magnets with very strong magnetic fields are employed, cooled by liquid helium. The spectra thus obtained are essentially simplified, since the value of the nuclear magnetic quantum number for ^{15}N (I) is $\frac{1}{2}$ (two lines of a hyperfine structure), and that of the magnetic moment of ^{2}H is much lower than that of ^{1}H. The spectrum lines become narrower and more intense; individual components related to various values of the g- and A-tensors become resolvable in simple cases.

In some cases it is necessary to record other parameters of the stationary spectrum rather than the absorption signal itself or its derivatives (Hyde & Dalton 1979; Thomas *et al.* 1986; Hyde *et al.* 1990). Such a situation arises when the saturation is transferred from one portion of the ESR spectrum to another (spectral diffusion) because of the rotational and translational processes or those of spin relaxation of nuclei. Saturation

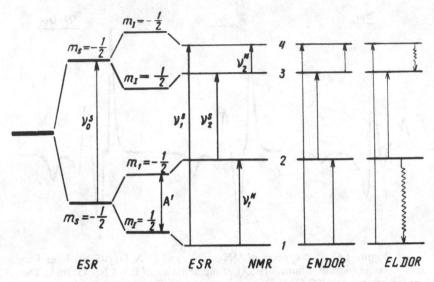

Figure 1.7. Transitions between the energy levels used in the ESR, NMR, ENDOR, and ELDOR methods: v_0^s, v_1^s, and v_2^s are the frequencies of transitions between the Zeeman electron levels; v_1^N and v_2^N are the frequencies of transitions between the Zeeman levels; A^1 is the HFI constant.

transfer affects in particular the dependence of the second harmonic of the absorption signal with a 90° phase shift (V'_2) on the field strength (H_z) (Figure 1.4).

The processes of saturation transfer provide the basis for the method of electron–electron double resonance (ELDOR) (Bendersky *et al.* 1968; Hyde *et al.* 1968; Hyde & Subczinski 1989; Hyde & Feix 1989). In this method, powerful MW radiation of the v_1 frequency saturates spins corresponding to a certain portion of the ESR spectrum (Figure 1.7). The influence of this effect on the intensity of the ESR signal of that spectrum portion is recorded at the other frequency (v_2). The efficiency of the saturation transfer due to competition between spin–lattice relaxation of the electron spin and such processes as spin nuclear relaxation, radical rotation, and spin exchange with other paramagnetics is characterized by the R parameter:

$$R = (A_0 - A_1)/A_0, \tag{1.6}$$

where A_0 and A_1 are the ESR signal amplitudes at the v_2 frequency before and after switching on the saturation power, respectively.

In the method of electron-nuclear double resonance (ENDOR), radioradiation saturates nuclear spins (Figure 1.7) (Isaacson & Feher 1990). Changes in the shape of the ESR spectrum line take place owing to the processes of saturation transfer caused by spin nuclear relaxation and spin electron–nucleus interactions. Analysis of ENDOR data makes it possible to determine the HFI constants (A_i) and the number of physical parameters dependent on the efficiency of electron–nucleus interactions.

1.4.2. Pulse methods

At present, approaches based on the action of short powerful pulses of high-frequency MW radiation followed by further monitoring of the corresponding responses of the system are being intensively studied (Lebedev & Muromtsev 1972; Beth & Robinson 1989; Hofer 1990). The simplest approach is that of pulse saturation, in which a sample put into a magnetic field is saturated with a short pulse for a period of a fraction of a microsecond (τ). The amplitude of the ESR signal first decreases, then increases under the influence of processes of spin–lattice relaxation.

In the simplest version of the electron spin-echo method (ESEM), a sample is placed in a constant magnetic field and is subjected to several pulses of MW radiation (Salikhov *et al.* 1976; Mims *et al.* 1977; Tsvetkov & Dikanov 1987; Mims & Peisach 1981; Kevan & Baglioni 1990). The first pulse, $\tau_1 = \tau/2\gamma H_1$, rotates the magnetization vector by 90°. Then during some time τ_2, individual spins precess in different local fields, and their phases diverge. The second pulse of 180°, with an amplitude twice as large as the first one, rotates each of the spins by the angle corresponding to the mirror reflection of its vector. The individual spin vectors begin converging and soon form a magnetization vector corresponding to an unstationary state. The return to the original magnetic state is accompanied by a burst of MW radiation that is called a spin-echo signal (SE).

Schemes for pulse succession have been proposed that allow the determination of parameters of radical spin relaxation. Thus, a series of 180° pulses with a period of τ_2 leads to a consecutive decay of the SE amplitude with the characteristic time T_{2e}, while another series of pulses gives the T_{1e} value.

The interactions between the electron spin and nuclear moments of the atoms are also manifested in the nuclear magnetic modulation effect (NMME). Nonmonotonic decay of the spin-echo signal is observed. The falloff modulation is described by the following equation:

$$v_m = (1 - \tfrac{1}{2}K) + \tfrac{1}{2}K[\cos\omega_\alpha t + \cos\omega_\beta t - \tfrac{1}{2}\cos(\omega_\alpha + \omega_\beta)t], \qquad (1.7)$$

where K is the parameter characterizing the extent of modulation, ω_α and ω_β are the values of hyperfine splitting expressed in the frequency units

$$\omega_{\alpha,\beta} = \{[\omega_1 \pm A'_{iso} \pm \tfrac{1}{2}F(1 - \cos^2\theta)]^2 + [\tfrac{3}{2}F\sin\theta\cos\theta]^2\}^{\frac{1}{2}},$$
$$\qquad (1.8)$$

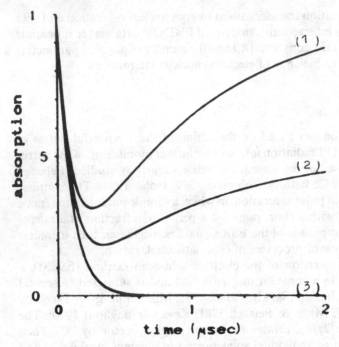

Figure 1.8. Theoretical saturation recovery absorption signals at three observed levels. Parameters used in the computations: $H_z = 3,300\,G$, $T_{1e} = 1.0\,\mu s$, $T_{2e} = 0.1\,\mu s$; pump time $T_p = 1\,s$, $\phi = 85°$; pump field $h_p = 1\,G$; observer field of low amplitude $h_0 = 0.01, 0.005$, and $0.0\,G$ for curves 1, 2, and 3, respectively (Beth & Robinson 1989).

where $\omega_1 = g_n\beta_n H_z/h$, θ is the angle between the H_z field direction and the line connecting spins, and

$$F = \mu_n g_n \mu_e g_e / h r^3. \tag{1.9}$$

The parameters K and F bear the most valuable information on the structure of the paramagnetic under study. As seen from equation (1.9), the latter parameter characterizes the efficiency of the electron–nuclear dipole interaction, which is r-dependent. Hence a quantitative analysis of the equations gives the distance r value. According to equations (1.7)–(1.9) the position of the harmonics' maximum $(\omega_\alpha + \omega_\beta)$ depends on the value of the HFI anisotropy, which allows its determination from the maximum shift relative to the main frequency of modulation.

The nuclear quadrupole interaction (NQI) should be taken into account for nuclei with magnetic quantum numbers $I_n \geqslant 1$, in particular, the values and orientations of tensors of HFI and NQI. The validity of the pertinent theory was demonstrated on the imidazolinoxide derivatives (Tsvetkov & Dikanov 1987).

The method of ESR saturation recovery (SR ESR) is based on the following procedure (Beth & Robinson 1989; Hyde & Subczinski 1989). An equilibrium spin system in a constant magnetic field under continuous MW radiation with a low amplitude, H_1^0, is irradiated with a MW pulse of high amplitude, H_1^1, for a time of τ_1. Then the field amplitude is decreased sharply down to H_1^0 and the response kinetics are recorded (e.g., the absorption amplitude). As seen from Figure 1.8, at first the amplitude decreases because of pulse relaxation for T_{2e}; then it increases due to saturation recovery in the weak MW field for a time T_{SR}.

The process of saturation recovery at a certain position of the ESR spectrum in the general case depends on the time of spin-lattice relaxation. T_{1e}, the spin nuclear relaxation time T_{2n}, and the rate of saturation transfer to other portions of the spectrum because of radical rotational diffusion. During the experiment, the parameters τ_1 and H_1^1 and the position of the signal being saturated in the spectrum are variable.

Figure 1.9 illustrates the principle of electron spin echo with magnetization transfer (ESE MT). The method is based on the transfer of magnetization of the system of spins initially placed in a magnetic field of strength H_z^0. The saturation pulse induces nonequilibrium magnetization of the system (Dzuba *et al.* 1984; Tsvetkov & Dikanov 1987; Dzuba 1988). The strength of the magnetic field changes by jumps to the H_z^1 value in time t, and the 90° and 180° pulse sequence is applied to initiate the spin-echo signal (Figure 1.9). Then the kinetics of the changes in the longitudinal magnetization of the magnetic field H_z^1 are recorded. Under these conditions, the magnetization transfer is mainly determined by the rotational mobility of the radicals.

The ELDOR method may be used in pulse modifications (Hyde & Feix 1989; Hyde *et al.* 1990). In this case, the saturation transfer from one portion to the other is monitored directly from the relative change in the amplitude of the ESR signal (parameter R). As stated earlier, the rate of saturation transfer of the radical is determined by the competition of the processes of spin and nuclear relaxation, radical rotation, and spin–spin interaction with other paramagnetics.

As follows from equation (1.1), resonance conditions can be attained at a certain strength of the magnetic field, H_z, and variable MW radiation. In Fourier ESR spectroscopy, all the spins of the system are irradiated by a short MW pulse, which rotates them all simultaneously. The resulting signal of free induction is Fourier-transformed. A certain component of the recovery corresponds to each resonance frequency of the spectrum. The ESR spectrum results from the inverse Fourier transform. The major advantages of Fourier spectroscopy are (1) that the whole range of the spectrum can be recorded after only one short pulse and (2) its ease of multiple accumulation in a computer. Thus, the time of the spectrum registration is reduced, and the signal–noise ratio is increased.

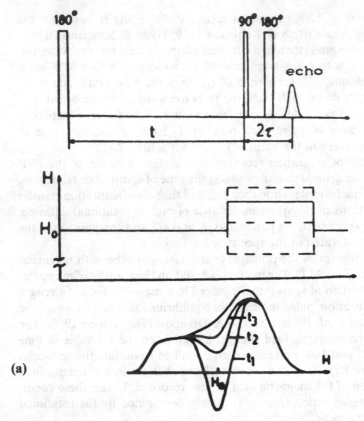

Figure 1.9. (a) Time sequence of HF pulses (top), change in the magnetic fields (middle) and form of the dip burned out by the first pulse (bottom). (b) Time evolution of the dip of the ESR spectra for two different portions of the burning-out pulse (arrows). The values of the orientation angles are given at the bottom. $T = 163K$, dibutylphthalate (Dzuba & Tsvetkov 1987).

Recently a number of new pulse methods have been developed, including the method of two-dimensional Fourier electron–electron double resonance (2D-ELDOR). The combination of a certain pulse sequence with the Fourier transform gives a high accuracy of measurement of the parameters of spin relaxation and dynamic processes (Bowman & Michalski 1990; Patyal *et al.* 1990; Shin *et al.* 1990). The method Fourier spectroscopy is also applicable to the modulation phenomena in electron spin-echo experiments.

Thus, modern techniques of ESR spectroscopy provide an opportunity to solve a vast range of dynamic and structural problems. However, the realization of this opportunity requires a theory to connect measured

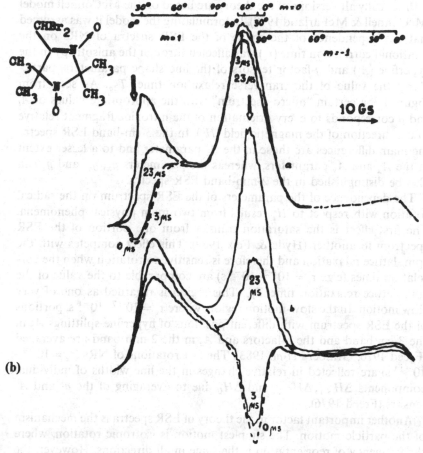

(b)

Figure 1.9 *(cont.)*

spectroscopic parameters with the physical properties of NRs. The following section deals with this problem.

1.5. Rotational diffusion of nitroxides

1.5.1. General

The parameters of the ESR signals of NRs depend on the mechanism and intensity of rotational diffusion, which in turn reflect conformational and rotational dynamics, microviscosity, and microstructure of biological and model systems. That is why many theoretical and experimental works have been dealing with the theory of the ESR spectra of NRs in solution.

Practically all versions of the theory are based on the McConnell model (McConnell & McFarland 1970). In formulating the model it was assumed that the dependence of the shape of the ESR spectra of NRs on the rotational correlation time (τ_c) was affected through the anisotropy in the hyperfine (A_i) and g-factor tensors of the line shape parameters, particularly the value of the transverse relaxation time, T_{2e}. As seen from Figure 1.3, a certain "micro spectrum" with the appropriate values of A_i and g corresponds to every orientation of the nitroxide fragment relative to the direction of the magnetic field (H_z). In the 3-cm-band ESR spectra the main differences are those in the A_z parameter and to a lesser extent in the A_x and A_y parameters, whereas the parameters g_x, g_y, and g_z can also be distinguished in the 2-mm-band ESR spectra.

The dependence of the parameters of the ESR spectrum on the radical rotation with respect to H_z results from two main physical phenomena. The first effect is the saturation transfer from one portion of the ESR spectrum to another (Hyde & Feix 1989). This effect competes with the spin–lattice relaxation and therefore is sensitive to rotation when the correlation times (e.g., $\tau_c = 10^{-3}$–10^{-6} s) are comparable to the value of the spin–lattice relaxation time T_{1e}. This region is regarded as one of very slow motion. In the slow-motion region, where $\tau_c = 10^{-7}$–10^{-8} s, portions of the ESR spectrum with different positions of hyperfine splittings A_z in the 3-cm band and the g-factors and A_z in the 2-mm band are averaged (Freed 1976; Grinberg *et al.* 1983). The fast rotations of NRs ($\tau_c = 10^{-9}$–10^{-11} s) are reflected in relative changes in the line widths of individual components ΔH_{+1}, ΔH_{-1}, and ΔH_0 due to averaging of the g- and A-tensors (Freed 1976).

Another important factor in the theory of ESR spectra is the mechanism of the particle motion. The simplest motion is isotropic rotation, where the frequency of reorientation is the same in all directions. However, the majority of NRs are asymmetric particles, and their rotation is anisotropic. The mechanism of the radical reorientation has a significant effect on the shape of the ESR spectrum as well. The following mechanisms usually are considered: (1) the Debye model of rotational Brownian motion, (2) the model of free diffusion, and (3) the rotation random-jump model.

1.5.2. Elements of the theory of the ESR spectra of rotating nitroxides

In the absence of radical rotation the theoretical ESR spectra (referred as "powder spectra") are calculated, taking into account all orientations characterized by the angle θ between the direction of the magnetic field and the radical axis. The g-factor value for each orientation of the radical is determined by the following formula, which is valid for radicals with axial anisotropy of the g-factor:

$$g(\theta) = (g_\perp^2 \sin^2 \theta + g_\parallel^2 \cos^2 \theta)^{\frac{1}{2}}, \tag{1.10}$$
$$A(\theta) = (A_\perp^2 \sin^2 \theta + A_\parallel^2 \cos^2 \theta)^{\frac{1}{2}}, \tag{1.11}$$

where $g_\perp = g_x \simeq g_y; g_\parallel = g_z; A_\perp = A_x \simeq A_y; A_\parallel = A_z$.

The stochastic Liouville equation is commonly used for describing the effects rotational diffusion on the line shape of ESR spectra (Freed 1976; Antsiferova & Lyubashevskaya 1986; Marsh 1989):

$$\delta\rho(\Omega, t)/\delta t = -i[H(\Omega), \beta(\Omega, t)] - \Gamma_\Omega \rho(\Omega, t), \tag{1.12}$$

where $\rho(\Omega, t)$ is the spin density matrix dependent on the time and dynamic variables (the polar angles Ω) of the radical, and Γ is the Markoff operand reflecting stochastic molecular motion. The operand is included in the equation

$$\delta P(\Omega, t)/\delta t = -\Gamma_\Omega P(\Omega, t), F_\Omega P_0(\Omega), \tag{1.13}$$

where $P(\Omega, t)$ is the probability of the radical being in the Ω state at the moment t, and $P_0(\Omega)$ is related to the equilibrium distribution. The operands of diffusion, say the Fokker-Planck operand, may serve as the Γ_Ω operand.

Numerical solution of the foregoing equations leads to theoretical spectra whose line shapes depend on Ω and t. For NRs containing ^{14}N, fast rotation ($\tau = 10^{-9}$ s) averages the anisotropy of A- and g-tensors and leads to spectra consisting of three separate lines (Figures 1.2 and 1.4). For the fast-motion region, the value of T_{2e} and consequently the homogeneous broadening of the Lorentzian line for each component can be calculated in the framework of time-dependent perturbation theory. The reason for the broadening is a modulation of A- and g-factor anisotropy by the radical rotation. The modulation process is expressed in terms of the autocorrelation of the Redfield function (Freed 1976; Marsh 1989, 1990):

$$G(\tau) = \langle H_1(t)H(t + \tau) \rangle, \tag{1.14}$$

where $H(\tau)$ is the time-dependent term of the spin Hamiltonian. The frequency dependence of the correlation time is expressed by the Fourier transforms of the spectral density function:

$$I(\omega) = \int_{-x}^{-x} G(\tau)e^{i\omega\tau}, \tag{1.15}$$

where ω is the angular frequency.

Rotational correlation times are eigenvalues of the diffusion equation, which is a function of time and of the coefficients of anisotropic rotational diffusion, $D_\parallel$ and $D_\perp$, corresponding to rotation along and perpendicular to the radical axis, respectively. The Hamiltonian $H(t)$ in the correlation function $G(\tau)$ includes the parameters of the anisotropy of HEI and the *g*-factor, which affects the Lorentzian line width:

$$\Delta H_{pp}(m_i) = A + B \cdot m_i + C m_i^2, \tag{1.16}$$

where ΔH_{pp} is the peak-to-peak width of the spectral line. In the case of isotropic rotation,

$$B = 1.81 \cdot 10^6 H_z (\Delta g \Delta A + 3 \delta g \delta A) j(0) \text{ Oe}$$
$$C = 1.81 \cdot 10^6 [(\Delta A)^2 + 3(\delta A)^2][J(0) - \tfrac{3}{8} j(\omega)] \text{ Oe}$$
$$\Delta A = A_{zz} - \tfrac{1}{2}(A_{xx} + A_{yy})$$
$$\delta A = \tfrac{1}{2}(A_{zz} - A_{yy})$$
$$\delta g = (\beta/h)[g_z - (g_x + g_y)]$$

In the case of isotropic liquids and exponential correlation functions,

$$j(\omega) = \tau_c/(1 + \omega^2 \tau_c^2) \simeq \tau_c. \tag{1.17}$$

The next section will consider the influence of the intensity and mechanism of the NR rotation on the parameters of ESR signals and methods of determination of those parameters.

1.5.3. *Very slow rotation*

The effects of saturation transfer between the components of the ESR spectra of NRs provide the basis for an approach to the study of very slow motions, with $\tau = 10^{-3}$–10^{-6} s. In the absence of spin–spin interactions the parameters of the saturation transfer depend on the electron spin–lattice relaxation time (T_{1e}) and nuclear spin–lattice relaxation time (T_{1n}). The relaxation processes remove saturation, and rotational motion competition transfers it. So the method of saturation transfer can be applied to rotational motion with characteristic times comparable to T_{1e} and T_{2n} (Beth & Robinson 1989; Hyde & Feix 1989). According to the experimental data for nitroxides, the value of T_{1e} at room temperature and in nonpolar solvents ranges from 3 to 5 μs.

A number of stationary and pulse techniques based on saturation of nitroxide spins by MW radiation of high power have been developed (Hyde & Dalton 1979; Watts 1985; Krinichny *et al.* 1987a,b; Beth & Robinson 1989). In the most common saturation transfer techniques the spectrum of the second harmonics of the absorption lines is often used with a 90° phase shift (V'_2) in the 3-cm and 2-mm ranges of MW radiation. The influence of the frequency and anisotropy of nitroxide rotation on the V'_2 spectrum is illustrated in Figures 1.10 and 1.11. Practically, the value of the rotational correlation time (τ_c) can be estimated by the use of calibration curves plotted for the systems with known rotational parameters, such as spin-labeled hemoglobin, or by theoretical calculation (Figure 1.11). In so doing, the dependence of the V'_2 spectral parameters on the spin-lattice relaxation time (T_{1e}) should be taken into account.

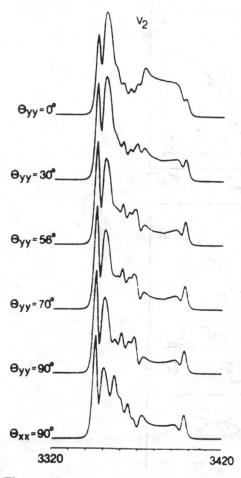

Figure 1.10. V_2' line-shape dependence on labeling geometry for a uni-axial rotational diffusion model. The computed line shapes for a constant-diffusion model are characterized by $\tau_\parallel = 10\,\mu s$ and $\tau_\perp = 0.1\,s$. The first five spectra were obtained by rotating the magnetic tensors by an angle θ_{yy} relative to the $\tau_\perp$ axis. The lower spectrum was obtained by rotating the magnetic tensors by the angle θ_{xx} (i.e., about the A_{xx} and D_{xx} axes) (Marsh 1989).

The resolution of the V_2' ESR spectrum was found to be markedly improved in the 2-mm band and for nitroxides enriched with ^{15}N and ^2H isotopes (Figures 1.5 and 1.12) (Krinichny *et al.* 1987a,b; Park & Trommer 1989). The line shape of the 2-mm-band ESR spectrum appeared to be especially sensitive to the anisotropy and mechanism of the radical rotation (Figure 1.12).

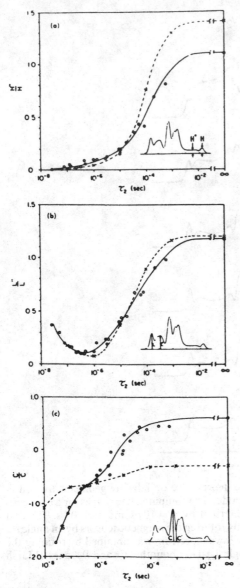

Figure 1.11. Dependence of V'_2 ratio parameters on τ_r for ^{14}N spin labels. Open circles are experimentally measured ratios from maleimide spin-labeled hemoglobin in glycerol buffer solution; × denotes the ratio measured from computer-generated spectra using axial A- and g-tensors (Thomas *et al.* 1976).

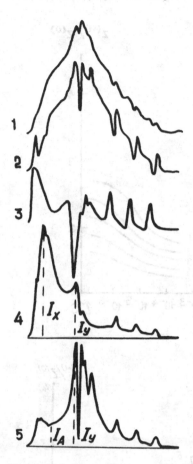

Figure 1.12. Theoretical V'_2 spectra in the 2-mm ESR band: 1 and 2 for $\tau_c = 10^{-3}$ s; 3–5 for $\tau_c = 10^{-7}$ s; 2 for $T_{1e} = 1\,\mu$s; others for $T_{1e} = 6.6\,\mu$s; 3, isotropic rotation; 4, anisotropic rotation around the x axis; 5, around the y axis ($\Delta H_0 = 0.6\,\text{mT}$, $g_x = 2.0089$; $g_y = 2.0061$; $g_z = 2.0027$; $A_x = A_y = 0.6\,\text{mT}$; $A_z = 3.2\,\text{mT}$ (Krinichny *et al.* 1987a).

The parameters of rotational diffusion of nitroxides (τ_c) in a region of very slow motion can be estimated by measuring the so-called saturation factor of an individual component of the "routine" differential ESR spectrum (Figure 1.13):

$$Z^m_{H_1} = A^m_{H_1} H^0_1 / A_0 H_1, \tag{1.18}$$

where $A^m_{H_1}$ is the amplitude of the mth component at a given MW power H_1, and A_0 and H^0_1 are the corresponding parameters in the absence of

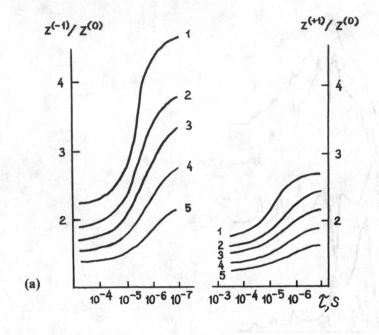

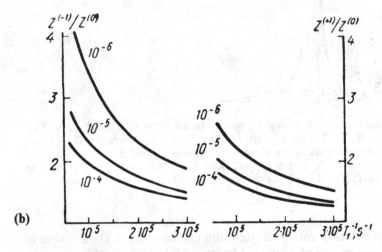

Figure 1.13. Theoretical dependence of the saturation factor Z^0 [equation (1.17)] (a) on the correlation time (τ_c) [$T_{1e} = 1.6 \cdot 10^{-5}$ (1), 10^{-5} (2), $0.75 \cdot 10^{-5}$ (3), $0.5 \cdot 10^{-5}$ (4), $0.33 \cdot 10^{-5}$ s (5)] and (b) on T_{1e} at different τ_c given in the figure ($H_1 = 0.32$ G, $1/\gamma T_2 = 3.1$ G, $A_1 = 34.4$ G, $A_2 = 6.9$ G, $\Delta g = 0.005$, $\nu_m = 20$ kHz) (Livshitz & Kuznetsov 1980).

saturation (Livshitz & Kuznetsov 1980). The T_{1e} and T_{2e} values are needed for the correct estimation of τ_c.

Stationary ELDOR was found to be an efficient approach to determination of the saturation transfer parameters, including the ones dependent on the rotational diffusion (Hyde & Feix 1989). The latter affects the value of ΔR (see Section 1.4.1), which is related to the ratios between the amplitudes of various components of the ESR spectrum with and without saturation.

Recently the high efficiency of pulse saturation techniques has been demonstrated. Monitoring of the saturation recovery after a short MW pulse allows estimation of the contribution made by one or several rotational exponential processes to the overall kinetics (Beth & Robinson 1989). In the pulse ELDOR method the rate of alteration of the parameter R (Section 1.4.2) directly reflects the radical rotation, if the latter is much faster than the electron and nuclear relaxations ($1/T_{1e}$ and $1/T_{2n}$) (Hyde & Feix 1989). Both techniques mentioned earlier require preliminary independent determination of the T_{1e} and T_{2n} values.

In the method of spin echo combined with magnetization transfer, the rotation rate of the NR is directly monitored by a time-dependent evolution of a dip in certain portions of the ESR spectrum related to different magnitudes of the magnetic field (H_z) (Dzuba *et al.* 1984) (Figure 1.9). Variation of the latter enables one to follow in detail the mechanism of rotation. In the case of Brownian diffusion, gradual rotation by small angles is expected to affect the ESR signal amplitudes of the sections of the spectrum that are adjacent to the dip position. In contrast, rotational jumps of high angles cause changes in the amplitude of the spectral sections that are remote from the dip position. The essential advantage of this approach is that the experimental values of the rotation rate are only slightly affected by the values of the electron and nuclear relaxation times.

1.5.4. Slow-motion regions

In the slow-motion region ($\tau_c = 10^{-7}$–10^{-8} s) the salient feature of nitroxide rotation is the averaging of the anisotropy of the A- and g-factor tensors; as a result, the hyperfine splitting (A_z) decreases in the 3-cm and 2-mm bands of the ESR spectrum. An additional shift observed in the 2-mm ESR line positions is related to the g-factor ($g_x, g_y,$ and g_z) (Figures 1.5, 1.10, and 1.14) (Freed 1976; Grinberg *et al.* 1983; Watts 1985; Marsh 1989, 1990). Another consequence of rotation is broadening of the ESR lines, which is formally analogous to the broadening from spin exchange between two spins with different g-factors (Figure 2.9).

For the model of isotropic rotation, the rotational correlation time (τ_c)

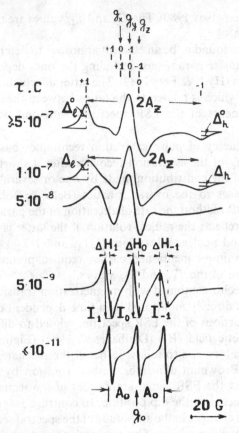

Figure 1.14. Schematic theoretical ESR spectra in the 3-cm band in slow and fast rotational regions (presented by Antsiferova & Lyubashevskaya).

can be calculated using the formula

$$\tau_c = a(1 - S)^b, \tag{1.19}$$

where $S = A'_z/A^0_z$, A'_z and A^0_z are the values of A_z for a given τ_c and $\tau_c \gg 10^{-7}$ s, respectively, and a and b are the coefficients derived from theoretical spectra. Theoretical calculation also gives

$$\tau_c = a'(\omega_i - 1)^{-b'}, \tag{1.20}$$

where $\omega_i = \Delta_i/\Delta^0_i$, Δ_i and Δ^0_i are the line widths for a given τ_c and $\tau_c \gg 10^{-7}$ s, and a' and b' are theoretical coefficients. The values for a, a', b, and b' are listed in Table 1.1.

The theoretical dependence of the parameters of the 2-mm band of the ESR spectrum on τ_c is depicted in Figure 1.15 for models of Brownian

Table 1.1. *Parameters for equations (1.17) (a and b) and (1.18) (a' and b')
for the low-field component of the NR spectrum calculated at the
individual line width of 0.3 mT*

Diffusion model	$a \cdot 10^{10}$ s	b	$a' \cdot 10^{10}$ s	b'
Isotropic Brownian diffusion	5.4	−1.36	1.15	0.943
Isotropic free diffusion	1.10	−1.01	1.29	1.033
Large-angle jumps	2.55	−0.615		
Brownian diffusion along the y axis	2.59	−1.39		
Fluctuating torques	2.60	−1.31		

Source: Data from Freed (1976) and Marsh (1989).

rotational diffusion (BRD) and jumps of large angles (JLA) (Antsiferova
& Lyubashevskaya 1986, 1990). As seen in the figure, various models
can be characterized by a specific dependence between the line width Δ
and the A_z shift (ΔR), with the ratio $\Delta/\Delta R$ being larger for the BRD model
than for the JLA model. Therefore the appropriate model can be chosen
experimentally. A more detailed analysis of the parameters of 2-mm-band
ESR spectra, including the line broadening and the shifts of various components, provides information not only on correlation times and mechanisms of the nitroxide rotation but also on the anisotropy of the rotation.

1.5.5. Fast-rotation regions

Rotation of a NR with $\tau_c = 10^{-9} - 10^{-11}$ s is efficient enough to average
the main parts of the hyperfine interactions in both the 3-cm and 2-cm
ESR bands and differences in the g-factors in the 2-mm ESR band. As a
result, three components are observed in the ESR spectrum. The key effect
of the rotation on the spectrum is narrowing of the individual components
due to modulation of the anisotropy of HFI and g-factors. In the case of
NRs, the theoretical treatment by Kivelson (1960) and Freed (1976) leads
to the following formula, which was derived from equations (1.16) and
(1.17):

$$T_{2e}(0)/T_{2e}(m) = 1 - (\tfrac{4}{5}b \cdot \Delta v H_z T_{2e}(0)m + \tfrac{1}{8}b^2 T_{2e}(0)m^2)\tau_c, \qquad (1.21)$$

where $m = 1, 0, -1$ is the magnetic quantum number of the nitrogen
nucleus, $b = \tfrac{4}{3}\pi(A_{\parallel} - A_{\perp})$, and $\Delta v = (\beta/h)[g_z - \tfrac{1}{2}(g_x + g_y)]$. This formula
was derived on the basis of the following assumptions: (1) the radical
motion is isotropic; (2) the magnetic anisotropy is low, $\Delta v \ll v_r$; (3) $\tau_c^{-2} \gg
\pi(A_{iso})^2$; (4) the HFI anisotropy is axially symmetric, i.e., $A_x = A_y = A_{\perp}$,

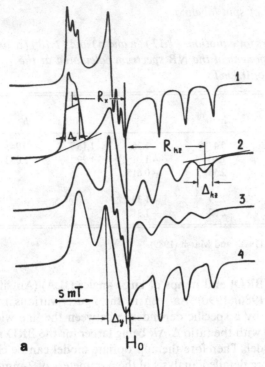

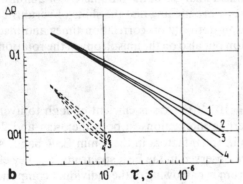

Figure 1.15. Theoretical 2-mm ESR spectrum of the NR (a) and theoretical dependence of the relative shift $\Delta R = (R_{hz} - R^*)/(R_h^* - H_0)$ on the isotropic correlation time T_{iso} (b): (a) 1, absence of rotation; 2–4, axisymmetric rotation with the correlation times $\tau_\perp = 2\cdot10^{-7}$ s, $\tau_\parallel = 2\cdot10^{-8}$ s, and orientations of the symmetry axis of fast rotations about the x, y, and z axes of the tensor, respectively. In the calculations, use was made of the following parameters: $g_{xy} = 2.0084$, $g_{yy} = 2.0060$, $g_{zz} = 2.0022$; $A_{xx} = A_{yy} = 6.73$ Hz, $A_{zz} = 3.725$ ml; the width of the spin packet was 0.3 mT, $H_0 = 4.9500$ T. (b) 1–4, values of $\Delta_{ax}^0 = 0.17$, 0.5, and 1.1 mT for the Lorentzian line shape of the spectrum; the solid lines correspond to the model of Brownian rotation; the dashed line represent the large-angle jump model.

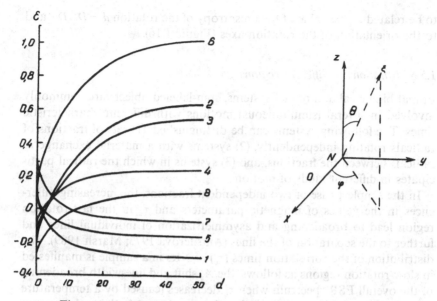

Figure 1.16. Dependence of the ε parameter on d (see text): 1, $\varphi = 0$, $\theta = n/2$; 2, $\varphi = n/2$, $\theta = n/2$; 3, $\varphi = n/3$, $\theta = n/2$; 4, $\varphi = \pi/2$, $\theta = n/3$; curves 5–8 were calculated with different values of A- and g-tensors (Vasserman et al. 1971; Likhtenshtein 1976a).

$A_z = A_{\parallel}$; (5) τ_c^{-1} is much less than the Larmor frequency (ω_L), though much larger than the anisotropy in HFI.

In practice, it is convenient to use equations obtained from (1.21), after substituting the numerical values for the coefficients and taking into account the fact that the value of T_{2e} for each hyperfine component is inversely proportional to the square of the intensity (height) of the ESR spectral line. In the case of $A_{\parallel} = 8.7 \cdot 10^6 \, \text{s}^{-1}$, $A_{\perp} = 14 \cdot 10^6 \, \text{s}^{-1}$, $g_x = 2.0089$, $g_y = 2.0061$, $g_z = 2.0027$:

$$v_{(+1)} = 1/\tau_{c(+1)} = 2 \cdot 10^8/(\sqrt{h_0/h_{+1}} - 1)\Delta H_0, \tag{1.22}$$

$$v_{(-1)} = 1/\tau_{c(-1)} = 3.6 \cdot 10^9/(\sqrt{h_0/h_{-1}} - 1)\Delta H_0, \tag{1.23}$$

$$v_{(\pm 1)} = 1.2 \cdot 10^{10}/(\sqrt{h_{+1}/h_{-1}} - 1)\Delta H_{(+1)}, \tag{1.24}$$

where ΔH_0 and $\Delta H_{(+1)}$ are the widths of the spectral components with $m = 0$ and $+1$, respectively, and h_0, h_{+1}, and h_{-1} are the intensities of the corresponding components.

For the model of anisotropic rotation, the theory of Kuznetsov (1974) predicted the value of the ε parameter,

$$\varepsilon = (\sqrt{h_0/h_{+1}} - 1)/(\sqrt{h_0/h_{-1}} - 1) \sim v_{(+1)}/v_{(-1)}, \tag{1.25}$$

to be related to the value of the anisotropy of the rotation $d = D_\parallel/D_\perp$ and to the orientation of the rotation axes (Figure 1.16).

1.5.6. *Rotations in different regions*

In real biological and model systems, spin-labeled objects are commonly involved in several (simultaneous) motions with different characteristic times. The following systems can be distinguished: (1) sets of fractions of radicals rotating independently, (2) systems with a material exchange of radicals between the fractions, and (3) systems in which the radical participates in different kinds of motion.

In the simplest case of two independent fractions, the increasing differences in the values of magnetic parameters and τ_c in the fast-rotation region lead to broadening and asymmetrization of individual lines and further to the separation of the lines (Antsiferova 1973; Marsh 1989). The distribution of the correlation times (τ_c) of NRs in a sample is manifested in slow-rotation regions as follows: the A_z shift and line-width broadening of the overall ESR spectrum when τ_c decreases (caused by a temperature increase, for example) will be delayed from that of individual fractions with low values of τ_c. Such properties of systems with independent fractions give rise to anomalously low values of the apparent activation energy of rotation. The methods for analysis of ESR spectra of heterogeneous systems have been described elsewhere (Antsiferova 1973; Antsiferova *et al.* 1989a).

The ESR spectrum is also affected when two fractions of radicals with correlation times τ_1 and τ_2, ranging within different rotational motion regions, undergo fairly fast exchange with a frequency v. If the value of v is comparable to or larger than the differences in the ESR spectral positions expressed in the same units, the spectral lines will broaden, followed by shifting of the line toward the central part of the spectrum (Marsh 1989; Antsiferova 1973).

Much experimental data can be considered in the framework of a model of nitroxide fast precession ($\tau_c \leqslant 10^{-9}$ s) around its long axis within a cone angle of 2θ (around the normal to the lipid bilayer, for example) (Griffith & Jost 1976; McConnell & McFarland 1970; Marsh 1989, 1990). In the 3-cm band of the ESR spectrum such a motion leads to a shift in A_z and can be characterized by the empirical order parameter

$$S = (A_\parallel - A_\perp)/[A_z^0 - \tfrac{1}{2}(A_x^0 + A_y^0)] = \tfrac{1}{2}(\cos\theta + \cos^2\theta). \qquad (1.26)$$

If the value of the precession angle is significantly higher, the ESR spectrum will reveal an internal extremum, $A'_\perp$ (Figure 1.17).

In the case of spin-labeled proteins in solution, one can determine the value of the correlation time for the protein globules' rotation as a whole

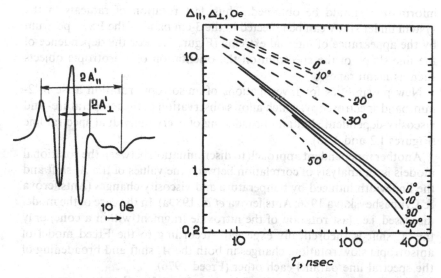

Figure 1.17. Dependence of the $\Delta_\parallel$ and $\Delta_\perp$ parameters (relative shifts) of the ESR spectra of NRs on the rotational correlation time for a macromolecule (τ) and the angle of fast rotation in a cone; solid and dashed lines correspond to the $\Delta_\parallel$ and $\Delta_\perp$ parameters, respectively (Dudich et al. 1977).

(τ_c) and the value of the order parameter (S) for fast precession of the nitroxide fragment. The procedure involves analysis of the experimental dependence of the A_z shifts of external ($A'_\parallel$) and internal ($A'_\perp$) spectral components on temperature and viscosity (η) (Shimchick & McConnell 1972; Dudich et al. 1977).

With the coordinates indicated in Figure 1.17 one can evaluate the molecular mass of the macromolecule or the molecular fragment moving together with the nitroxide group. Extrapolation of the curve to the region of high viscosity allows estimation of the order parameter S for the nitroxide rotation relative to the macromolecule.

The experimental dependence of the parameters of the spin-labeled proteins in solution and nitroxide derivatives of lipids in membranes can be described in the framework of an alternative model as well (Freed 1976). This model also suggests that the radical is involved in two types of mobility, though it is assumed that both motions occur within the slow-rotation region. According to the model, extrapolation of the dependence between the A_z shift and T/η to infinite viscosity leads to the value A_z^ω, which is related to the hindered rotation of the nitroxide fragment with respect to the macromolecule.

To distinguish between the previously mentioned models, additional information should be obtained. Thus, fast rotation of radicals in the system under study can be reflected in the 3-cm band of the ESR spectrum by the appearance of internal extrema (Figure 1.17) or the dependence of the line shape of the spectrum on the orientation of anisotropic objects such as membranes.

New possibilities for investigations of anisotropic rotation arise in 2-mm-band spectroscopy, which allows observation of the temperature- and viscosity-dependent shift and broadening of every individual spectral line (Figures 1.2 and 1.15).

Another independent approach to discrimination between the rotational models is an analysis of correlation between the values of the A_z shift and the line width induced by temperature and viscosity changes (Antsiferova & Lyubashevskaya 1986; Antsiferova *et al.* 1989a). In the case of the model discussed, i.e., fast rotation of the nitroxide fragment within a cone, only the A_z shift is theoretically expected. According to the Freed model of anisotropic slow rotation, changes in both the A_z shift and broadening of the spectral line parallel each other (Freed 1976).

In the case of radicals involved in two types of motion, slow and very slow, the values of the correlation times (τ_c) can be determined by a special approach (Squier & Thomas 1989). This approach requires taking separate ESR spectra of the nitroxide under saturating and under nonsaturating MW power.

1.5.7. *High-frequency low-amplitude dynamics*

When the atomic and molecular motions speed up because of changes in temperature, viscosity, or humidity of the medium, the changes also affect the spin relaxation parameters and line shapes of the ESR signals of NRs.

One such effect is narrowing of the spectral line initially broadened as a result of inhomogeneous spin dipole–dipole interactions between the unpaired electron and protons of the nitroxides and the medium (Johnson 1979; Marupov *et al.* 1981; Likhtenshtein *et al.* 1986a). A relatively low amplitude of local wobbling of the nitroxide fragment with the characteristic time of $\tau_c \leqslant 10^{-7}$ s averages the interaction and leads to narrowing of the ESR line.

Another such effect is a small (0.1–0.2 mT) gradual A_z shift in the ESR spectra of spin-labeled proteins as the temperature increases from 77K to 200K (Frolov *et al.* 1974). The cause of the shift appears to be high-frequency low-amplitude wobbling of the radical. The value of the wobbling angle derived from formulas (1.25) and (1.26) was on the order of several degrees.

According to theory, spin–lattice $(1/T_{1e})$ and transverse $(1/T_{2e})$ relaxation rates of radicals depend significantly on the intensity of phonon dynamics in the medium (Lebedev & Muromtsev 1972; Alexander *et al.* 1986). The phonon oscillations are accompanied by electric field oscillations within a wide range of frequencies. If the distribution of the electric oscillation frequencies partially overlaps the MW resonance frequency, the radical magnetization energy will be transferred to the energy of the phonon oscillations through the spin–orbital interaction.

In the case of a one-quantum direct transfer, the spin–lattice relaxation rate is proportional to the temperature: $1/T_{1e} \sim T$.

The Raman mechanism of the transfer is predicted to give the expression

$$1/T_{1e} \sim T^n, \tag{1.27}$$

where $n = 9$ for three-dimensional Kramer transfer and $n = 7$ for non-Kramer transfer. Rotational vibrations of radicals with anisotropy of the g-factor and hyperfine splitting A can also accelerate the spin–lattice relaxation (Lebedev & Muromtsev 1972).

An analysis of the temperature dependence of the spin–lattice relaxation rates $(1/T_{1e})$ of radicals and paramagnetic ions allows an investigation of collective phonon oscillations (fractions) in proteins and other biological objects (Alexander *et al.* 1986). For example, a two-fraction mechanism in the case of the three-dimensional Kramer transfer is predicted to give $n = 2.63$, with the value of n for non-Kramer transition being 4.63.

Thus, the fine mechanisms of phonon processes in biological systems can be easily studied by measuring the temperature dependence of the relaxation rate of spin labels.

1.5.8. Superslow motion

In 1981, Lazarev and Lebedev developed a procedure to measure superslow $(\tau_c = 10^{-6}\text{--}10^{-2}\,\text{s})$ reorientation of radicals. In their procedure, randomly oriented radicals with anisotropic magnetic parameters are irradiated with intense, photochemically active polarized light. The polarized radiation destroys radicals whose electron transition moment coincides with the polarization vector of the incident light. The light thus causes a dip in the ESR spectrum. The time of recovery of the dip corresponds approximately to the rotational correlation time. Radical pairs induced by radiation of complexes of quinones and phenols, or triplet carbenes, can serve as the oriented probes. NRs, which in contrast to the aforementioned radicals are chemically stable, would be ideal oriented probes.

Superslow motions $(\tau_c = 10\text{--}10^6\,\text{s})$ in a condensed medium can be recorded by studying the recombination of radical pairs by the ESR method (Grinberg *et al.* 1983; Likhtenshtein *et al.* 1983). Radical pairs can be easily produced by the introduction of certain compounds, such as tetraphenyl hydrazine, into the samples under study, with further dissociation of these compounds by radiation.

1.6. Nitroxides as dielectric, pH, and redox probes

The magnetic parameters of NRs, namely, the g-factor and hyperfine splitting constants (A), appear to be sensitive not only to the structure of radicals but also to the properties of the medium (Figure 1.18; see also Figure 1.6). According to theory, deviations of the g-factor value from $g_e = 2.0023$ for free electrons depend on the spin density (ρ), the spin-orbital coupling constant (λ'), and the value of the energy gap between the orbital of the unpaired electron residence and the nearest vacant orbital (ΔE) (Buchachenko & Wasserman 1973; Griffith & Jost 1976). Thus, for the x-component of an NR ESR spectrum,

$$\Delta g_x = \lambda_O \rho_O^\pi / \Delta E(n\pi^*), \tag{1.28}$$

where ρ_O^π is the spin density on the oxygen atom, $\Delta E(n\pi^*)$ is the energy of the $n \to x^*$ transition in the N$\div$O fragment, and λ'_O is the spin-orbital coupling constant for oxygen. For piperidine radicals, $\rho_O = 0.58$ and $\rho_N = 0.42$. The value of the HFI constant is a function of spin density on the s-orbitals of a nitrogen atom:

$$A = \tfrac{8}{3}\pi h^2 \gamma_e \gamma_n \rho_N, \tag{1.29}$$

where γ_e and γ_n are electron and nuclear gyromagnetic ratios.

The anisotropic components of the HFI constants, $A_{z,x,y}$, are also affected by dipolar electron–nucleus interactions. For the component of the A-tensor oriented along the magnetic field (H_z),

$$A_z = \tfrac{4}{5} h^2 \gamma_e \gamma_n \langle r_{en}^{-3} \rangle, \tag{1.30}$$

where $\langle r_{en}^{-3} \rangle$ is the inverse cube of the average distance between the electron and the nucleus.

The dependence of the g-factor and A for nitroxides on spin density and ΔE allows them to be used as probes to provide information on local polarity, pH, and the ability to form H-bonds with the N$\div$O group. The physical reason for such a dependence is that the contribution of resonance structure II,

$$\begin{array}{cc} \mathord{>}\text{N}\div\text{O} \leftrightarrow \mathord{>}\dot{\text{N}}^+ \!-\!\text{O}^- \\ \text{(I)} \qquad\qquad \text{(II)} \end{array}$$

causes an increase in spin density on the nitrogen atom and therefore increases the A_{iso} value. At the same time, the A_z values are larger for the ionic structure because of the reduction of the $\langle r_{en}^{-3} \rangle$ distance.

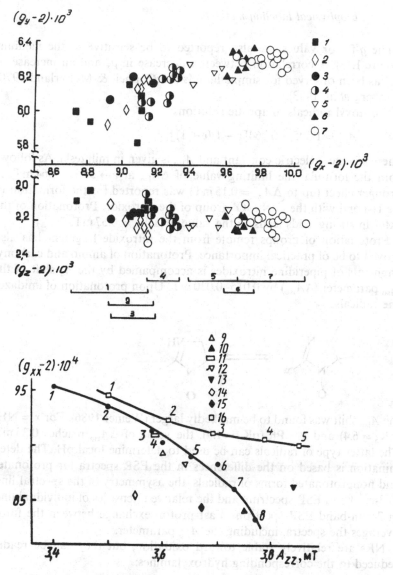

Figure 1.18. (a) Parameters of the g-tensor of NR derivatives of (1) imidazolidine, (2) imidazoline, (3) pyrrolidine, (4) 3-imidazolin-3-oxide, (5) imidazolidine, (6) 3-piperidine, and (7) piperidine (Grinberg *et al.* 1983). (b) Correlation between g_{zz} and Λ_{zz} at $T - 140$K. Radical SLXIV in toluene (1), isobutanol (2), ethanol (3), a water-ethanol mixture with H_2O content of 0.25 (4), 0.50 (5), and 0.66 (6), methanol (7), a water-glycerol mixture (0.50) (8), egg lysozyme (16) with relative humidity $P = 0.04$ (1), 0.35 (2), 0.60 (3), 0.80 (4), and 0.95 (5); human albumin with $P = 0.04$ (9) and 0.96 (10); α-chymotrypsin with $P = 0.04$ (11); radicals SLXIII (12, 13) and SLXIX (14, 15) in serum albumin with $P = 0.04$ and 0.96, respectively (Krinichny 1990).

The g-factor values are also reported to be sensitive to the medium (Figure 1.18). A correlation between a decrease in g_x and an increase in A_z has been observed for simple liquids (McConnell & McFarland 1970; Grinberg *et al.* 1983).

For doxyl radicals in aprotic solutions,

$$A_{iso} = 1.395 + 0.064[\varepsilon - 1/(\varepsilon + 1)],$$

where ε is the dielectric constant and A_{iso} is given in millitesla. As follows from the formula, the limiting value of ΔA_{iso} at $\varepsilon \to \infty$ is 0.064 mT. A stronger effect (up to $\Delta A_{iso} = 0.15$ mT) was reported for the formation of the H-bond with the $>$N—O$^\cdot$ group of the nitroxide. Protonation of the latter in strong acids leads to the value of $\Delta A_{iso} = 0.57$ mT.

Protonation of groups remote from the nitroxide fragment has also proved to be of practical importance. Protonation of amino and carbonyl fragments of piperidine nitroxides is accompanied by the A_z shift of the A_{iso} parameter (ΔA_{iso}) by 0.015–0.030 mT. Upon protonation of imidazoline radicals,

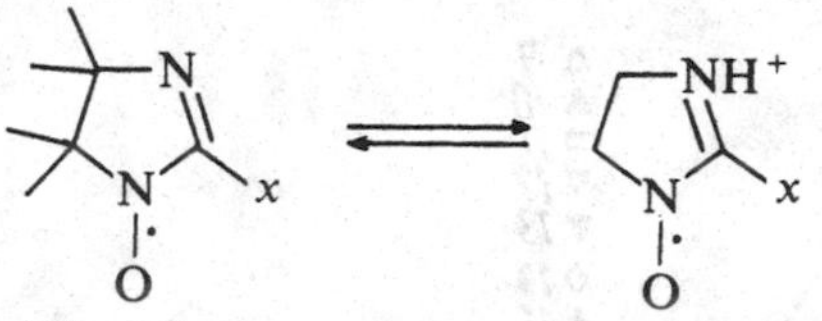

the A_{iso} shift was found to be markedly larger (Weiner 1986). For $x = NH_2$ (p$K_a = 6.4$) and $x = Ph$ (p$K_a = 1.9$), the value of ΔA_{iso} reaches 0.35 mT. The latter type of radicals can be used to determine local pH. The determination is based on the differences in the ESR spectra for protonated and nonprotonated forms of radicals: the asymmetry of the spectral lines in 3-cm-band ESR spectrum and the relative intensities of individual lines in 2-mm-band ESR spectrum. Fast proton exchange between the forms averages the spectra, including the A_{iso} parameter.

NRs are relatively stable toward oxidation, but they can be readily reduced to the corresponding hydroxylamines:

$$>N—O^\cdot \underset{}{\overset{e, H^+}{\rightleftharpoons}} >N—OH$$

The redox potential of a nitroxide (~ 0.2 eV for piperidine, for example) is high enough to oxidize such biological compounds as ascorbic acid, semiquinones, and superoxide radicals (Meisel & Czapski 1975). Taking advantage of this potential, two approaches have been developed to use nitroxides as redox probes. In the first, nitroxides are reduced by a standard reductant (e.g., ascorbate), and the reduction rate is measured. The reduction rate then indicates the location and steric hindrance of the nitroxide fragment. For example, it is relatively fast in the aqueous phase, but very slow for radicals that are embedded in the lipid phase of membranes. The second approach consists of recording the reduction rate of the nitroxide by a product of metabolic reactions, say semiquinones.

The values of the nitroxide redox potentials depend on their chemical structures. Thus, pyrrolidine nitroxides have been found to be more stable toward reduction by ascorbate in neutral solutions than are piperidine nitroxides. Doxyl radicals are reduced markedly faster than proxyl and acetoxyl radicals, but much slower than four-member ring nitroxides (Keana 1984). Special experiments on electrochemical reduction of nitroxides have shown that the reduction rate decreases in the following series: imidazolines < imidazoline oxides ≈ piperidines (Volodarsky 1988).

The ability of hydroxyl derivatives of nitroxides to be oxidized by dioxygen in the presence of transition metal ions (Fe^{3+}, Cu^{2+}) and directly by superoxide radicals is used for regeneration of the reduced spin labels and for analytical purposes (Rashba *et al.* 1986; Rosen *et al.* 1982).

Thus, there are experimental and theoretical grounds for quantitative studies of local physical and chemical properties of biological systems by the spin-labeling method.

1.7. Nitroxides in ESR tomography

ESR tomography is a new approach to the study of spatial distributions of paramagnetic centers. Owing to their chemical stability, the high intensity of their ESR signals, and their potential for specific labeling, NRs appear to be promising sources of paramagnetism for ESR tomography of biological and model systems. ESR tomography can be used for studying the spatial distribution and dynamics of spin-labeled metabolites, drugs, water, dioxygen, etc., in normal and pathological materials, with resolution of $10\ \mu m$. ESR tomography can be used to examine single grains of roe-corns of fish, fragments of tissue, and even the organs of small animals.

In most commonly used version of ESR tomography, the cavity of the ESR spectrometer is placed in an inhomogeneous magnetic field (Sueki *et al.* 1990; Ohno 1985; Eaton & Eaton 1987a; Smirnov *et al.* 1988; Lurie *et al.* 1990). The inhomogeneous field makes each location in the sample correspond to a slightly different magnetic field intensity and thus slightly different conditions for resonance. In a sample with chemically homogeneous radicals, the intensity of the ESR signal at a given magnetic field intensity is then directly proportional to the radical concentration at the corresponding sample location (Figure 1.19). A similar result can also be obtained in the case of radicals with anisotropic magnetic parameters, if the magnetic field inhomogeneity in the sample does not exceed the difference in the position of the ESR spectral lines.

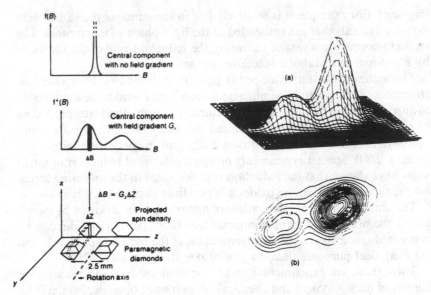

Figure 1.19. X-band ESR image of two diamonds. The left part of the figure represents the experimental arrangement and the method of constructing the image. In the right part of the figure (a, side view; b, top view), two displays of the spin density distribution in the diamond are depicted.

A number of techniques for creating the magnetic field gradient has been proposed. A one-dimensional gradient can be obtained with the aid of (1) a ferromagnetic wedge, (2) induction coils in the anti-Helmholtz configuration or (3) twisted coils of the main magnet. In such a way a gradient of up to 20 mT/cm can be produced. A temperature gradient can induce a magnetic field gradient of up to 300 mT/cm. In two-dimensional ESR tomography, spatial effects are attained by rotation of the sample in the cavity or by a special disposition of additional coils (Ohno 1985; Jansen *et al.* 1987).

Commonly, ESR tomograms are obtained in 3-cm-band ESR spectrometers. Nevertheless, the use of promising S-band techniques with a superficial coil has been reported (Berliner & Fujii 1985). The latter approach allows significant enlargement of the sample bulk.

Another approach to ESR tomography involves induction of the intensity gradient in the MW radiation or modulation of the amplitude of the radiation. Two new prospective pulse methods should also be mentioned. The first one involves determination of the spin–lattice relaxation parameter $(1/T_{1e})$ for radicals in various parts of the sample space. Figure 1.20 shows the sequence of the MW pulses and the dynamics of intensity

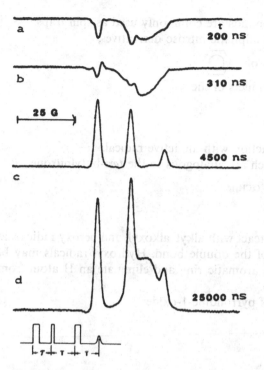

Figure 1.20. Spin-echo imaging experiments (Eaton & Eaton 1987a).

changes in the ESR signals of the nitroxide in the model heterogeneous system (Eaton & Eaton 1987a; Sueki *et al.* 1990). In the second method, the magnetic field gradient is created by short pulses, followed by Fourier analysis of the dynamics of the ESR signals. By such a technique a three-dimensional picture of the spatial distribution of radicals was obtained (Ewert & Freed 1990).

There is good reason to believe that certain difficulties in ESR tomography related to loss of sensitivity is samples of small size, dielectric adsorption of the MW field, and the complexity of theoretical analysis of three-dimensional objects, particularly for radicals with anisotropic magnetic parameters, will be overcome in the near future.

1.8. Spin traps

A spin trap is a compound that is converted to a stable NR after reaction with an active radical. Analysis of ESR spectrum of the nitroxide formed allows identification of the structure and amount of the active radical.

The following compounds are commonly used as spin traps:

1. Aromatic and aliphatic nitroso derivatives,

$(CH_3)_3CNO$ or   ⬡NO

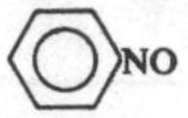

which yield adducts of the

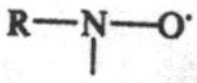

R—N—O˙
　|

type upon reaction with an active radical.

2. Nitrones, which are analogous to the *tert*-butylnitrone

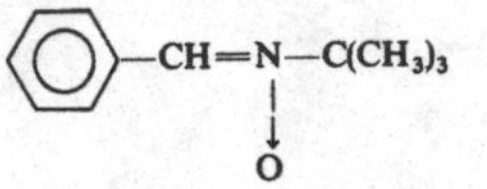

⬡—CH=N—C(CH₃)₃
　　　　|
　　　　↓
　　　　O

The nitrones react with alkyl, alkoxyl, and peroxy radicals at the carbon side of the double bond. Hydroxyl radicals may be attached to the aromatic ring and eliminate an H atom from the methyl group.

3. Derivatives of pyrrolidine-1-oxide:

which accept alkyl and alkoxyl radicals.

4. Methylenenitrones:

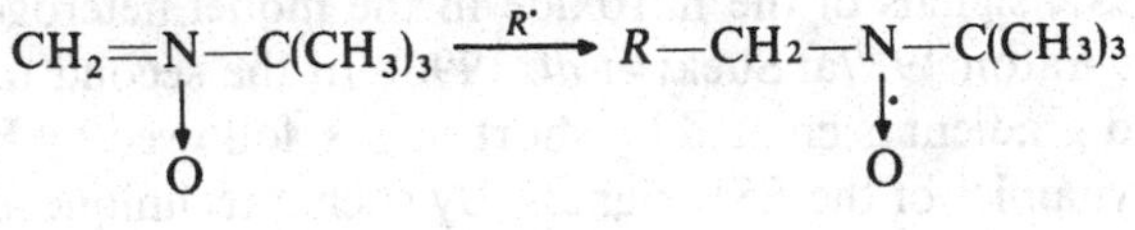

CH_2=N—C(CH₃)₃ $\xrightarrow{R˙}$ R—CH_2—N—C(CH₃)₃
　　|　　　　　　　　　　　　　　　　|
　　↓　　　　　　　　　　　　　　　　↓
　　O　　　　　　　　　　　　　　　　O

5. Bifunctional compounds, say

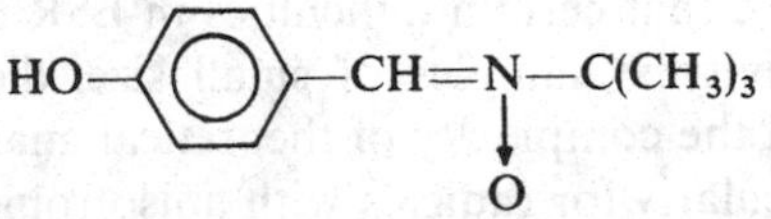

HO—⬡—CH=N—C(CH₃)₃
　　　　　　　|
　　　　　　　↓
　　　　　　　O

which can react with active radicals via two schemes: (1) by attaching to the methylene carbon atom or (2) by eliminating an H atom from the aromatic hydroxyl.

The resolution capacity of this approach is enhanced when spin traps with ^{15}N and ^{2}H isotopes are used, for example, for nitrone (Keana *et al.* 1990):

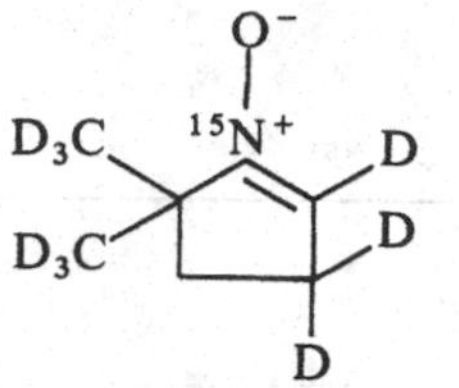

$$
\underset{\underset{\text{D}}{|}}{\overset{\overset{\text{O}^-}{|}}{\text{N}^+}}
$$

D$_3$C, ^{15}N$^+$ D
D$_3$C D
D

The rate constants for the reactions between active radicals and spin traps have been found to be fairly high (up to 10^7 M s^{-1}).

From the practical viewpoint, chemical stability of the adducts is particularly important. According to Zubarev (1984), the stability diminishes in the following series: (alkyl or aryl)-N-2,6-nitrones > (alkyl or aryl)-N-*tert*-alkyl nitrones > 2,4-substituted nitroso compounds > *tert*-nitroso compounds. It should be noted that in some cases nitroxide products can be formed from corresponding spin traps in reactions with non-radical compounds, e.g., hydroperoxides. It seems conceivable that spin traps could be used to generate nitroxide spin labels and probes in certain portions of such complicated biological systems as membranes.

In this chapter we have presented evidence that the spin-labeling method can be used to attack many structural and dynamic problems of molecular biology. Applications of the method to biological and model objects of various complexities will be considered in Chapters 6–8.

2

Double-labeling techniques

2.1. General

In solving a number of important problems of enzyme catalysis it is necessary to know the distances between certain functional groups of enzymes and other bioobjects. Therefore, one must first decipher the arrangement of the groups present at the active site. One must also know the distance between the active site and the site of an allosteric transition in order to study allosteric transitions, the depth of immersion of paramagnetic centers, and other long-range effects. Experimental investigations of the structures of multicenter metal enzymes containing several metal atoms per macromolecule are beset with serious difficulties.

A technique making use of double paramagnetic labels has been proposed for deciphering the structures of biological materials, primarily enzymes (Likhtenshtein 1968, 1976a; Likhtenshtein & Bobodzhanov 1968; Taylor *et al.* 1969). The principle of this method is that various residues of protein molecules can be modified by spin labels of an identical or different structure. Under certain conditions, an analysis of the parameters of the ESR signals of spin-labeled proteins will allow determination of the distance between the modified groups. This method has been further developed by other investigators. The paramagnetic complexes of transition metals, Cu^{2+}, Mn^{2+}, Fe^{3+}, Co^{2+}, etc., which either give ESR signals or affect the ESR parameters of other paramagnetics, can serve as labels as well.

Under certain circumstances, the ESR spectra of paramagnetic centers will respond suitably to the approach of other centers. Two types of interaction can be distinguished (Abragam 1961): (1) dipole–dipole interaction associated with the fact that the magnetic dipole of one paramagnetic group induces a local magnetic field at the site of another paramagnetic group; (2) exchange interaction caused by overlap of the orbitals of unpaired electrons as paramagnetic particles approach each other (Figure 2.1).

The magnetic field strength induced by a dipole of magnetic moment μ_m at distance r is given by

$$\Delta H_d \sim \mu/r^3. \tag{2.1}$$

46

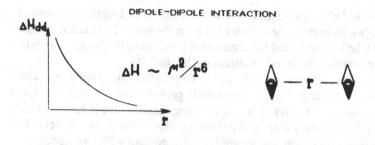

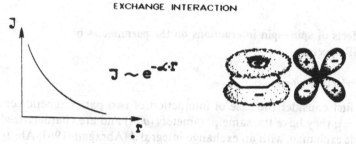

Figure 2.1. Schematic representation of the dependence of dipole–dipole and exchange interactions on the distance between the paramagnetic centers. ΔH_{dd} is the dipole–dipole contribution, J is the exchange integral, and μ is the magnetic dipole momentum.

Dipole–dipole interaction is characterized by relatively weak attenuation as the distance between paramagnetic centers increases. For paramagnetics with the spin value $S = \frac{1}{2}$, $\Delta H_d \approx 10\,mT$ at $r \approx 0.7\,nm$, and $\Delta H_d \approx 0.1\,mT$ at $r \approx 2\,nm$. For a pair of dipoles, the following formula can be used for magnetic interaction:

$$\Delta H_{dd} = \bar{\mu}_1 \bar{\mu}_2 / r_{12}^3 - (\bar{\mu}_1 F_{12})(\bar{\mu}_2 r_{12}) / r_{12}^5, \tag{2.2}$$

where $\bar{\mu}_1$ and $\bar{\mu}_2$ are magnetic moments, and r_{12} is the distance between the spins.

According to quantum mechanics, the exchange interaction between pairs of spins leads to energy splitting, with the formation of two states: (1) a singlet state with antiparallel disposition of spins and (2) a triplet state with parallel disposition. The energy gap between the states is equal to $2J$, where J is an exchange integral.

The Hamiltonian of the exchange interaction H_S between spins with operators $\hat{S}_1$ and $\hat{S}_2$ is described by the formula

$$H_S = -2J\hat{S}_1\hat{S}_2 \tag{2.3}$$

The value of the exchange integral J is quantitatively characterized by the degree of overlap of the electron orbitals with unpaired electrons. The value $J \approx 10^{15}\,\mathrm{s}^{-1}$ at a distance of a chemical-bond length decreases exponentially with increasing distance between the spins. The exchange interaction can be strengthened by conductive (conjugated) bridges. The latter type of interaction is called "indirect exchange."

Because of the differences in the parameters of the spin–spin interactions at distance r, exchange interaction usually prevails over the dipole–dipole interaction at $r \leqslant 0.8$–$1.0\,\mathrm{nm}$. Thus, exchange interaction is more suitable for the study of closely disposed paramagnetics in the active centers. The dipole–dipole interaction is more applicable to systems with remote centers.

2.2. Effects of spin–spin interactions on the parameters of ESR spectra

2.2.1. Principal effects

Let us first consider the case of interaction of two paramagnetic centers with $S = \frac{1}{2}$; they have the same parameters (g, A) and are characterized by isotropic exchange, with an exchange integral J (Abragam 1961; Abragam & Bleaney 1970). If the spins are brought closer together or the temperature (T) is changed, the position and total line intensity of the ESR spectrum will not change down to the T value at which J becomes comparable to $K_B T$ (Figure 2.2). As the temperature continues to drop, the intensity of the spectrum will begin to fall if the spins in the paired (singlet) state have lower energy than in the unpaired (triplet) state. If the anisotropic exchange or dipolar interaction has the value J', the line will split into two, each shifted by $3J'/4$.

When spins approach each other, the interaction (J) is enhanced. For different ions with different resonance energies (Δv), the following sequence of changes in the spectrum must be observed (Figure 2.2): (1) splitting of each level by an amount approximately equal to J (at $\Delta E_v \gg J$); (2) diminishing intensity of the extreme external components, followed by their moving apart from each other, with the internal components increasing in intensity and coming closer to each other $(\Delta E_v \sim J)$; (3) almost complete disappearance of the external pair of components, and merging of the internal components to be observed as a single line $(\Delta E_v \ll J)$.

The dipole–dipole interaction of a given spin and other spins will depend on the distance between them and on the angle θ between the direction of the magnetic field and the line connecting the given pair of spins with the isotropic g-factors, g_1 and g_2:

$$\Delta H_{dd} = g_1 g_2 \beta^2 (1 - 3\cos^2 \theta)/r^3. \tag{2.4}$$

This interaction is revealed as dipole splitting or broadening of the ESR spectrum line. If the local fields are of a random value, the ESR spectrum line will have a Gaussian form.

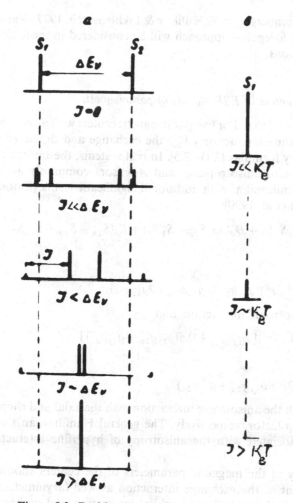

Figure 2.2. Positions and intensities of the ESR lines of two spins, S_1 and S_2, upon their approach as functions of the magnitude of the spin–spin interaction (J), energy splitting between the lines (ΔE_v), and thermal energy $K_B T$: (a) $\Delta E_v \neq 0$; (b) $\Delta E_v = 0$.

Upon intensive exchange at a frequency of $v_{ex} = J/h$, the central part of the line will assume a Lorentzian shape, with a half-width of

$$\Delta v_{\frac{1}{2}} = \langle \Delta v_0^2 \rangle / v_{ex}, \qquad (2.5)$$

where $\langle \Delta v_0^2 \rangle$ is the value of the second moment of the parent Gaussian line.

The determination of the contribution made by the dipole–dipole inter-action to the values of T_{1e} and T_{2e} for the spins observed (the former being more sensitive to weak interactions) is of paramount importance for prac-tical problems of estimating the distance between the paramagnetic centers (Solomon & Blombergen 1956; Kulikov & Likhtenshtein 1977). Various applications of the foregoing approach will be considered in more detail in subsequent sections.

2.2.2. On the parameters of ESR signals of paramagnetics

Spin–spin interaction effects. For two paramagnetic centers with an isotropic g-factor and hyperfine interaction (A_{iso}) the exchange and dipole effects can be described by formulas (2.1)–(2.5). In real systems, the interactions of paramagnetics with anisotropic g- and A-tensors commonly are described by the Hamiltonian, with additional significant terms (Eaton & Eaton 1989; More *et al.* 1990):

$$H_d = -J_d S_1 S_2 + D_d(3S_z S_{2z} - S_1 S_2) + E_d(S_{1x} - S_{2x} - S_{1y}S_{2y}),$$
$$(2.6)$$

where

$$J_d = -(\beta^2/3r^3)(g_{1x}g_{2x} + g_{1y}g_{2y} - 2g_{1z}g_{2z})$$

is related to the isotropic interaction, and

$$D_d = (-\beta^2/3r^3)[2g_{1z}g_{2z} + 0.5(g_{1x}g_{2x} + g_{1y}g_{2y})]$$

and

$$E_d = (\beta^2/3r^3)(g_{1x}g_{2x} + g_{1y}g_{2y})$$

are associated with the anisotropic interaction with the axial and rhombic symmetry of the g-factor, respectively. The general Hamiltonian should include terms associated with the anisotropy of hyperfine interactions (HFI).

The asymmetry of the magnetic parameters of the centers should be taken into account in the exchange interaction as well (asymmetric exchange). Systems with marked spin–orbital interaction and without an inversion center show so-called antisymmetric exchange. In general, anisotropic exchange and dipole interactions can be described by the following Hamiltonian:

$$H_* = S_1 J S_2,$$

where

$$J = \begin{vmatrix} -J & 0 & 0 \\ 0 & -J & 0 \\ 0 & 0 & -J \end{vmatrix} + \begin{vmatrix} 0 & d_z & -d_y \\ -d_z & 0 & d_x \\ d_y & -d_x & 0 \end{vmatrix} + \begin{vmatrix} D_{xx} & D_{xy} & D_{xz} \\ D_{yx} & D_{yy} & D_{yz} \\ D_{zx} & D_{zy} & D_{zz} \end{vmatrix}$$
$$\quad\quad\text{I}\quad\quad\quad\quad\quad\quad\text{II}\quad\quad\quad\quad\quad\quad\text{III}$$

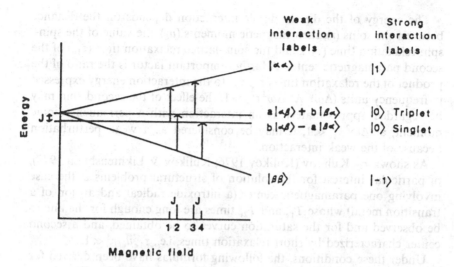

Figure 2.3. Sketch of an energy-level diagram for two interacting $S = \frac{1}{2}$ centers in fluid solution for the case of antiferromagnetic interaction and the exchange integral J comparable to the energy separation due to the g-value differences or nuclear hyperfine splitting (Eaton & Eaton 1989).

Terms I, II, and III reflect isotropic, antisymmetric, and anisotropic interactions, respectively.

If the value of J is less than the energy of the microwave (MW) quantum ($\sim 0.3 \, \mathrm{cm}^{-1} \approx 10^{12} \, \mathrm{s}^{-1}$), the anisotropic exchange does not practically contribute to the ESR spectrum in the X-band.

Figure 2.3 shows a diagram of the energy level for a two-spin system with isotropic interaction J in the magnetic field. As can be seen, the energy of the allowed transition ($\Delta m_s = 1$) depends on the value of the splitting at zero field (J), that of the forbidden transition ($\Delta m_s = 2$) being J-independent. The value of the relative intensity of the allowed (I_1) and forbidden (I_2) transitions is

$$I_1/I_2 = (1630 + 870 \cdot \Delta g)/v_r^2 r^6, \tag{2.7}$$

where r is the distance between the spins, and v_r is the resonance frequency. As follows from equation (2.7), the lower the value of v_r, the higher I_1/I_2 is expected to be, in the S-band for example.

For the spectra of solid powder systems, the position of the low-field edge of the half-field signal is given by

$$H_{\min} = \{(hv_r)^2 - 4[D^2/3 + E^2]\}^{\frac{1}{2}}/2g\beta, \tag{2.8}$$

where D and E are the parameters of anisotropic interaction [equation (2.6)].

The energy of the dipole–dipole interaction depends on the distance between the spins (r), their magnetic moments (μ_e), the value of the spin–spin relaxation time (τ_{2e}), and the spin–lattice relaxation time (τ_{1e}) of the second paramagnetic center. A further important factor is the ratio of the product of the relaxation times $\tau_{1e}\cdot\tau_{2e}$ to the interaction energy expressed in frequency units ($\Delta\omega$). At $\Delta\omega^2\tau_{1e}^2 \gg 1$, the effect of the second spin may be regarded approximately as an interaction with a permanent dipole moment; at $\Delta\omega^2\tau_{1e}^2 \ll 1$, it may be considered as a weak perturbation because of the weak interaction.

As shown by Kulikov (Kulikov 1976; Kulikov & Likhtenshtein 1977), of particular interest for the solution of structural problems is the case involving one paramagnetic center (a nitroxide radical and an ion of a transition metal) whose T_{1e} and T_{2e} times are long enough for the line to be observed and for the saturation curves to be obtained, and a second center characterized by short relaxation times, i.e., $\tau_{1e}\mu_e/r^3 \ll 1$.

Under these conditions, the following formulas have been derived for the dipolar contribution of the spin under study to the relaxation time:

$$
\frac{1}{T_{1e}} = \frac{\mu^2\gamma^2}{6r^6}\left[(1 - 3\cos^2\theta)^2 \frac{\tau_{2e}}{1 + (\omega - \omega_i^2)^2\tau_{2e}^2} \right.
$$
$$
\left. + \frac{9}{2}\sin^2 2\theta \frac{\tau_{1e}}{1 + \omega^2\tau_{1e}^2} + 9\sin^4\theta \frac{\tau_{2e}}{1 + (\omega + \omega_i)^2\tau_{2e}^2} \right], \tag{2.9}
$$

$$
\frac{1}{T_{2e}} = \frac{\mu^2\gamma^2}{6r^6}\left[2(1 - 3\cos^2\theta)^2\tau_{1e} + \frac{9}{2}\sin^2 2\theta \frac{\tau_{2e}}{1 + \omega^2\tau_{2e}^2} \right.
$$
$$
+ \frac{1}{2}(1 - \cos^2\theta)^2 \frac{\tau_{2e}}{(\omega - \omega_i^2)^2\tau_{2e}^2} + \frac{9}{4}\sin^2 2\theta \frac{\tau_{1e}}{1 + \omega^2\tau_{1e}^2}
$$
$$
\left. + \frac{9}{2}\sin^4 2\theta \frac{\tau_{2e}}{1 + (\omega + \omega_i)^2\tau_{2e}^2} \right], \tag{2.10}
$$

where ω and ω_i are the resonance frequencies of the radical and the paramagnetic ion, r is the distance between the spins, μ is the magnetic moment of the ion, and θ is the angle between the direction of the magnetic field and the line connecting interacting spins.

According to these formulas, it is possible to calculate the distance r using the experimentally measured values of $\Delta(1/T_{1e})$ and $\Delta(1/T_{2e})$ and the known values of μ, τ_{1e}, and $\Delta\omega$. Under the most favorable conditions, when $\tau_{1e} \sim \Delta\omega_i^{-1}$, the maximum distances that could be determined to the second paramagnetic center with spin S_2 were found to be as follows: $r_{max} = 10\,\text{nm}$ for $S_2 = \frac{1}{2}$, and $r_{max} = 15\,\text{nm}$ for $S_2 = \frac{5}{2}$.

For the cases in which r is known, e.g., from X-ray data, equations (2.9) and (2.10) may be used to determine the parameters of the second unobserved center (μ_e or $\tau_{1,2e}$).

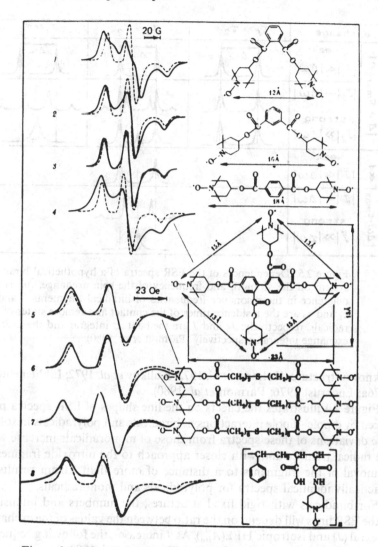

Figure 2.4. Chemical formulas of biradicals (1–3, 5–7) and polyradicals (4, 8) and their respective ESR spectra: chloroform, radical concentration 10^{-4} M, 77K (rigid glassy matrices); solid lines spectra of polyradicals; dashed lines spectra of monoradical R (Likhtenshtein 1976a).

2.2.3. Spin–spin interactions in biradicals and polyradicals and paramagnetic complexes of metals with nitroxide ligands

The known structures of biradicals and polyradicals and the relative simplicity of their ESR spectra make them convenient objects for verifying the theory of ESR spectra and for deciphering the spectra of systems of

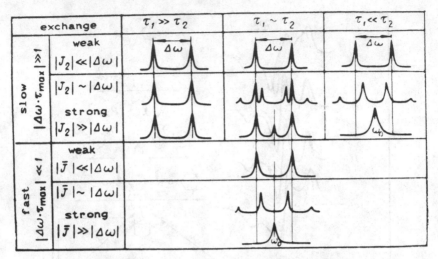

Figure 2.5. Dependences of the ESR spectra of a hypothetical biradical system on the strength and frequency of the spin exchange; $\Delta\omega$ is the difference in the resonance frequencies of biradical fragments 1 and 2; τ_1 and τ_2 are the residence times of the contact and remote states of the radicals, respectively; J_2 and J are the contact integral and the average exchange integral, respectively (Parmon *et al.* 1980).

unknown structure (Kokorin *et al.* 1972; Kulikov *et al.* 1972; Likhtenshtein, 1976a; Luckrust 1976; Parmon *et al.* 1980).

Figure 2.4 illustrates the effects on the line shapes of ESR spectra produced by dipole–dipole interactions in biradicals and polyradicals in solids. The deviations of these spectra from those of monoradicals increase with the radical number and with closer approach to the nitroxide fragments. Removal of the fragments to a distance of more than 2.0 nm results in practically identical spectra for polyradicals and monoradicals.

For biradicals with rigid fixed structures, the numbers and intensities of the ESR lines will depend on the ratio between the values of the exchange integral (J) and isotropic HFI (A_{iso}). As J increases, the following sequence of spectral changes may be observed (Parmon *et al.* 1980):

1. At $J/A_{iso} < 1$, each line is split into five components, with one of them retaining the initial position, and the other four appearing in pairs in the high and low fields.
2. A further increase in J/A_{iso} induces the line pairs to fall apart; the intensity of the inner lines increases, whereas that of the outer lines decreases.
3. At the limit of strong exchange, $J/A_{iso} \gg 1$, five lines remain, with the splitting between the lines equal to $A_{iso}/2$ and an intensity distribution of 1:2:3:2:1.

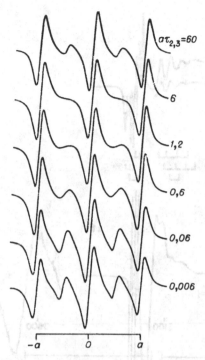

Figure 2.6. Effects of nitroxide fragment mobility on the ESR spectrum of a biradical with these conformations of fraction P: $P_1 = 0.5$; $P_2 = P_3 = 0.25$; $a\tau_1 = 60$; $\tau_2 = \tau_3$; $J_1 = J_2 = 0$; $J_3 = 300a$; $a(T_2)_0 = 9$; a is the isotropic constant; τ_i is the residence time in a given conformation (Parmon *et al.* 1980).

If the mobility of radical fragments allows the nitroxides to collide, the experimental spectra can be analyzed in the framework of this model, which suggests several fixed conformations with different values of lifetime (τ_i) and exchange integral (J_i). Figure 2.5 depicts the transformations in the ESR spectra of biradicals that result from alterations of the values of the lifetimes (τ_1 and τ_2), resonance integrals (J_1 and J_2), and the difference in the resonance frequencies ($\Delta\omega$). According to the physical essence of the model, weak, slow exchange features two initial lines, while fast, strong exchange leads to the merging of the lines into a single one with an intermediate frequency.

For biradicals with two conformations, the effect of exchange frequency enhancement on the line shape is presented in Figure 2.6. The comparison of experimental and theoretical spectra allows calculation of values of J and dynamic parameters for flexible nitroxide fragments. Such an analysis of the ESR spectrum of spin-labeled systems of an unknown structure

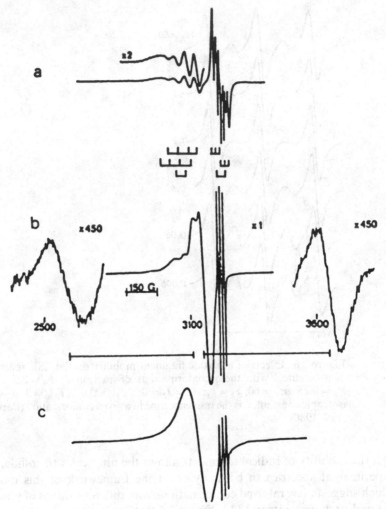

Figure 2.7. X-band ESR spectra of nitroxide pyridine coordinated to Cu(II)bis(hexafluoroacetylacetonate) in fluid solution at room temperature: (a) the value of J is comparable to that of the copper nuclear hyperfine splitting; (b) the J value is about 10 times greater than the separation between the copper and nitroxide energy levels ($\Delta\omega$); (c) $J \gg \Delta\omega$, and an averaged signal is observed (Eaton & Eaton 1989).

permits the study of mutual dispositions of the modified groups, microrelief, and microviscosity in the region of the spin-label location.

Figure 2.7 illustrates the effects on the ESR spectrum of the exchange interaction in a complex containing Cu(II) and nitroxide fragments. As the exchange interaction increases, the lines merge.

In general, the effects of spin–spin interactions on the ESR spectra of transition metals with paramagnetic ligands will depend on the values of the electron spin relaxation time, the number of spins, the spin-orbital constant, the distances and mutual orientations of the metal atoms and nitroxide fragments, and the character of the connecting bridge (Table 2.1).

2.3. Determination of the distance between spins

As noted in Section 2.2, the ESR spectra of paramagnetic centers exhibit a subtle response to the approach of other paramagnetics under certain conditions. The values of the exchange integral J, which are quantitative measures of the exchange interaction, are strongly dependent on the distance between the centers (r) and their mutual orientation. The theory (Marcus & Sutin 1985) predicts an exponential dependence of J on r. If the distance is fixed, the value of J for the parallel orientation ($J_{\parallel}$) is on the same order as the value of J for the perpendicular orientation ($J_{\perp}$), with $J_{\parallel}$ being positive and $J_{\perp}$ negative. A rough approximation of such a calculation shows that the dependence of J on r is insufficient to be used for structural purposes. This conclusion appears to be particularly valid in the case of indirect exchange.

Nevertheless, the experimental data available permit us to plot appropriate calibration curves. As can be seen from Figure 2.8, for the dependence of the quantitative exchange parameters, the exchange integrals (J), and the rate constants (k_{ex}) of triplet–triplet transitions and triplet–singlet conversion (intersystem crossing, see Section 2.4) on the distance r between the centers, the experimental data lie on two approximately exponential curves described by the following equation:

$$J, k_{ex} \simeq 10^{16} \exp(-2r/a)\,\mathrm{s}^{-1}. \tag{2.11}$$

For systems in which the exchanging centers are separated by a non-conducting medium (molecules or groups with saturated chemical bonds), $a = 0.078\,\mathrm{nm}^{-1}$. For systems in which the centers are linked by conjugated bonds, $a = 0.691\,\mathrm{nm}^{-1}$. In nonconducting systems, the exchange intensity is dampened by about a factor of 10 when the centers are at a distance of 0.1 nm from each other. In conducting systems, the effect can be reduced by a factor of 10 when the centers are separated by a distance of 0.7 nm.

The empirical equation (2.11) and the data of Table 2.1 offer interesting possibilities for solution of a number of theoretical and structural problems. They can be used to determine the distance between the centers, the depth of immersion of paramagnetic centers, and transmission coefficients; they can also be used to study the ability of protein macromolecules to conduct spin density, etc.

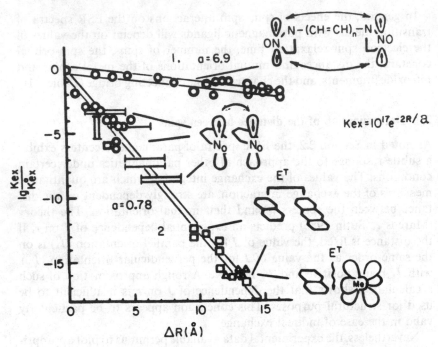

Figure 2.8. Dependences of logarithms of relative parameters of the exchange interaction on the distance between the interacting centers (R); $k_{ex} = J, k_{TT}$; J is the exchange integral of the interaction between nitroxide radicals; k_{TT} is the rate constant for the triplet–triplet energy transfer of the interconversion (denoted as E^T); 1 and 2 are conductive and non-conducting media (Likhtenshtein *et al.* 1986a,b; Likhtenshtein 1988a,b).

Analysis of ESR spectra is most conveniently performed under conditions in which the molecular motion of spins is either practically stopped (frozen solutions) or slowed down to such an extent that it no longer affects the ESR spectrum (when the centers are rigidly bound to very large macromolecules or are in highly viscous media).

The following methods have been used to calculate the distances between paramagnetic centers from the magnitude of the dipole–dipole interaction: (1) empirical calibration; (2) the method of second moments; (3) measurement of the dipole–dipole splitting; (4) analysis of the broadening of single spectral components; (5) analysis of relative intensities of forbidden and allowed transitions; (6) comparison of experimental and calculated spectra.

The method of empirical calibration is based on comparison of the shape and parameters of the ESR spectrum of the object under study with the spectrum of a sample with a known arrangement of spins. The data

Table 2.1. *Attenuation of the exchange interaction between nitroxide radical R˙ and a paramagnetic P (nitroxide, transition metal ion) in structure R-X-P*

Group	γ^a	Group	γ
—CH₂—	6–8	$\overset{\displaystyle O}{\underset{\parallel}{-C-}}$	6.8
—CH₂— in cycle	>5	—S—	3.3
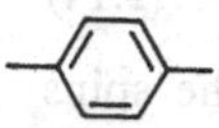 (phenyl ring)	6	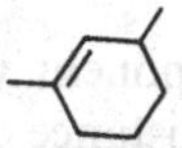 (cyclohexene ring)	>25
—O—	2.5	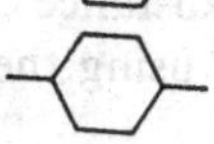 (methylcyclohexene ring)	>100
—NH—	3		
—C≡C—	1.0	—SO—	3.2
H H ⟍ ⟋ —C=C—	1.7	—SO₂	3
⟍ —Si— ⟋	2.5	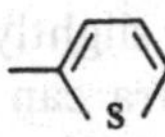 (thiophene, S)	3.3
—N=N—	1.1	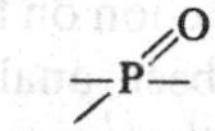 ⟍ —P=O ⟋	3.5
—CO—NH—	13	H····O ⟍ —C⤳	140
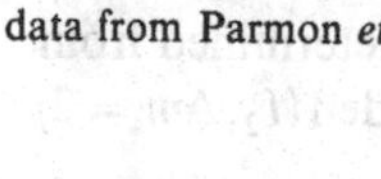 ⤳C⟋ ··O=C—N—H·· ⤳C—	36	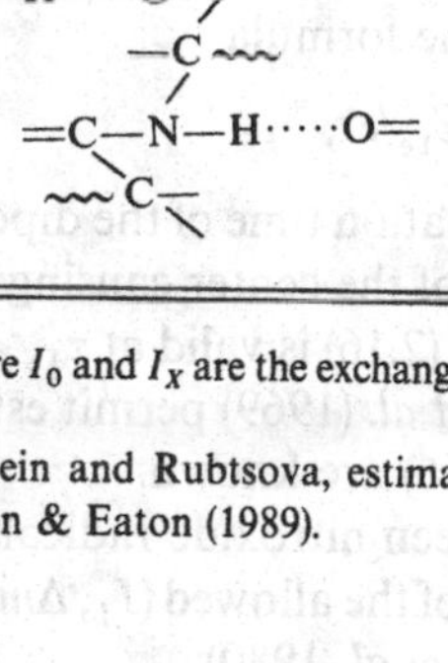 =C—N—H····O= ⤳C⟍	5·10³

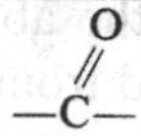

^{a}The attenuation parameter $\gamma = I_X/I_0$, where I_0 and I_X are the exchange integrals for —C≡C— and a given group X.

Source: Unpublished data of Likhtenshtein and Rubtsova, estimated from an analysis of data from Parmon *et al.* (1980) and Eaton & Eaton (1989).

on solid solutions of radicals and complexes or chemical compounds of known structure, such as polyradicals, may be used as standard data.

In the method of second moments, use is made of the well-known van Vleck equation, which relates the dipole–dipole component of the second moment in the line (ΔM_2) to the parameters of the magnetic moments of several spins (N) and the distance (r_{ik}) between them:

$$\Delta M_2 = a_2 \gamma_e^2 h^2 S(S+1) N^{-1} \Sigma r_{ik}^{-\frac{1}{6}}, \tag{2.12}$$

where γ_e is the gyromagnetic ratio, S is the spin of the interacting particles, a_2 is a coefficient, and N is the number of spins. For a pair of radicals the following formula can be used:

$$r = 1.08 \Delta M_2^{-\frac{1}{6}}. \tag{2.13}$$

(Here and later, all line-width parameters are expressed in millitesla, and r is in nanometers.)

In magnetically dilute polycrystalline samples (biradicals on macro-molecules, the active sites of enzymes, radical pairs obtained as a result of radiolysis, etc.) the dipole–dipole interaction may lead to the splitting of the line by an amount D_d. In the absence of exchange, the distance between the pairs can be calculated from the formula

$$r = 1.41 D_d^{-\frac{1}{3}}. \tag{2.14}$$

If the spectrum shows a component whose width varies as the spins are brought closer together, the distance can be estimated from the magnitude of the broadening ΔH_d by using the following formula:

$$r = 1.63 \Delta H_d^{-\frac{1}{3}}. \tag{2.15}$$

In principle, this formula is valid only for the cubic lattice. The numerical coefficient, however, depends only slightly on the arrangement of the spins. In a number of cases, ESR spectra can be calculated theoretically. The effect of the dipole–dipole interaction on the line shape under the influence of a paramagnetic group has been analyzed by Taylor *et al.* (1969). In their study, the contribution of the interaction to the line width was taken into account using the formula

$$\Delta H_d = g\beta\mu^2\tau_{1e}/r^6, \tag{2.16}$$

where τ_{1e} is the correlation time of the dipolar interaction, e.g., the electron spin relaxation time of the center causing the "broadening" [at a distance of $r = 1$ nm, equation (2.16) is valid at $\tau_{1e} < 10^{-9}$ s]. The theoretical curves obtained by Taylor *et al.* (1969) permit estimation of r if the other parameters in equation (2.16) are known.

The distance between nitroxide radicals (NRs) can be determined from the relative intensity of the allowed (I_1, $\Delta m_s = 1$) and forbidden (I_2, $\Delta m_s = 2$) transitions (Parmon *et al.* 1980):

$$r = 0.18(I_1/I_2)^{\frac{1}{6}}. \tag{2.17}$$

It should be emphasized that the ratio I_1/I_2 depends on the dipole–dipole interaction only and is not affected by the exchange interaction (Eaton & Eaton 1989) (see Figure 2.3). Equation (2.17) makes it possible to determine r; the latter can be used for calculation of the dipole–dipole contribution to the ESR spectral line shape in the region of the allowed transition. Comparison of the calculated and experimental spectra then permits determination of the value of the exchange integral J. It is noteworthy that the forbidden transitions are manifested at a distance of $r \leqslant 1.0$ nm, e.g., under conditions of noticeable contribution from the exchange interaction. Observation of the forbidden transitions can thus be useful in cases of close disposition of the centers.

For a rigid solution of radicals of axial symmetry, $r = 1.42 D_{as}^{-\frac{1}{3}}$, where D_{as} is the parameter of the asymmetric interaction.

At distances ranging from 1.3 to 3.0 nm, it is convenient to use the Kokorin–Zamaraev parameter d_1/d (Figure 2.16), which is connected with r by the following empirical relation:

$$r = 0.93 + 0.077/(d_1/d + 0.036 A_{zz}^0 - 1.76).$$

The value of A_{zz}^0 can be determined from the same ESR spectrum with rather high accuracy (0.1–0.2 nm) because of the strong dependence of the line shape on the parameters measured. However, the validity of the results may be significantly lowered by the uncertainty in the orientation of nitroxide fragments and the possibility of the existence of several conformations of the labels. The efficiency of the method can be increased by combined analysis of experimental and calculated spectra, taking into account the possible variants of interaction.

As shown by Kulikov (Kulikov 1976; Kulikov & Likhtenshtein 1977), interesting possibilities are suggested by investigation of the paramagnetic contribution to the value of the spin–lattice relaxation rate $1/T_{1e}$ (Chapter 1).

These possibilities are especially strongly manifested in the case of NRs, for which the values of $1/T_{1e}^0$ and $1/T_{2e}^0$ differ significantly in the absence of foreign paramagnetic groups, with the value of $1/T_{1e}^0$ being much lower than of $1/T_{2e}^0$. For example, they are equal to $2.7 \cdot 10^3 \, \text{s}^{-1}$ and $\approx 10^7 \, \text{s}^{-1}$, respectively, at 77K in a water–glycerin mixture (1:1). From the approximately equal contributions of the dipole–dipole interactions to $1/T_{1e}^0$ and $1/T_{2e}^0$ at small τ_{1e} [equations (2.9) and (2.10)] it follows that the value of $1/T_{1e}^0$ is much more sensitive to the interaction than is $1/T_{2e}^0$, which is associated with the line shape:

$$\Delta(1/T_{1e}) = \frac{\mu^2 \gamma_e^2}{6r^6 \tau_{1e}} = \left[\frac{4}{5(\omega_1 - \omega_2)^2} + \frac{24}{5(\omega_1 + \omega_2)^2} + \frac{12}{5\omega_1^2} \right], \quad (2.18)$$

$$\Delta(1/T_{2e}) = \frac{4}{15} \frac{\mu^2 \gamma_e^2}{6r^6} \tau_{1e}, \quad (2.19)$$

where ω_1 and ω_2 are the resonance frequencies of the interacting spins, τ_{1e} is the relaxation time of the second paramagnetic group, μ is the magnetic moment of the paramagnetic group, and γ_e is the gyromagnetic ratio for the electron.

The formulas (2.18) and (2.19) were derived on the assumption that $(\omega_1 - \omega_2)^2 \tau_{1e}$ is much smaller than unity; they have been averaged over the angle between the line connecting the centers and the direction of the magnetic field. Application of this method requires knowledge of the value of the electron spin relaxation time τ_{1e} of the second paramagnetic group. However, some approaches have been suggested that allow one to avoid this condition. Thus, for example, the τ_{1e} value may be omitted from the product of $1/T_{1e}$ and $1/T_{2e}$, a circumstance that can be used for calculations. An ingenious method of finding τ_{1e} is based on the experimental dependence of $1/T_{1e}$ on τ_{1e}. Since for ions in solids the τ_{1e} value decreases monotonically with the temperature, one can find experimentally the temperature at which $\tau_{1e} \sim \Delta\omega^{-1}$, where $\Delta\omega$ is the difference of the resonance frequencies of the spectrum. It can be easily shown that under these conditions the maximum dipolar contribution is equal to

$$[\Delta(1/T_1)]_{max} \approx \mu^2\gamma_e^2/15r^6\Delta\omega. \tag{2.20}$$

These methods have been tested in a system where the location of the spin label is known, i.e., hemoglobin samples with the β-93 SH group modified by various radicals. The values of r thus obtained agree well with the X-ray diffraction analysis.

2.4. The spin label–spin probe method

2.4.1. General

In this section we shall analyze new possibilities offered by the spin probes, which are chemically inert paramagnetic species capable of diffusing freely in solution. The spin probe technique has been successfully employed for studying the exchange relaxation and the magnetic and other properties of paramagnetic complexes in solution (Zamaraev *et al.* 1981). It has been suggested (Likhtenshtein *et al.* 1970; Hyde *et al.* 1979; Likhtenshtein 1976a) that this method can be used for probing paramagnetic active centers of enzymes and also for investigating microstructures of proteins in the vicinity of the added spin labels. More recently, versions of the method have been developed to allow determination to the depth of immersion of the paramagnetic center into the biological matrix and to facilitate measurements of dioxygen concentration.

NRs attached to definite parts of proteins and membranes are commonly used as spin labels, whereas nitroxides, aqueous and other complexes of transition metals, and molecules of oxygen are used as spin probes. Two versions of the spin label–spin probe method usually are employed in laboratory practice. One of them is based on analysis of the effect of a probe on the line shape of the spin label. The other approach involves determination of the contribution of spin interaction to the spin relaxation rates of the spin labels.

When a stable nitroxide radical in solution encounters another paramagnetic species that is within the concentration range $C = 6 \cdot 10^{-3}$ to 10^{-1} M, the magnitude of broadening of the Lorentzian line of the NR is related to the rate constant of the exchange relaxation (k_{ex}) by the following simple equation:

$$1/T_{2e} \sim \Delta H_L \simeq 6.5 \cdot 10^{-9} k_{ex} \cdot C, \tag{2.21}$$

where ΔH_L is expressed in mT, C in M, and k_{ex} in $M^{-1} s^{-1}$. This equation allows one to determine the value of k_{ex}. In the case of inhomogeneous broadening of the initial line, adequate correction should be made for the value of the experimentally observed broadening (Marsh 1989). In some cases the formula can be formally applied to the inhomogeneously broadened line. For example, for a nitroxide radical at the rigid limit, the theoretically calculated width of the central component of the ESR spectrum is proportional to ΔH_L (Parmon *et al.* 1980). However, in general, inhomogeneous effects must be taken into account. Thus, for a NR in solution, the first effect of the exchange interaction with a paramagnetic is narrowing of the radical line because of averaging of the dipole contribution from protons. In liquids with dissolved dioxygen, the exchange narrowing is usually complete.

A further increase in the radical concentration leads to shifting of the components of the ESR spectrum to the center and then their merging into a single line (Figure 2.9). Comparison of the theoretical and experimental spectra makes it possible to determine the diffusion coefficient for model and biological systems, e.g., biomembranes (McConnell & McFarland 1970; Marsh 1989).

A method has recently been worked out for determination of k_{ex} from the extent of the change in the rate of the spin–lattice relaxation of the radical in the presence of a paramagnetic species:

$$\Delta(1/T_{1e}) \approx k_{ex} \cdot C. \tag{2.22}$$

Since the $1/T_{1e}$ parameter is more sensitive than $1/T_{2e}$, in particular in the region of slow rotation, this method allows one to widen the range of the k_{ex} values being determined. (Methods for determining $1/T_{1e}$ may be found in Sections 1.4.1 and 1.4.2.) It is because of the sensitive response of $1/T_{1e}$ in the slow rotation region that the effect of saturation transfer from one type of radical to another depends on the frequency of collisions between the radicals. Another method of particular interest is an elegant approach based on ENDOR using two types of nitroxides, ^{15}N and ^{14}N, characterized by nonoverlapping positions of the ESR lines (Hyde & Feix 1989).

In principle, both dipole–dipole and exchange interactions can occur in solutions. The authors who have studied the behavior of paramagnetic particles in detail (Likhtenshtein *et al.* 1974; Zamaraev *et al.* 1981) have come to the conclusion that in a medium of relatively low viscosity, the major contribution to the ESR line width is made by exchange interactions during encounters. The probability of relaxation during the encounter of

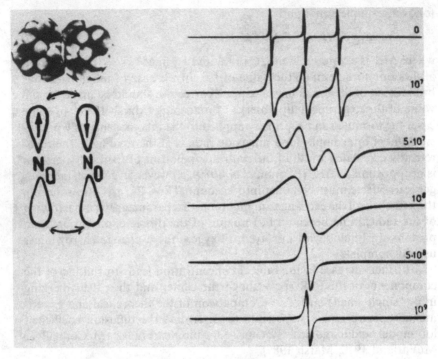

Figure 2.9. Schematic representation of the effects of dynamic spin exchange on collisions of two nitroxide fragments in the ESR spectrum, with the different frequencies of collisions (in s^{-1}) as noted in the figure (Marsh 1989).

paramagnetic particles is determined, in particular, by the magnitude of the exchange integral J. According to the available data (Parmon *et al.* 1980), J depends on the densities of unpaired electrons at the sites of direct contact between the colliding particles. The value of J is of the order of $10^{11} s^{-1}$, i.e., it is close to the inverse of the lifetime of the colliding complexes in nonviscous solutions (τ_{col}). The probability of relaxation (P_{ex}) upon biomolecular collisions is given by the formula

$$P_{ex} = f_{ns} J^2 \tau_{col}^2 / (1 + J^2 \tau_{col}^2), \tag{2.23}$$

where f_{ns} is the nuclear statistical factor, equal to $\frac{2}{3}$ for NRs. P_{ex} is defined by the following equation:

$$k_{ex} = P_{ex} K_d, \tag{2.24}$$

where K_d is the number of encounters in solution, which can be estimated by using well-known formulas for diffusion processes. By this definition, P_{ex} incorporates a factor (unknown in the general case) that characterizes the geometric steric factor (f_g).

Table 2.2. *Probabilities of spin relaxation of radicals (P_1) and complexes (P_2) during encounters for different ratios among the exchange integral (J), the spin–lattice relaxation time (T_{1e}), and the duration of the encounters (τ_c)*

Relationship between J and δ	$\delta\tau_c$	$J\tau_c$	JT_{1e}	$J^2 \cdot T_{1e} \cdot \tau_c$	P_1	P_2
				$\tau_c \ll T_{1e}$	$\dfrac{J^2\tau_c^2\Sigma m^2}{2S_2+1}$	$\tfrac{1}{2}J^2\tau_c^2$
For $J \ll \delta$	$\delta\tau_c \lesssim 1$	$J\tau_c \ll 1$		—	1 (even S_2)	$\tfrac{1}{2}J^2\tau_c^2$
	$\delta\tau_c \gg 1$	$J\tau_c \gg 1$		—	$\dfrac{2S_2}{2S_2+1}$ (odd S_2)	
For $J \gg \delta$	$\delta\tau_c \gg 1$	$J\tau_c \gg 1$	$JT_{1e} \ll 1$	$\tau_c \gg T_{1e}$	$\dfrac{1}{\tfrac{2}{3}S_2(S_2+1)+1}J^2\tau_c^2$	$\dfrac{1}{2(S_2\tau_c^{\frac{1}{2}})^2}$
	$\delta\tau_c \ll 1$	$J\tau_c \ll 1$	$JT_{1e} \ll 1$	$J^2T_{1e}\tau_c \ll 1$	$\dfrac{2S_2(S_2+1)}{3}\dfrac{1}{(S_2+\tfrac{1}{2})^2}$	
		$J\tau_c \gg 1$	$JT_{1e} \gg 1$	$J^2T_{1e}\tau_c \gg 1$	1	1
				$J^2T_{1e}\tau_c \gg 1$	1	1

Source: Salikhov et al. (1971).

A more general theory has been developed for exchange relaxation during encounters between paramagnetic particles in solution (Zamaraev *et al.* 1981). This theory takes into account the influence of such factors as the paramagnetic relaxation time of electron spins (T_1), the values of the magnetic spin numbers of paramagnetic particles $(S_1$ and $S_2)$, and the difference between the resonance frequencies of the spins (δ) (Table 2.2).

According to the theory, the relaxation probability (P_{ex}) during an encounter is mainly determined by the relationship among the time T_1, τ_{col}, and the exchange integral J. Experimental criteria have been found for strong $(P_{ex} \sim 1)$ and weak $(P_{ex} \ll 1)$ exchanges. For example, in the case of strong exchange, the temperature dependence of k_{ex} must be determined by the temperature dependence of the number of encounters, i.e., by the dependence of K_d on the viscosity of the medium. In the case of weak exchange, the value of k_{ex} must be weakly temperature-dependent, because the dependences of K_d and τ_{col} on the viscosity are opposed.

In cases of low values of the geometric steric factor $(f_g \ll 1)$, e.g., for a spin-labeled macromolecule or paramagnetic active center, k_{ex} can be small despite the high value of the total frequency of collisions. Such a mechanism may reasonably be called a pseudostrong exchange.

2.4.2. Selection of spin probes

The main criteria in selecting spin probes may be stated in the following way. Depending on the objective, the probe must be readily soluble in aqueous or hydrophobic phase or capable of being absorbed on the surface of the matrices under study. The probes must be chemically inert toward functional groups of the biological systems and NRs so as not to produce a concomitant ESR signal. Highly efficient exchange relaxation during the encounters is an important property of the probe.

Values of the exchange rate constant are listed in Table 2.3 for interactions in water and in various other solvents between a NR and several paramagnetic complexes that are potential spin probes. According to these data, the value of k_{ex} strongly depends on the nature of the central atom and the ligand environment.

The most suitable probes for the study of biological matrices in aqueous solution are potassium ferrocyanide (PFC) and dibenzene chromium iodide (DBC). They have a high relaxation probability $(P_{ex} \sim 0.2)$, form very stable, chemically inert ligand shells, and have short electron relaxation times $(T_{1e} \ll \tau_c)$. As shown in one experiment (Likhtenshtein 1976a), the value of k_{ex} is inversely proportional to the viscosity of a water–glycerol mixture, indicating that the value of P_{ex} is τ_c-independent.

It can also be seen from Table 2.3 that the k_{ex} values of probes soluble in nonpolar media are close to the frequency of collisions after the geo-

Table 2.3. *Values of the rate constants for spin exchange,* k_{ex} ($M^{-1}s^{-1}$), *between the NRs of the piperidine row and some complexes of paramagnetic metals at 20°C*

Complex[a]	Solvent	$k_{ex} \cdot 10^{-8}$
$CuCl_2$	H_2O	15
$NiCl_2$	H_2O	12.7
$Co(SO_4)_2$	H_2O	9.7
$MnCl_2$	H_2O	20.0
$FeCl_3$	$0.05N\ H_2SO_4$	20.5
$K_3Fe(CN)_6$	H_2O	12.2
$VOSO_4$	H_2O	8.8
$Cu(SCN)_2(v\text{-}mPy)_4$	Py	28.7
$Ni(SCN)_2(v\text{-}mPy)_4$	Py	21.0
$Co(SCN)_2(v\text{-}mPy)_4$	Py	28.7
$Mn(SCN)_2(v\text{-}mPy)_4$	Py	38.5
$Cu(AA)_2$	Cl	22
$Mn(AA)_2$	Cl	30
$Fe(AA)_2$	Cl	32
$Cr(AA)_2$	Cl	19
$Cu^{2+}[(CH_3)_2NCH_2CH_2N(CH_3)_2]_2$	H_2O	35.0
Hemin	Py	24.0
Etioporphyrin-Fe	Py	25.7
Etioporphyrin-Cu	Py	15.3
Mono-CET-TMP-Fe	Py	32.8

[a] Py, pyridine; Cl, chloroform; AA, acetylacetonate; CET-TMP, carboethoxytetramethyl porphyrin.

metric steric factor of the NR ($f_g \approx 0.07$) is taken into account (Berdnikov & Makarshin 1982). Examples of such spin probes include acetylacetonate complexes of transition metals and hemin and its derivatives.

2.4.3. Investigation of steric, electrostatic, and exchange effects

For a particular pair of paramagnetic species the value of k_{ex} depends on the electronic structure, microviscosity, steric hindrances, and distribution of electrostatic charges in the region of encounters (Figure 2.10). Hence, after preliminary empirical calibration, the method of examining k_{ex} and its variation can be used for experimental study of these factors in biological and other objects.

The diffusion coefficient (D_{tr}) is inversely proportional to the radius (R) of colliding particles (Likhtenshtein 1976a). Therefore, as a first approximation, it is expedient to use the value $\alpha' = k_w/2k_p$ ("microviscosity") as an accessibility parameter that characterizes efficiently the increase in viscosity and steric hindrance in the region of the encounter between the

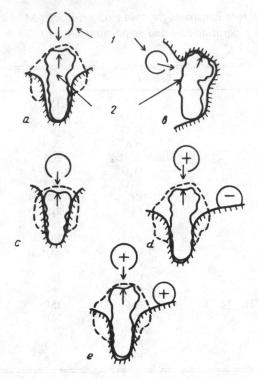

Figure 2.10. The spin probe (1)–spin label (2) approach: effects of micro-relief (a–c) and charge distribution (d, e) on the nature of the label–probe interaction (Likhtenshtein 1976a).

probe and the radical on the protein (k_w and k_p are the k_{ex} constants for the radical in water and the spin label on the protein, respectively).

Experiments have been carried out using labels and probes with various electrostatic charges; the results are presented in Figure 2.11. The sign of the electrostatic charge on the substituent and its distance from the nitroxide group have significant effects on the values of the k_w and k_p constants. The experimental data (Figure 2.12) have been shown to be described well by the Debye electrostatic equation with $\varepsilon_0 = 80$ and the relevant theory regarding the effect of ionic strength. Therefore, these equations can be used for approximate estimation of the distance between the nitroxide group of the spin label on the protein and the nearest charged group, provided that this distance does not exceed 1.0–1.2 nm.

New possibilities have been reported for studying the effect of exchange on the rate of spin–lattice relaxation of spin labels (Kulikov & Likhtenshtein 1977). This parameter is markedly more sensitive to much weaker exchange interactions than are the parameters of the line shape. Such an approach

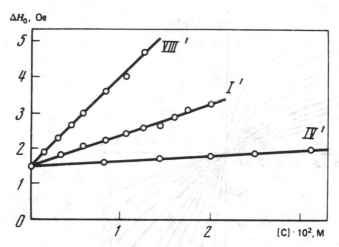

Figure 2.11. Experimental broadening (ΔH_0) of the ESR spectra of charged radicals in solution as a function of the ferricyanide concentration: I', IV', and $VIII'$ are piperidine nitroxides with $=C=O$, $=CH_2COO^-$, and $-NH_3^-$ groups in the fourth position of the ring, respectively (Likhtenshtein 1976a).

allowed the first measurements of the collisions between paramagnetic macromolecules in solution (Yudanova & Kulikov 1984).

In the first version of the approach, the rate constant of the exchange interaction at the collision of two nitroxide spin labels attached to macromolecules is determined using the formula (Hyde *et al.* 1968; Thomas *et al.* 1976)

$$1/T_{1e} = (b + 3b' + 1)(3b + 3b' + 1)/T_{1e}^0[(1 + 3b)(1 + b') + (b + 4b + 2)],$$

$$(2.25)$$

where $b' = T_{1e}^0 k_{ex}[R^\cdot]$ and $b = 1.3 \cdot 10^{10}\tau_c/(1 + \omega_N^2\tau_R^2)$. ($[R^\cdot]$ is the radical concentration, τ_R is the rotation correlation time of the radical, and ω_N is the frequency of ralaxation of the radical nucleus, ^{14}N or ^{15}N in this case.) If $[R^\cdot] \to \infty$, $1/T_{1e} \to 3T_{1e}^0$. The physical meaning of such a dependence is that the spin exchange at collisions between two radicals leads to saturation transfer between the ESR spectral components, just as nuclear relaxation does. Formula (2.25) presupposes the dependence of $1/T_{1e}$ on $[R^\cdot]$ to have an inflection point. At fast rotation ($\tau_R^{-1} \gg \omega_N$), in particular, the $k_{ex} = ([R^\cdot] \cdot T_{1e}^0)^{-1}$ value is just at that point.

The second version of the spin–lattice relaxation approach requires that one kind of micromolecules (e.g., lysozyme) be modified by a nitroxide spin label, whereas the other one (e.g., heme protein) must have a paramagnetic ion center. In the latter case, the value of k_{ex} can be determined from a simple equation similar to equation (2.24).

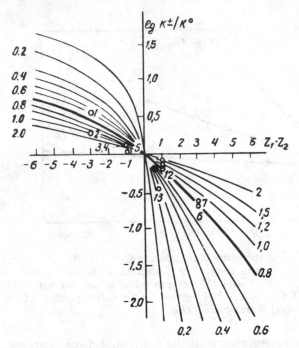

Figure 2.12. Variations in the relative relaxation rate constants (k^+/k_0) for nitroxide radicals in solution as a function of the charges $(z_1 \cdot z_2)$ of the colliding particles, being calculated from the Debye equation for the distances between the charges, r, equal to the values denoted in the figure; 1–13 are experimental data (Likhtenshtein 1976a).

2.4.4. Determination of the immersion depth of a radical center

The phenomenon of spin relaxation of NRs and other centers, under the influence of other paramagnetic species, can be used to establish the depth of immersion of the centers into biological matrices, e.g., proteins and membranes. For this purpose, a quantitative study should be undertaken to evaluate the effects that the dipole–dipole interactions between the centers and the paramagnetic ions randomly distributed in the bulk of the vitrified sample have on the relaxation parameters of the centers, e.g., on the value of T_{1e}, which is determined by the continuous-wave saturation method (Kulikov *et al.* 1981; Likhtenshtein *et al.* 1986a).

A method has been developed for determining the nearest distance (r_{min}) between a stable radical [R˙] and an ion-relaxator (IR). It is obvious that in the case of R˙ penetration into an impermeable matrix, the r_{min} value is equal to the depth of immersion r_{im}. If the center resides at a sufficient depth, $r_{min} > \bar{r}$, where the latter value is the average distance between the stable radical R˙ and the IR. As has been shown, in this case the contribution of the dipolar interaction to the rate of the R$^{•}$ relaxation is expressed by the following formula:

$$[\Delta(1/T_{1e})]_d = A_d\mu^2\gamma^2\tau_{1e}C/r_{min}^3,\qquad(2.26)$$

where C is the ion concentration in the solid solution, and A_d is a factor that depends on the geometry of the surface and impedes the approach of the radical and the ion. For example, if the surface is flat, $A_d = 0.7$. If R$^{•}$ is at a distance r_{im} from the surface of the spheric sample of radius R,

$$A_d = 16\pi/9(2 - r_{min}/R)^3.\qquad(2.27)$$

Figure 2.13 shows the results obtained by testing the method in the case of NR immersion at different depths in the membranes of chromophores. The linear dependences of $[\Delta(1/T_{1e})]_d$ on C are seen (Figure 2.13) to correspond to equation (2.26). The values of r_{im} for probes SLXX ($m = 3$, $n = 12$) and SLXXI ($m = 3, n = 12$) are equal to 2.2 and 2 nm, respectively, which agree well with the values that might be expected from the structures of these compounds.

An ion-relaxator can be included into superficial portions of the membranes of interest (e.g., Co^{2+} acetylacetonate) or into the surface of the membranes (e.g., Co^{2+}). Such an approach allows precise determination of the location of the radical in the biological matrix (Figure 2.14).

2.4.5. NRs in oxymetry

Dioxygen is a paramagnetic molecule, and therefore it can affect the spin–lattice ($1/T_{1e}$) and transverse ($1/T_{2e}$) relaxation rates of NRs. This property has been used for determination of the $k_{ex}\cdot[O_2]$ product, which is an important characteristic of the oxidative capacity of molecular oxygen at the location of the spin label (Subczynski & Hyde 1981; Yudanova & Kulikov 1984; Likhtenshtein *et al.* 1986a; Hyde *et al.* 1990; Halpern & Peric 1990; Likhtenshtein 1990a). Two versions of oxymetry have been developed.

One of them is based on analysis of the line shape of the NR, which is sensitive to the presence of dioxygen when the value of $k_{ex}\cdot[O_2] \geqslant 10^7\,s^{-1}$ (Backer *et al.* 1977). Paramagnetic O_2 blurs the superfine structure of inhomogeneously broadened ESR lines and increases the Lorentzian line width (ΔH_L) of the NR in proportion to the concentration of the radical. In aqueous solutions of nitroxide the ΔH_L value is directly proportional to the dioxygen concentration starting from $\Delta H_L \approx 0.15\,mT$.

Methods of oxymetry based on the effects of dioxygen on the spin–lattice relaxation parameters of NRs allow determination of the $k_{ex}\cdot[O_2]$ product as approximately equal to $10^5\,s^{-1}$ (Subczynski & Hyde 1981; Yudanova & Kulikov 1984). All three of the commonly used techniques (continuous

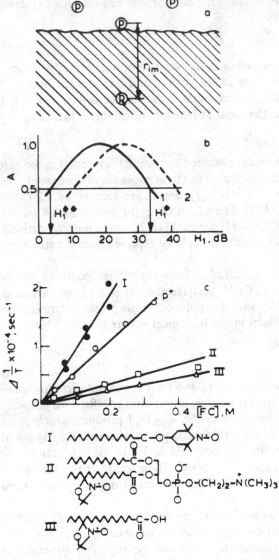

Figure 2.13. Determination of the depth of immersion, r_{im}, of radical R˙ in biological membranes in chromatophores from *Rhodospirillum rubrum*: (a) Diagram illustrating the principle of the method. FC is ferricyanide; (b) saturation curves of the ESR spectrum of a spin probe in chromatophores, (1) [FC] = 0.45 M, (2) [FC] = 0; water:glycerol = 1:1; 77K; (c) dependence of the dipole contribution $\Delta(1/T)$ on the concentration of ferricyanide. Below are the formulas of spin probes; P⁺ is the bacteriochlorophyll cation of the reaction center (Likhtenshtein *et al.* 1986a; Likhtenshtein 1988a).

(a)

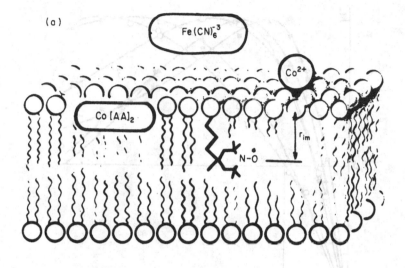

(b)

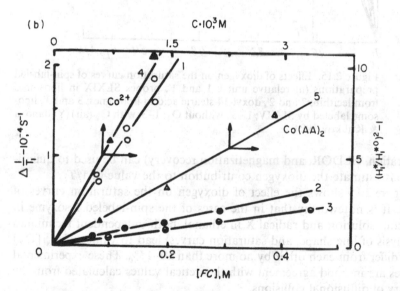

Figure 2.14. Determination of the locations of nitroxide radicals in membranes: (a) Diagram illustrating the principle of the method. (b) Dependences of the relaxation parameters of nitroxides on the concentration of the ion-relaxators ferricyanide (FC) (1–3) and Co^{2+}-acetylacetone (5); 1–3 relate to the membrane of chromatophores from *R. rubrum* (probes I–III, Figure 2.13); 4 and 5 relate to the membrane of *M. capsulatus* (4 is 5-doxylundecanoic acid, 5 is probe III) (Cherepanova *et al.* 1990).

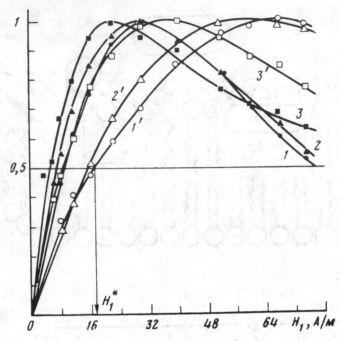

Figure 2.15. Effects of dioxygen on the saturation curves of spin-labeled preparations (in relative units): 1 and 1′, probes SLXIX in liposomes from lecithin; 2 and 2′, doxyl-14 stearic acid in liposomes; 3 and 3′, liposome labeled by SLXIV; 1′–3′, without O_2; 1–3, with O_2 (air) (Yudanova & Kulikov 1984).

saturation, ELDOR, and magnetization recovery) can be used to quantitatively estimate the dioxygen contribution to the value of $1/T_{1e}$.

Figure 2.15 shows the effect of dioxygen on the saturation curves of NRs. It is noteworthy that in the cases of the spin-labeled lysozyme in aqueous solution and radical X in ethanol, both independent techniques (analysis of line shape, and saturation curves) lead to values of $k_{ex} \cdot [O_2]$ that differ from each other by no more than 10–15%. These experimental values are in good agreement with theoretical values calculated from the theory of diffusional collisions.

Considering that the reactions of O_2 evolution and consumption are processes of paramount importance in living nature, it is difficult to overestimate the potential of the new oxymetry techniques described here.

2.5. Nuclear magnetic resonance of paramagnetic systems

Nuclear magnetic resonance (NMR) is considered here only in connection with the problem of the interaction between nuclear spins and electron spins of paramagnetic ions and radicals. Analysis of this interaction may be useful in investigations of metal enzymes and metal carriers. The dipolar and contact (Fermi) interactions are reflected both in the positions of NMR lines and in the relaxation parameters T_{1n} and T_{2n} of the NMR spectrum (Lamer *et al.* 1974; Dwek 1977; Bradbury & Nicolini 1986; Wütrich, 1986).

In cases involving radicals and ions of transition metals from the first and second rows of the Mendeleyev table, the factor responsible for a paramagnetic shift usually is the isotropic contact term of the Fermi interaction:

$$\delta_{NC} = \Delta H/H_0 = -[a_h g_e^2 \beta_e^2 S(S+1)]/3g_N \beta_N k_B T, \qquad (2.28)$$

where a_h is the hyperfine interaction constant, which is proportional to the spin electron density at a given nucleus.

In those cases in which the population of the levels is changed with a change in the temperature, e.g., upon transition of the spin system from the antiferromagnetic to the paramagnetic state, equation (2.28) predicts that the observed deviation will be $\delta_{NC} \sim T^{-1}$. The a_h value usually varies within the range of 10^8–10^9 s^{-1}, which leads to shifts exceeding the shifts from diamagnetic molecules. This makes it possible to study the nuclear environment of the ligands of the active sites of metal enzymes.

Another mechanism for a paramagnetic shift, called a pseudo-contact shift, is due to the dipole–dipole interaction of the nuclear spin with the electron spin of the complex. The magnitude of the pseudo-contact shift depends on the magnetic anisotropy of the ligand field and on the angle θ between the radius connecting the spins and the principal axis of the complex. For molecules with axial symmetry,

$$\delta_{Npc} = \Delta H_0/H_0 = -[\beta_e^2 S(S+1)/k_B T](g_\parallel^2 - g_\perp^2)[(3\cos^2\theta - 1)/r^3]. \quad (2.29)$$

This expression may, in principle, be used to determine r (up to 1.5 nm) for systems with strong anisotropy of the g-factor, e.g., for some lanthanides, Mn^{2+}, Cd^{3+}, Mn^{2+}-EDTA, etc.

In the presence of unpaired electrons, the nuclear relaxation parameters are substantially affected by the following factors: the distance between the spins (r), the magnitudes of the spin quantum numbers of electrons (S) and nuclei (I_n), the lifetime of the complex (τ_M), the correlation time of the dipolar electron–nucleus interaction (τ_c'), the unpaired-electron density at the nucleus, and the intensity of the chemical exchange between the nuclei. Here we shall consider the dipole–dipole interactions, since they are most widely used in studying the behavior of substrates at the active sites of metal enzymes.

In a rapid chemical exchange of ligands at paramagnetic active sites ($\tau_M \leqslant 10^{-4}$ s), the experimentally measured rates of relaxation depend on the proton relaxation time of the electron spin (τ_c'), the ratio of the proton concentration in the paramagnetic complex to the total concentration of protons (f_M), the contribution of diamagnetic interactions, the effects that arise upon direct contact with paramagnetic particles without formation of a complex with a metal, and the ratio of resonance frequencies of protons before and after the complex formation. The complete mathematical expression that includes all these factors has been analyzed in the literature (Wien *et al.* 1972).

If the complex formation is not accompanied by a strong chemical shift, the T_{2n} value is directly related to the structure and molecular motion of the paramagnetic complex and is calculated from the experimental data using the following equation:

$$\Delta\Delta H_{\frac{1}{2}} = (f_M/\pi)(1/T_{2n}), \tag{2.30}$$

where $\Delta\Delta H_{\frac{1}{2}}$ is the broadening of the resonance line of protons caused by the formation of a paramagnetic complex. The $\Delta\Delta H_{\frac{1}{2}}$ value is defined as the width of a Lorentzian line at half-height expressed in hertz. The contribution of the paramagnetic complex to the increase in the relaxation rate is determined from the difference between the rates in the presence and absence of the complex.

According to the theory developed by Solomon and Blombergen (1956),

$$1/T_{2n} = \frac{1}{15} \frac{S(S+1)\gamma_I^2 g^2 \beta_e^2}{r^6} \left[4\tau_c' + \frac{3\tau_c'}{1+\omega_I^2\tau_c'} \right], \tag{2.31}$$

$$1/T_{1n} = \frac{2}{15} \frac{S(S+1)\gamma_I^2 g^2 \beta_e^2}{r^6} \left[\frac{3\tau_c'}{1+\omega_I^2\tau_c'^2} \right], \tag{2.32}$$

where τ_c' is the correlation time of the dipole–dipole interaction (the preceding formulas hold for $\tau_c' > 2 \cdot 10^{-10}$ s, which corresponds, for example, to Mn^{2+} and Cu^{2+} ions and NRs), ω_I is the frequency of the Larmor precession of the nuclear spin, γ_I is the gyromagnetic ratio, and g is the g-factor. The functions in brackets will be denoted as f_1 and f_2.

In the general case, the efficiency of the dipole–dipole interaction can be determined by three parallel processes: (1) molecular rotation (with the correlation time τ_r), which averages the magnetic field created by electron spins at the point where the proton resides; (2) relaxation of the electron spins, which averages the interaction with the efficient correlation time τ_{1e}; (3) movement of the spins away from one another as a result of dissociation (with the time τ_M). In accordance with the equation for parallel processes,

$$\tau_c^{-1} = \tau_r^{-1} + \tau_{1e}^{-1} + \tau_M^{-1}, \tag{2.33}$$

the efficient correlation time is determined by the shortest time of one of the processes mentioned earlier. Equations (2.31)–(2.33) provide unique possibilities for studying the fine structure of paramagnetic complexes through measurement of the T_{1e} and T_{2e} times. Indeed, as follows from equations (2.31)–(2.33), these values depend on the three most important parameters: the electron spin (S), the separation between the electrons and the nucleus of interest (r), and the correlation time τ'_c.

In cases where the NMR parameters are being measured for rotating groups, it is necessary to include a term for the rotation correlation time (τ_{sp}). For a rotating methyl group, for example,

$$\tau_{sp}^{-1} = (3/8)b^2 T_{1n},$$

$$b = 4h\gamma_c^2\gamma_D^2 T_{1n}^c/3\pi r^6 T_{1n}^D, \tag{2.34}$$

where b is the constant of the quadruple splitting, γ_c and γ_D are the gyromagnetic ratios for the ^{12}C and ^2H nuclei, T_{1n}^c and T_{1n}^D are the nuclear relaxation times of ^{13}C and ^2H, respectively, and r is the length of the C—H bond. The τ_{sp} value for the rotating CH_3-group is estimated to be $(3-4)\cdot10^{-10}$ s.

Most often it appears necessary to determine the separation between the various protons of the ligand and the paramagnetics at an enzyme. Since the S value is known as a rule, the task of calculating r is reduced to finding out which of the three processes (electron spin relaxation, rotation of the macromolecule, dissociation of the complex) leading to the disturbance of the dipole–dipole spin–nuclear interactions is the most rapid [see equation (2.35)]. For metal enzymes the value of τ_{1e} lies within the range of 10^{-6}–10^{-11} s, $\tau_r = 10^{-8}$–10^{-6} s, and $\tau_M = 10^{-5}$–10 s (for the enzyme-substrate and enzyme-inhibitor complexes); therefore, the term containing τ_M is usually disregarded.

To calculate the separation r, given the extent of paramagnetic broadening of the ESR line, the following formula can be derived from equations (2.31) and (2.33):

$$r = 46\left[S(S+1)f_2(\tau'_c)\frac{f_M}{\Delta\Delta H_{\frac{1}{2}}}\right]^{\frac{1}{6}}nm, \tag{2.35}$$

where $\Delta\Delta H_{\frac{1}{2}}$ is expressed in hertz, and τ'_c is expressed in seconds. For instance, according to data reported in the literature (Wien *et al.* 1972), for a spin label at a relatively rapidly rotating lysozyme molecule ($\tau'_c = 10^{-8}$ s) in aqueous solution at room temperature ($a_3 = 2.3$),

$$r = a_3[f_M/\Delta\Delta H_{\frac{1}{2}}]^{\frac{1}{6}}nm. \tag{2.36}$$

For the spin labels of macromolecules with molecular masses equal to or exceeding 500,000, or for very viscous media, the value of τ'_c is equal

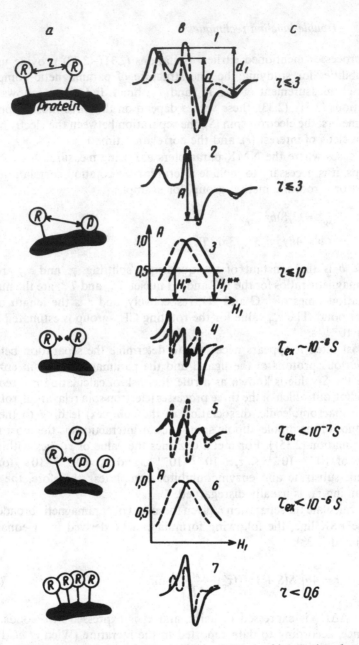

Figure 2.16. Double-labeling techniques. Types of interactions between paramagnetic centers: (a) corresponding changes in the ESR spectra and saturation curves and (b) the range of distances measured (r, nm) between the centers of the range of efficient times of spin exchange (τ_{ex}). R is a nitroxide radical; P is a paramagnetic center. The broken lines represent the noninteracting centers; 1–3, dipole–dipole interaction; 4–6, dynamic exchange interaction; 7, static exchange interaction (Likhtenshtein 1988b).

to $2 \cdot 10^{-7}$ s, and an analogous calculation gives $a_3 = 4.2$ for a macromolecule with a τ_r value close to that of τ_{1e} of the spin label, $a_3 = 3.9$. For complexes of macromolecules with Mn^{2+}, one can use the value $a_3 = 2.3$, which was calculated assuming that $\tau'_c = \tau_{1e} \approx 4 \cdot 10^{-10}$ s. For rapidly relaxing ions that are not observable by ESR (e.g., for ions with $\tau_{1e} = 10^{-11}$ s and $a_3 = 1.1$), paramagnetic broadening is difficult to observe even in those cases in which the protons are near the paramagnetic center.

Thus, the NMR method, especially when used in combination with ESR spectroscopy and spin-label techniques, is a very efficient tool for investigating the structures of enzyme-substrate complexes.

Figure 2.16 illustrates all the types of interactions used for structural investigations in paramagnetic systems, the corresponding changes in the ESR spectra, and the range of measured distances between the paramagnetic centers. The double spin-label and spin label–paramagnetic probe techniques permit determination of the distances between the labels within 0.3–8.0 nm and the depth of immersion up to 4 nm, which practically covers the sizes of individual enzymes and biomembranes. Examples of applications of the method described here will be given in Chapters 6–8.

3

Fluorescent labeling methods

Luminescent dyes have been routinely used in biochemistry as indicators of acidity, in biochemical assays, in histometry, and in other areas. For the past 30–40 years luminescent labels have found wide application in the solution of a number of structural and dynamic problems in enzyme and membrane systems (Vladimirov & Dobretsov 1980; Lakowicz 1983; Hoffman & Restall 1984; Waggoner 1986; Dobretsov 1989).

The areas of application for luminescent labels are as wide-ranging as are those for spin labels, e.g., determination of local acidity and polarity of objects under study and estimation of the parameters of microviscosity and molecular dynamics from data on the rotational and translational diffusion of labels. The methods of luminescent labels and probes allow measurements of transmembrane and surface potentials, studies of localization of charged and hydrophobic portions, and estimation of the distances between discrete centers and the depth of immersion into the biological matrix. Luminescent chromophores, like spin labels, can serve as "microscopic seismic stations," following subtle changes in the local conformation during the functioning of biological objects.

In this chapter the physicochemical principles and the scope of the method of luminescent labels and probes will be briefly considered. Examples of their application to studies of biological objects will be given in Chapters 6–8.

3.1. General

3.1.1. Absorption spectra

The phenomenon of absorption of electromagnetic radiation in the ultraviolet (UV) and visible regions of optical spectra results from the discreteness of the energy levels in molecules. The main parameters of an absorption spectrum are its maximum frequency (v_{max}^{ab}) and its wavelength (λ_{max}^{ab}), reflecting the value of the energy gap between the individual energy levels. Other parameters characteristic of a spectrum are the width and the shape of the spectral line, which are determined by a set of oscillatory

80

and rotational levels of the molecule and by its interaction with the environment.

The possibility of observing an absorption spectrum is determined by the magnitude of the line intensity, characterized by the molar extinction coefficient (ε). The larger the change in the electron dipole moment (μ_{el}) at a given electron transition, the higher is the ε value. The ε values of readily absorbing molecules can reach as high as $10^5\,M^{-1}\,cm^{-1}$, with the value of μ_{el} equal to 3–7 Debye (D) in the ground state and 10–22 D in the excited state. The transition probability is another parameter of an absorption spectrum:

$$k_f \sim |\bar{M}|^2 = 2{,}900n^2 v_{max}^{ab} \int \varepsilon \cdot dv, \tag{3.1}$$

where $\bar{M}$ is the transition dipole moment, n is the refraction index, and $\int \varepsilon \cdot dv$ is the appropriate integral.

Organic molecules may undergo the following transitions (Parker 1968) (excited orbitals are indicated by an asterisk):

$$n \to \pi^*, \quad \pi \to \pi^*, \quad n \to \sigma^*, \quad \pi \to \sigma^*, \quad \sigma \to \sigma^*, \quad l_a \to \pi^*.$$

The latter is the transition of an unpaired electron from a heteroatom into the π-conjugation.

Charge transfer transitions are the most intense ones, due to a strong change in the dipole moment upon absorption of a quantum.

According to the Franck–Condon principle, nuclei remain practically fixed as they undergo electronic transitions (vertical transitions between terms, Figure 3.1), since the transition time ($\tau \sim 10^{-16}$ s) is considerably shorter than the characteristic time of the nuclear displacement ($\tau_n = 10^{-12}$–10^{-14} s). The lifetimes of the lower oscillatory singlet state usually range within 10^{-8}–10^{-11} s.

The sensitivity of the nonspecific electrostatic properties of the environment allows the use of chromophores as indicators of polarity (Parker 1968; Bakhshiev 1972, 1989; Gorodisky 1989). The Onsager model is used for quantitative consideration of the medium. According to the model, a molecule with a dipole moment (μ_{el}) forms a homogeneous, spherically closed cavity of radius a. The medium beyond the cavity is characterized by uniform values of the dielectric constant (ε_0) and the refraction index (n). The dipole moment of the molecule induces dipole moments in the environment. Within the framework of the Onsager model, the value of the shift of the absorption spectrum at the transition from the gas phase to the given medium can be described by the following formula:

$$\Delta v_{max}^{ab} = 5.035[2\mu_{el}^g(\mu_{el}^g - \mu_{el}^e)/a^3 - 2(\mu_{el}^g - \mu_{el}^e)^2/a^3]$$
$$\cdot [(\varepsilon_0 - 1)/(\varepsilon_0 + 2) - (n^2 - 1)/(n^2 + 2)], \tag{3.2}$$

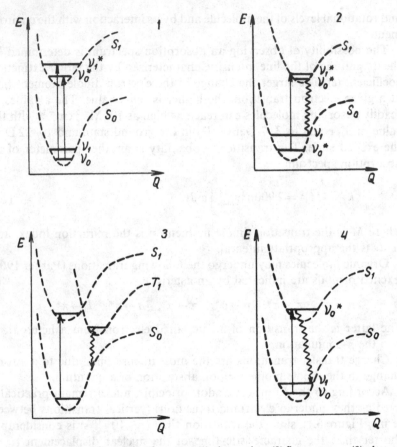

Figure 3.1. The origin of spectra: (1) absorption; (2) fluorescence; (3) phosphorescence; (4) resonance Raman scattering.

where μ_{el}^g and μ_{el}^e are the dipole moments of the ground and excited states (in Debye units), respectively; a and $\Delta\nu_{max}^{ab}$ are given in nm and cm^{-1}, respectively.

The Onsager model obviously gives only a rough picture of the effects in the system. Nevertheless, in some cases this model appears to reflect the main properties of nonspecific solvation and can be a tool to approximate the characteristics of the solvato-chromic effects in the system under investigation.

It should be emphasized, however, that specific donor–acceptor interactions, including H-bond formation, can contribute significantly to the value of solvato-chromic shifts.

3.1.2. Fluorescence and phosphorescence

Upon absorption of light, a molecular system undergoes a transition from the ground state (S_0) to the electronically excited state (S_1). The reverse transition may be accompanied by secondary emission (fluorescence), or it may take place by a nonradiative mechanism (Figure 3.1). On giving up a part of its energy to the environment, the system may pass over from the S_1 state to an excited triplet state (T_1). The latter may be inactivated in two ways: by emission (phosphorescence) or by a nonradiative mechanism (Parker 1968; Lakowicz 1983; Kapinus 1988).

A significant difference between fluorescence and phosphorescence is that the former process is an $S_1 \rightarrow S_0$ transition allowed by quantum mechanics, since it does not change the multiplicity of the system. The latter transition, $T_1 \rightarrow S_0$, involves a change in the multiplicity, and hence it is forbidden. Therefore the lifetimes of the excited states during fluorescence ($\tau_f^* = 10^{-8} - 10^{-9}$ s) are much shorter than phosphorescence lifetimes ($\tau_{ph}^* = 10 - 10^{-6}$ s).

The main parameters of both types of luminescence are their intensity, I, the frequency of the maxima of the fluorescence and phosphorescence spectra, $v_{max}^{f,ph}$, the line shape of the spectrum, the quantum yield, φ_0, and the lifetime of the excited state, τ^*.

The following characteristics of chromophores provide high quantum yields: (1) the π-electron system, (2) ortho-, para-substituents, (3) high rigidity of the molecule, and (4) the $\pi^* \rightarrow \pi$ and $l_a \rightarrow \pi$ transitions from the lowest singlet excited state. The following factors reduce the value of the quantum yield: (1) the $\pi^* \rightarrow n$ transition, (2) meta-substitutents, and (3) heavy and/or paramagnetic atoms. External factors such as interactions with quenching molecules by the mechanism of energy or electron transfer reduce the φ_0 value as well.

The excitation of a chromophore group is accompanied by a change in the electron dipole moment of the molecule. This involves a change in the energy of interaction with the surrounding molecules. If the characteristic time of reorganization of the dipoles (τ_r) in the medium is much longer than τ^*, the dipoles have no chance to follow the change in the light-induced electric field, and the transition from the "unsolvated" level to the ground state takes place. In another limiting case, where $\tau_r \ll \tau^*$, the interaction between the dipoles and the excited molecule lowers the energy of the system, and the emission is effected from the "solvated" level. A decrease in the transition energy with a decrease in τ_r is observed experimentally in a change in v_{max} (a relaxation shift) (Bakhshiev 1972, 1989; Mazurenko & Bakhshiev 1970; Mazurenko 1989):

$$v_{max}(\tau) = v_{max}(\infty) + [v_{max}(0) - v_{max}(\infty)] \exp(-\tau/\tau_r), \qquad (3.3)$$

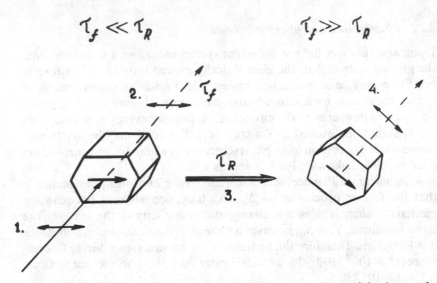

Figure 3.2. Schematic representation of the dependence of the degree of fluorescence depolarization of a chromophore on the rotational correlation time (τ_R) and the excited-state lifetime (τ_f): (1) irradiation by polarized incident light; (2) polarized fluorescence from "immobile" chromophore; (3) rotation of the chromophore; (4) depolarized fluorescence from mobile chromophore.

where the indices τ, ∞, and 0 are related to the ν_{max} of the time-resolved emission spectrum at a given moment, $\tau \to \infty$ and $\tau \to 0$, respectively.

For a normal emission spectrum,

$$\nu_{max} = \nu_{max}(\infty) + [\nu_{max}(0) - \nu_{max}(\infty)]\tau^*/(\tau^* + \tau_r). \tag{3.4}$$

In real systems (viscous lipids, polymers, proteins, membranes, etc.) there is, as a rule, a set of energy levels, τ_r values, and other parameters. Analysis of relaxation shifts in such systems requires special approaches (*vide infra*).

Chromophores that are being used as luminescent labels and probes usually are asymmetric and consequently are sensitive to the polarization of incident light. If the rotation time (τ_R) of the chromophore is comparable to its excited-state lifetime, the emitted light will be depolarized, provided that the directions of the dipole transition moment and the vector of the incident-light polarization are oriented parallel to each other (Figure 3.2). The efficiency of energy migration between chromophores also depends strongly on the mutual reorientation of the dipole transition moments of the chromophores.

Inorganic cations (Tl^+, La^{3+}, Tb^{3+}, and Eu^{3+}) are also used as fluorescent probes. In lanthanides, optical transition occurs within the inner $4f$ shell, which is shielded by $5s^2$ and $5p^6$ electrons. The intensity of absorption of luminescent spectra of lanthanides is low. However, the intensity of fluorescence can be increased as a result of energy migration from other chromophores.

3.2. Chemical properties of fluorescent labels and probes

As can be seen from the chemical formulas, the most commonly used fluorescent probes contain a hydrophobic part and an electrostatic charge (Vladimirov & Dobretsov 1980; Waggoner 1986; Dobretsov 1989):

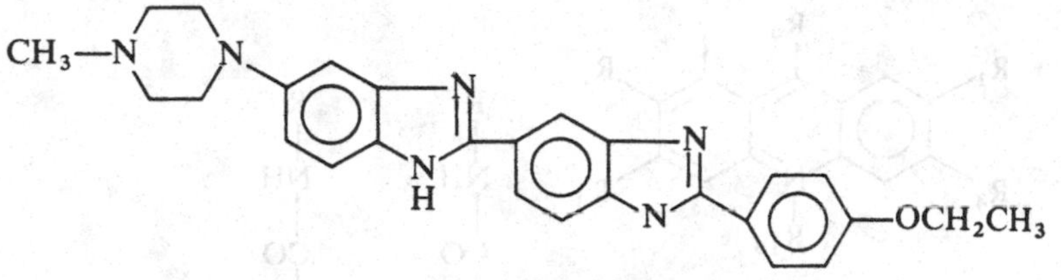

HOECHST 33342

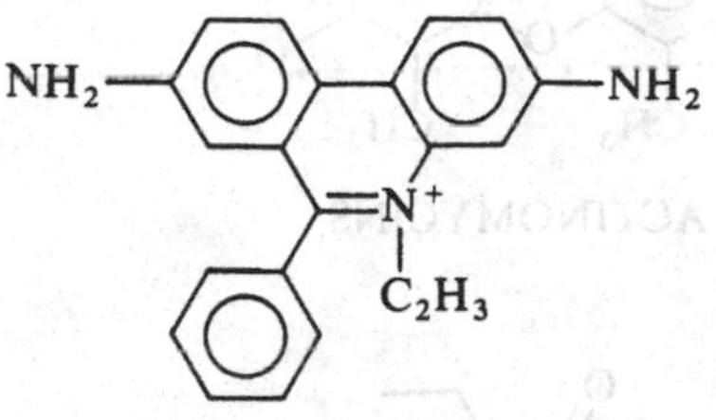

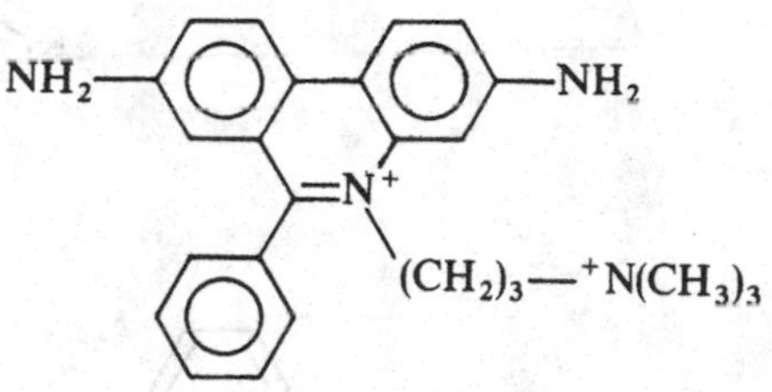

ETHIDIUM   PROPIDIUM

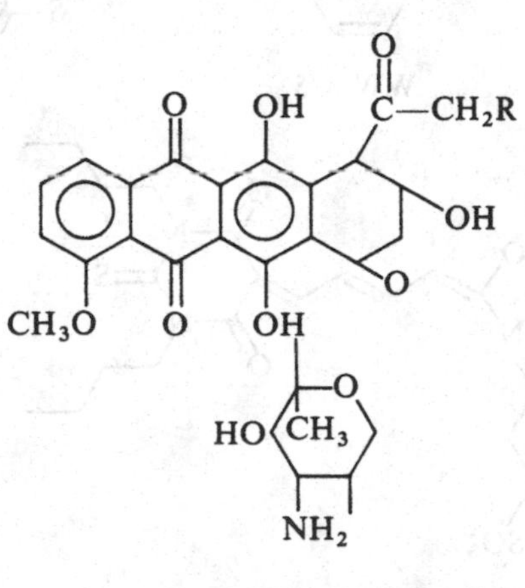

ANTHRACYCLINE DYES

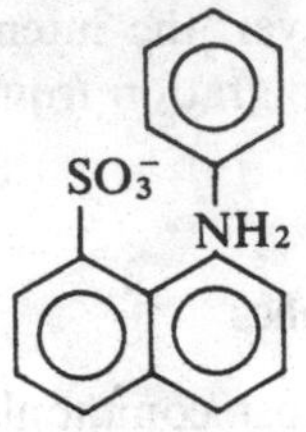

ANS

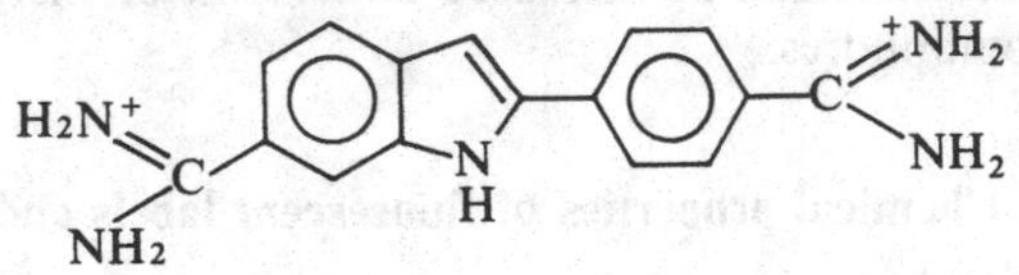

DAPI

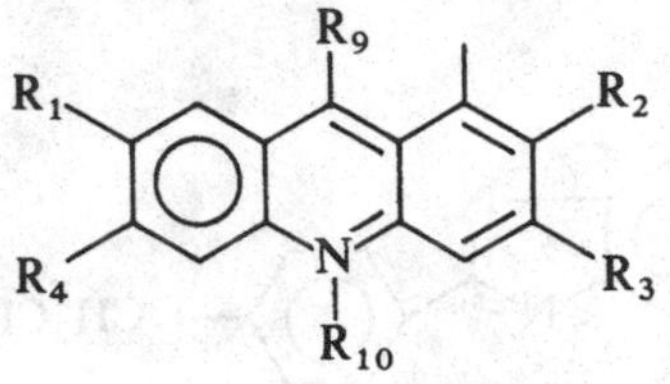

ACRIDINE DYES

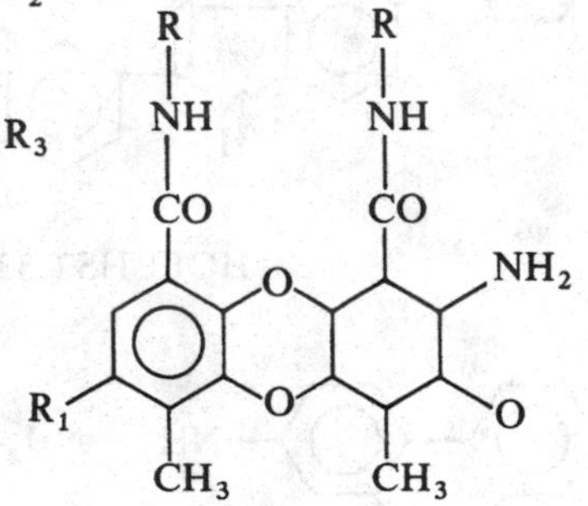

ACTINOMYCINS

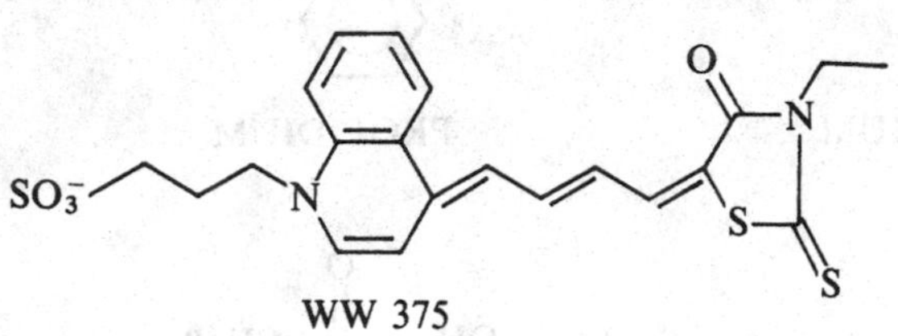

WW 375

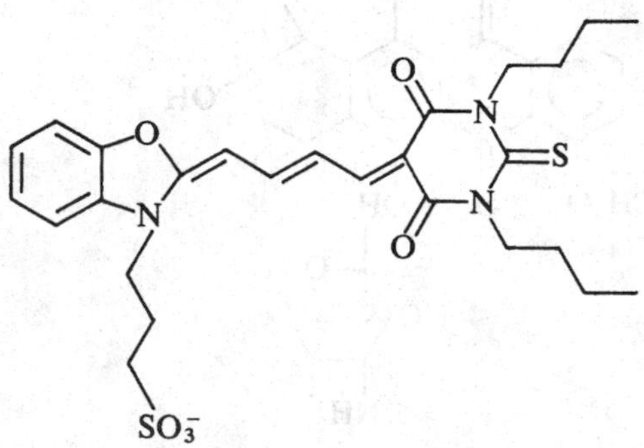

MEROCYANINE 540

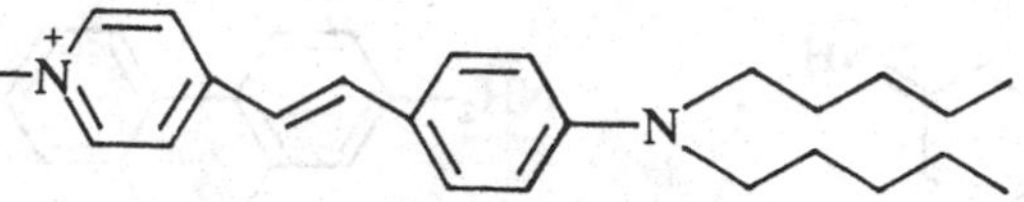

di-C$_5$-ASP

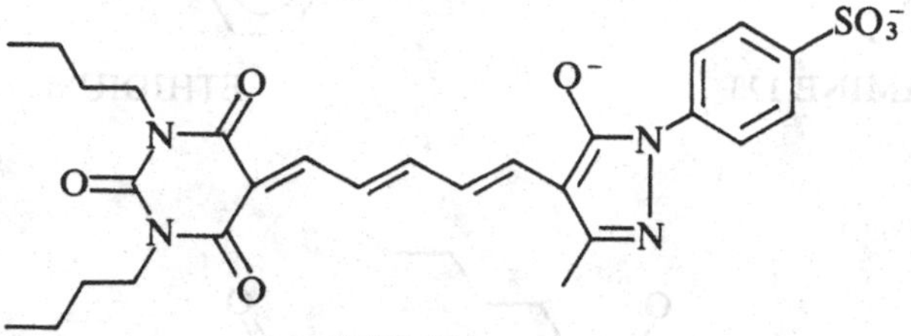

WW 781

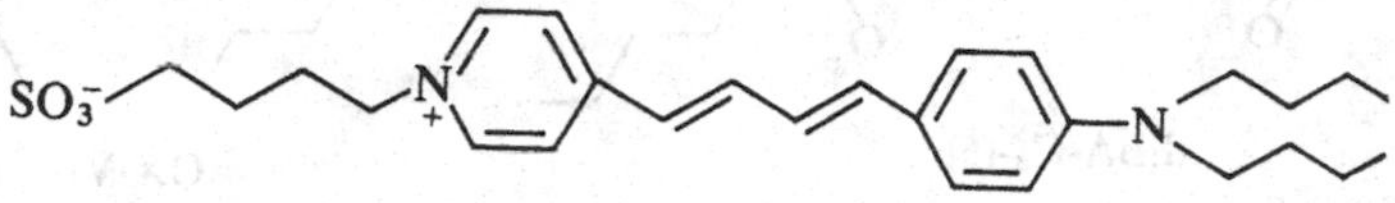

RH 160

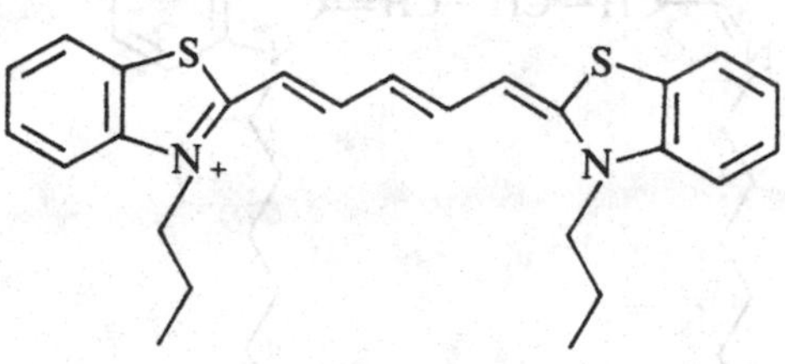

diS-C$_3$-(5)

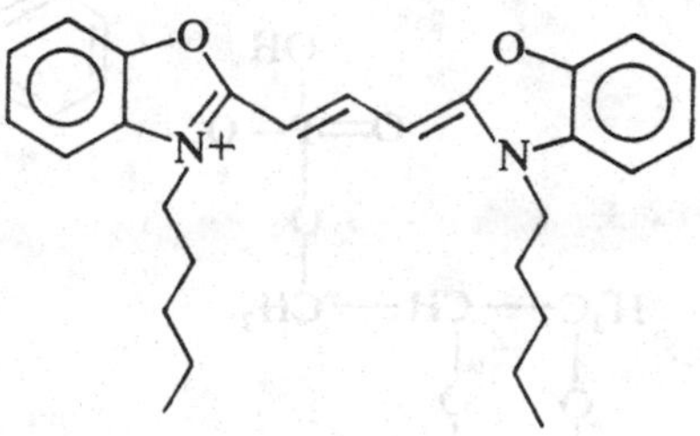

diO-C$_5$-(3)

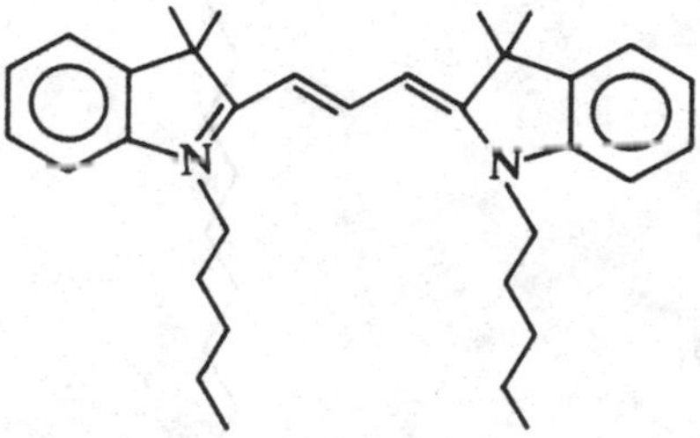

diI-C$_5$-(3)

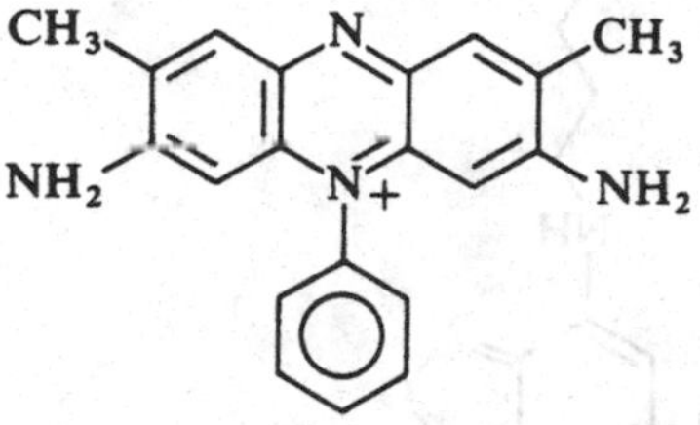

SAFRANIN

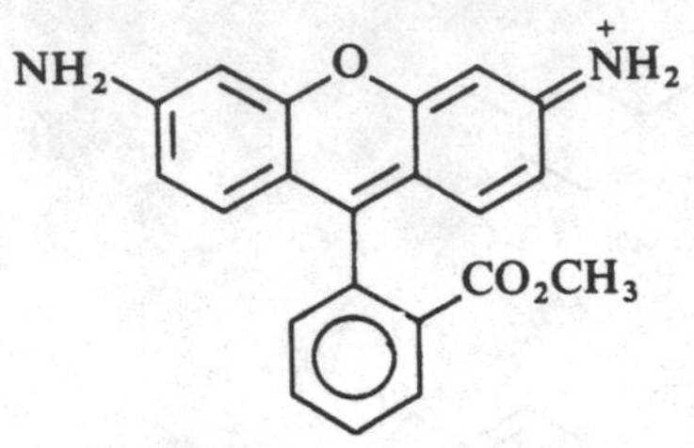

RHODAMINE 123

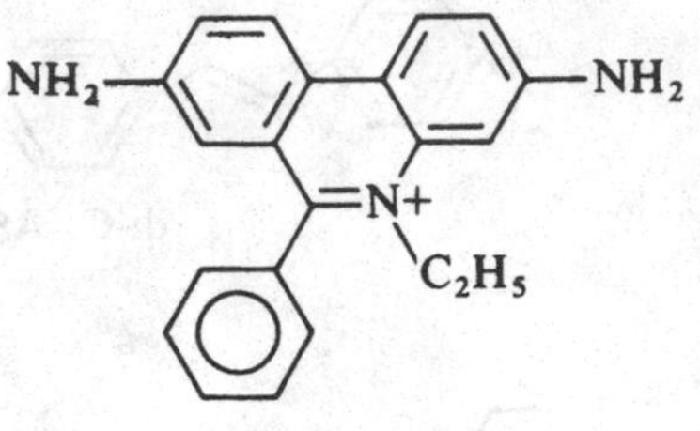

ETHIDIUM

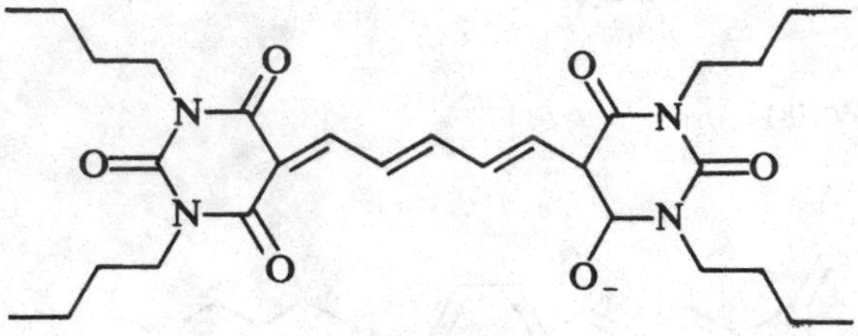

diBA-C$_4$-(5)

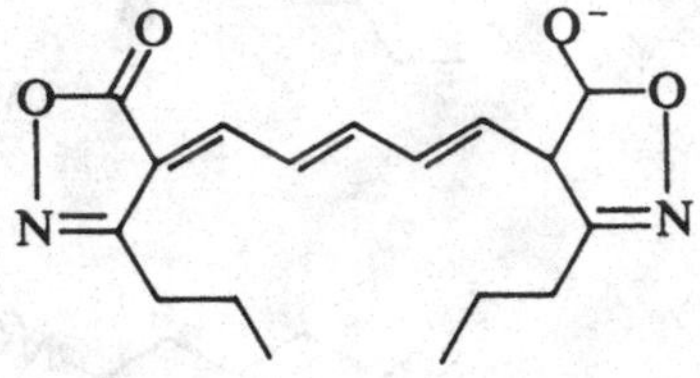

OX-VI

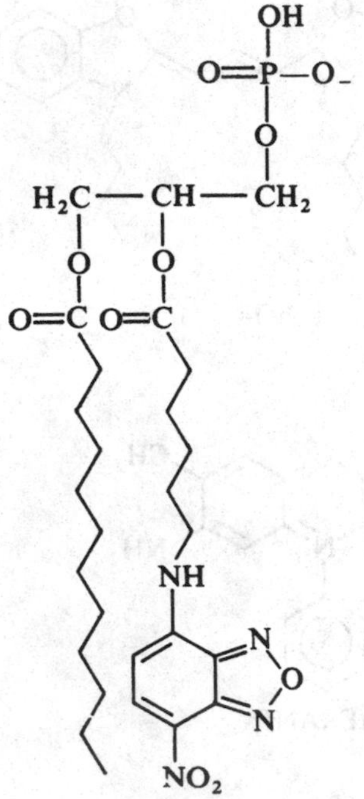

C$_6$-NBD-PA

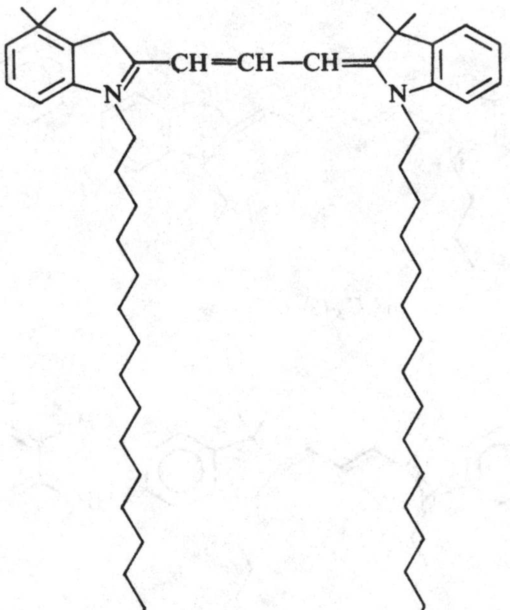

diI-C$_{18}$-(3)

TRANS-PARANARIC ACID

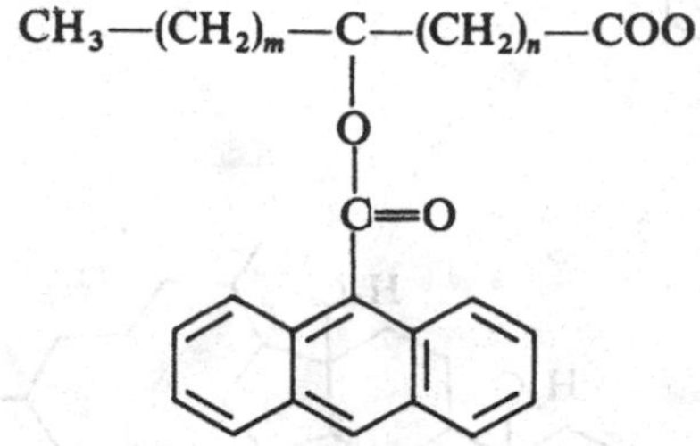

(*m,n*)-ANTHROYLSTEARATE

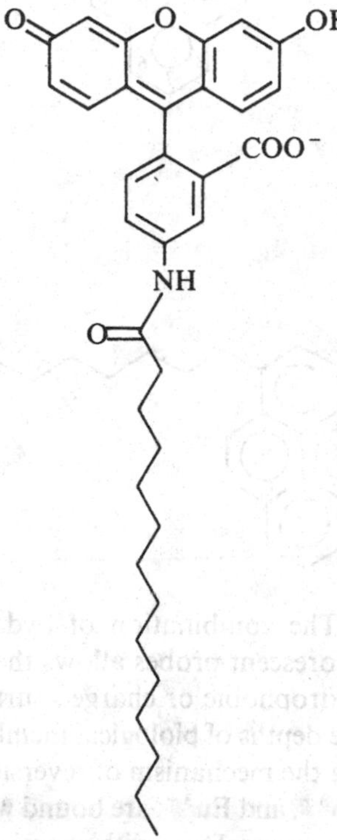

ACYL AMINOFLUORESCEIN

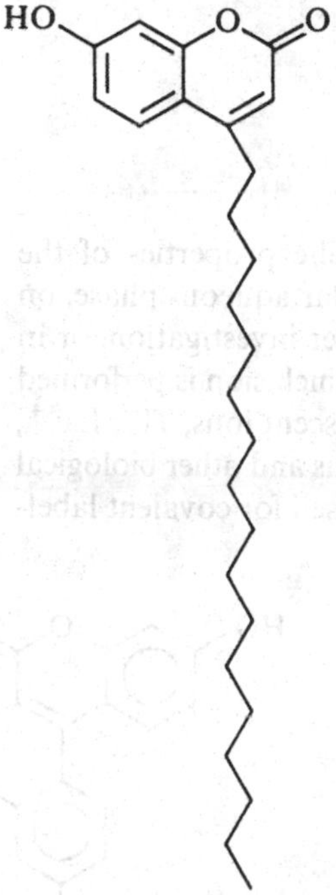

4-ALKYL UMBELLIFERONE

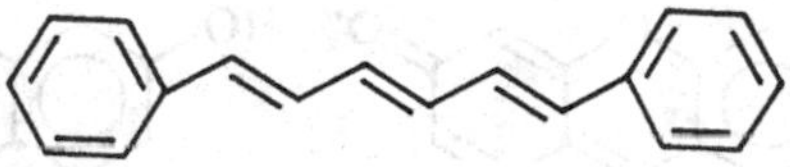

DPH

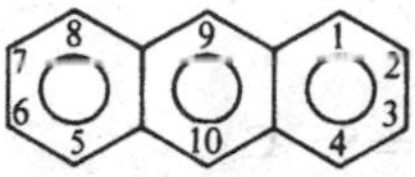

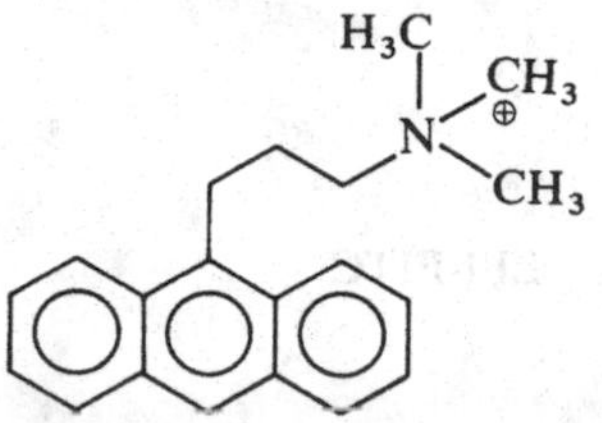

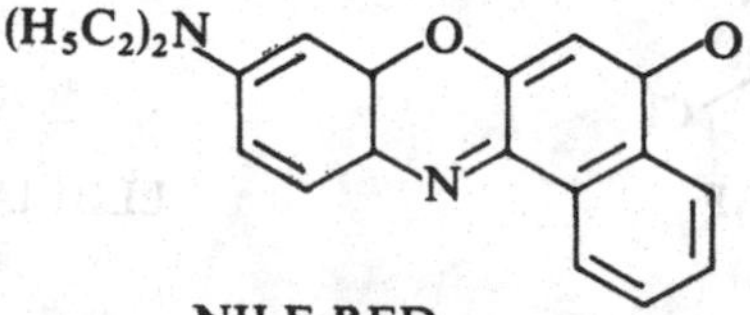

NILE RED

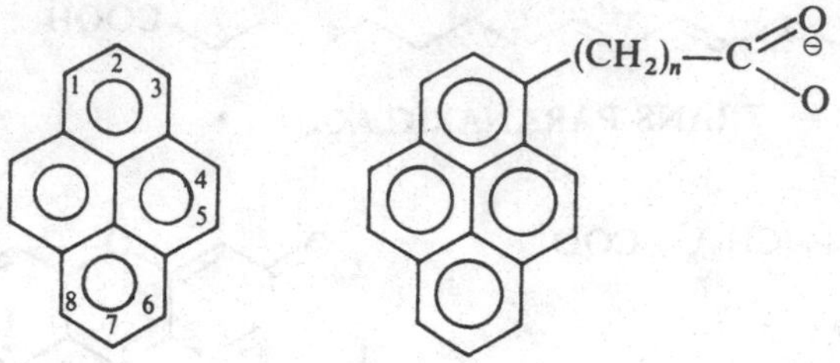

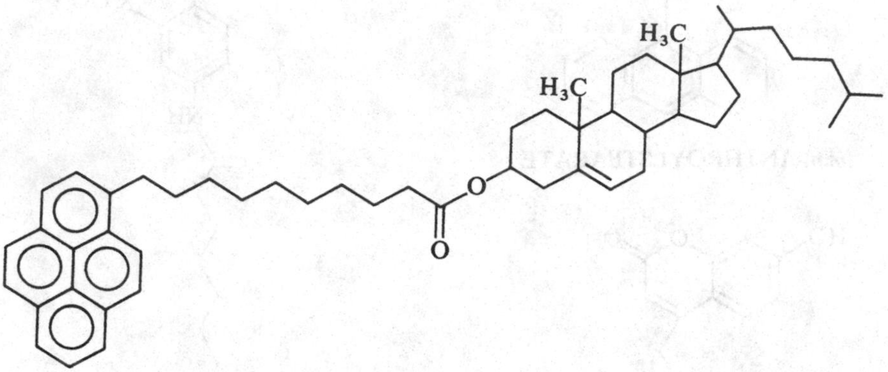

The combination of hydrophobic and hydrophilic properties of the fluorescent probes allows them to be used in the polar aqueous phase, on hydrophobic or charged surfaces of the objects under investigation, or in the depths of biological membranes (Figure 3.3). The inclusion is performed via the mechanism of reversible adsorption. Luminescent ions, Tl^+, La^{3+}, Tb^{3+}, and Eu^{3+}, are bound with the ligands of proteins and other biological structures. Dyes with chemically active groups are used for covalent label-

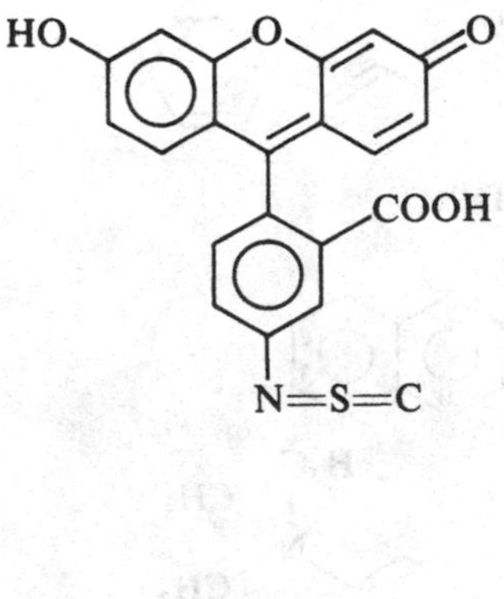

LLI-FITC 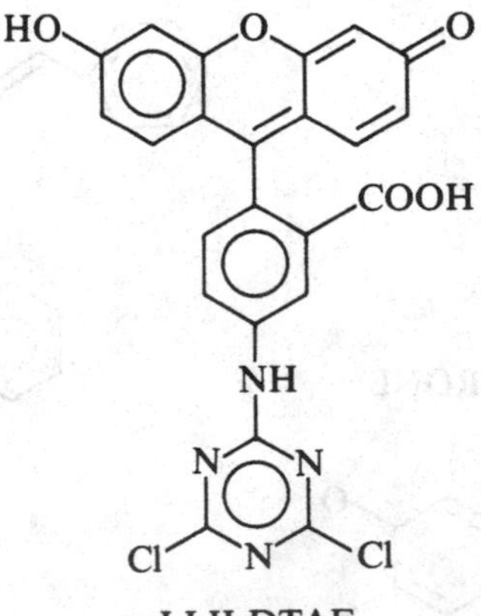   LLII-DTAF 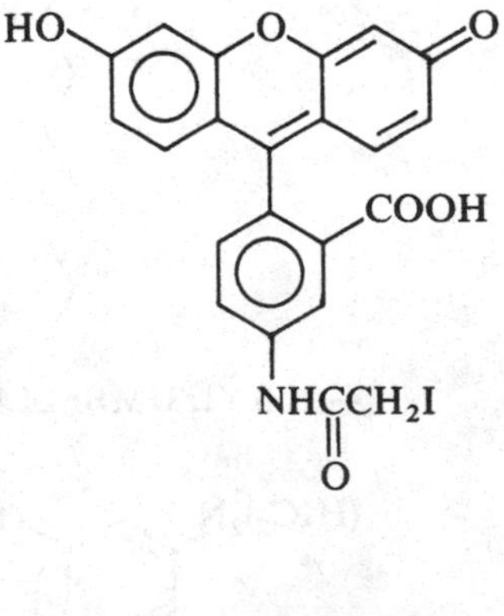   LLIII-IAF

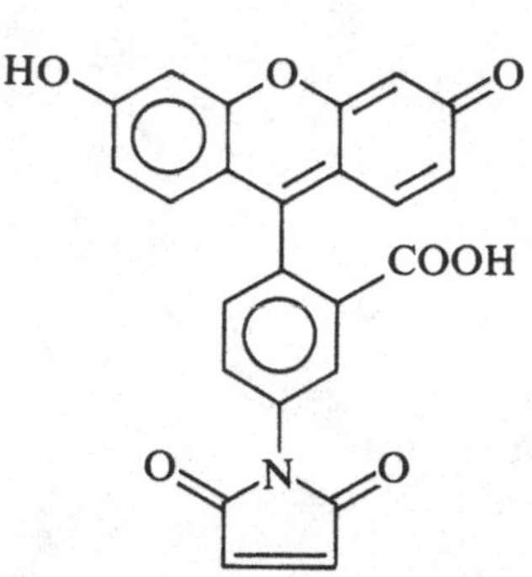

LLIV-MALF

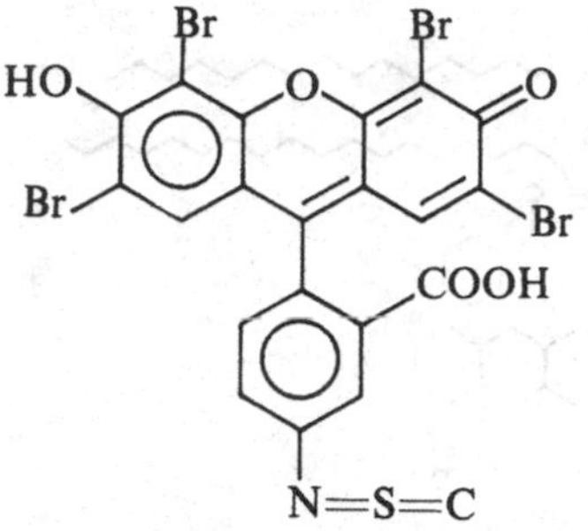

LLV-EYITC

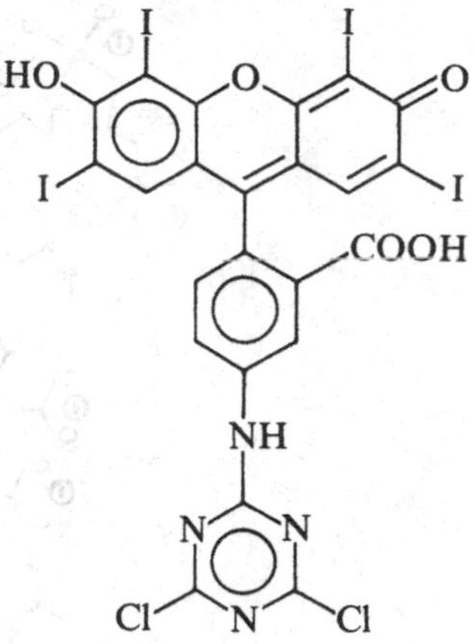

LLVI-DTAEry

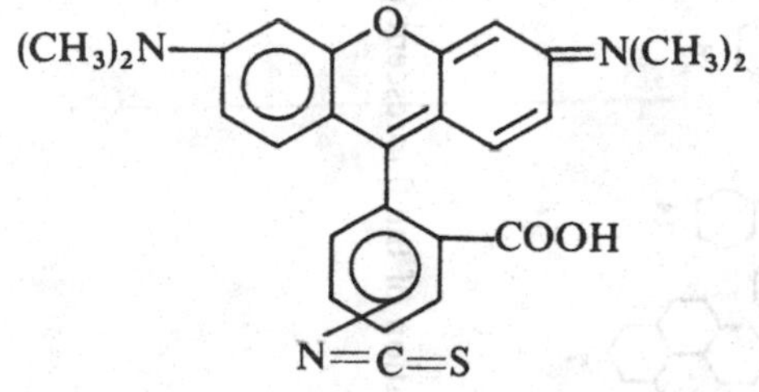

LLVII-TRITC

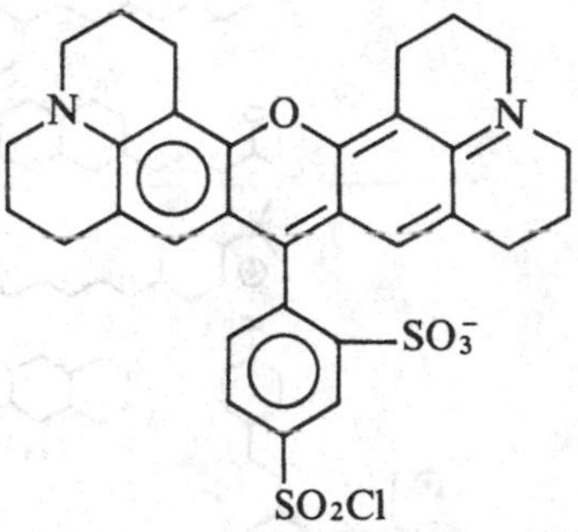

LL-TEXAS RED

ing of the chosen functional residues of proteins, nucleic acids, membranes, hydrocarbons, etc. These groups commonly react with nucleophilic portions of modified molecules (Waggoner 1986; Dobretsov 1989):

Labels that fluoresce after covalent binding have also been synthesized:

$$R-HgO-\overset{\overset{\displaystyle O}{\parallel}}{C}-CH_3 \text{ (FLX)}, \quad R-SO_2-CH=CH_2 \text{ (FLXI)},$$
$$R-SO_2NH-NH_2 \text{ (FLXII)},$$

where R is a luminescent chromophore.

Labels that are fluorescent only after covalent binding have also been synthesized: 2-aminobiphenyl, *N*-(1-anilinenaphthyl-4-maleimide, and phthalaldehyde, for example. The latter reacts with amines

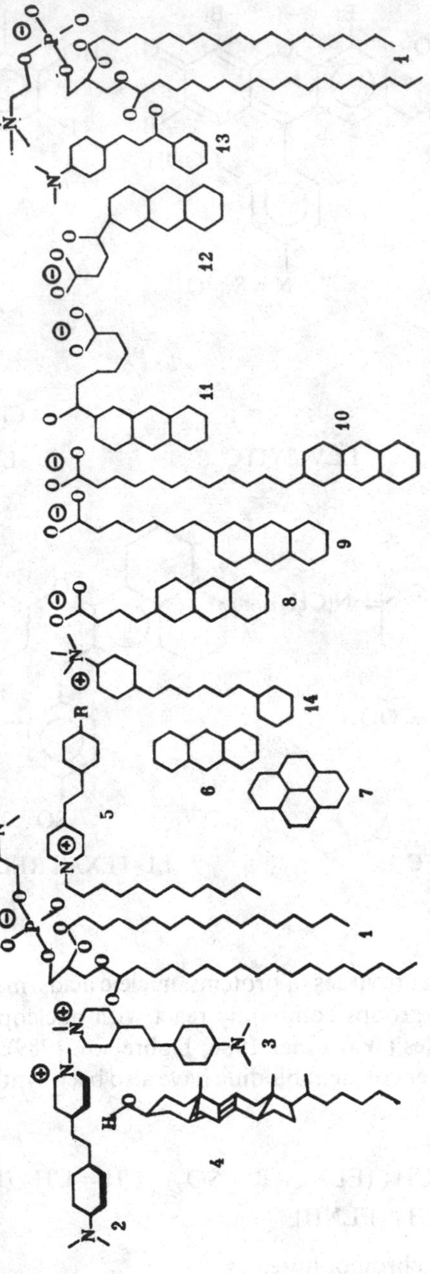

Figure 3.3. Schematic representation of the fluorescent probe's location in the membrane (Dobretsov 1989).

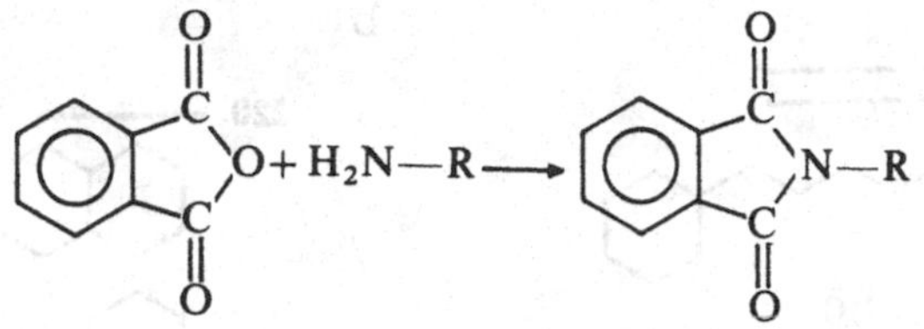

Binding of fluorescent labels to the functional groups is accomplished according to the following schemes (R* is the fluorescent part of the label; R is a residue of the modified object):

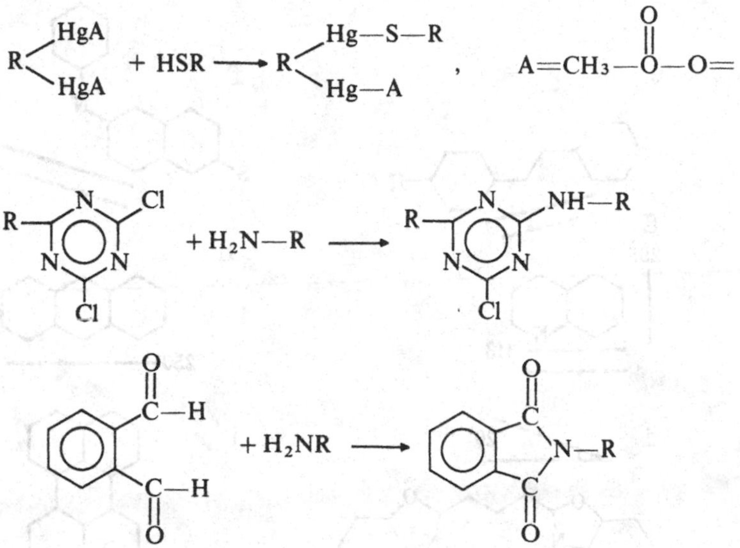

Thus, modern organic and inorganic chemistry offers a wide collection of probes and labels providing a great variety of both luminescent properties and targets for the proposed labeling.

3.3. Rotational diffusion of fluorescent chromophores

3.3.1. Depolarization of fluorescence

Asymmetry of chromophore molecules suggests anisotropy of their transition dipole moments. Figure 3.4 shows the direction of the transition dipole moments of light absorption and emission for several luminescent probes.

The rotation of an asymmetric chromophore with a correlation time of τ_R comparable to the lifetime of the molecular excited state, τ^*, may

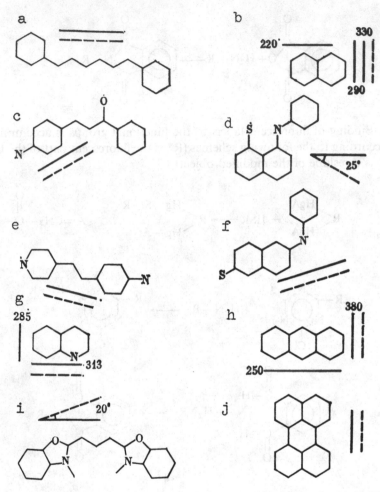

Figure 3.4. Directions of transition moments for chromophores. Solid and dashed lines relate to absorption and emission processes, respectively; numerals denote the absorption line width (nm) (Dobretsov 1989).

be accompanied by depolarization (Figure 3.2). The value of τ_R depends on the microviscosity and microstructure of the medium.

The requirements for depolarization are, as seen from Figures 3.2 and 3.4, anisotropy of the transition dipole moment for light absorption, subsequent emission, and closeness of the τ_R and τ^* values. The theory of the process and the experimental data have been described in a number of papers (Cherry 1979; Beddard 1986; Waggoner 1986; Edidin 1987; Matko et al. 1989; Dobretsov 1989; Lakowicz et al. 1989; Weber 1989; Libertini

& Small 1990). The degree of polarization may be presented as

$$P = (I_{\parallel} - I_{\perp})/(I_{\parallel} + I_{\perp}), \tag{3.5}$$

where $I_{\parallel}$ and $I_{\perp}$ are the intensities of emitted light passing the polarizer with the electric axis directed parallel or perpendicular to the polarization of the incident light, respectively.

For a spherical molecule,

$$P_0/P = 1 + \tau_f^*/\tau_R = 1 + 3RT\tau_f^*/4\pi b^3 N_A \eta, \tag{3.6}$$

where P_0 is the polarization for a chromophore without rotation, b is the radius of rotation, and η is the viscosity of the medium. Measurement of the dependence of P_0/P on T/η allows determination of the value of τ_R and the radius of rotation of the chromophore, if τ_f^* is known.

The τ_R value can be measured directly from the change in the degree of depolarization after pulse radiation of the sample under study.

For ellipsoidal rotation, with the $D_{\parallel}$ and $D_{\perp}$ diffusion coefficients, the transition dipole moment is perpendicular to the main rotation axis, and the value of the degree of polarization is

$$P = 0.1[(10P_0 - 1)/(1 + 4D_{\perp}\tau_f^* + 2D_{\parallel}\tau_f^*) + 1/(1 + 6D_{\perp}\tau_f^*)]. \tag{3.7}$$

In such a case the polarization kinetics can be described by two exponents.

For a system with the transition dipole moment parallel to the main rotation axis, the kinetics follows a monoexponential law with $D_{\perp}$. In the general case of three tensors of rotational diffusion, the theory predicts the superposition of five exponents.

For a chromophore with high-frequency wobbling of the characteristic time $\tau_w \ll \tau_f^*$ and the angle θ,

$$P/P_0 = [\tfrac{1}{2}\cos\theta(1 + \cos\theta)]^2. \tag{3.8}$$

If a chromophore is bound to a macromolecule and is involved in two movements, i.e., a relatively slow rotation along with the macromolecule with a rotational correlation time τ_R, and a high-frequency wobbling with the angle θ, then the values of τ_R and θ can be determined by measuring the dependence of polarization on the viscosity of the medium [formulas (3.6)–(3.8)].

The aforementioned techniques are concerned with fast rotational diffusion with $\tau_R \leqslant \tau_f^* \approx 10^{-8}-10^{-9}$ s. An alternative method for study of rotational and translational diffusion with a wider range of correlation times is fluorescence correlation spectroscopy (Magde *et al.* 1974; Aragon & Record 1975). Rotational diffusion is accompanied by fluctuation of the luminescence intensity. According to the theory of Brownian diffusion, spherical rotators can be described by means of the time-resolved autocorrelation function for the fluorescent intensity fluctuation. The general expression includes three relaxation times; two of them are related to rotational diffusion, and the third is equal to the lifetime of the chromophore excited state, τ_f^*. This technique makes it possible to study rotational processes that are much slower than the excited-state deactivation.

3.4. Fluorescence and molecular dynamics of the medium

Abrupt changes in the electronic state of a chromophore cause a nonequilibrium state in the chromophore–medium system. There are two competitive processes: the deactivation of the excited state with τ_f^* and the transition of the particles in the medium to a new equilibrium state (relaxation) with characteristic time τ_r (Figure 3.5). Relaxation can be monitored by two techniques (Section 3.1.2): (1) time-resolved spectroscopy, measuring the kinetics of the shift of the maximum of the fluorescent spectrum $v_{max}^f(\tau)$, and (2) routine spectroscopy, measuring the dependence of v_{max}^f on temperature.

Equations (3.3) and (3.4) allow estimation of the value of τ_r, which is related to the characteristic time of dielectric relaxation τ_D (Bakhshiev 1972, 1989):

$$\tau_r \approx \tau_D(n^2 + 2)/(\varepsilon_0 + 2). \tag{3.9}$$

For polar media, τ_D is 10–20 times higher than τ_r. The physical explanation for such a difference is that dielectric relaxation occurs under the action of a homogeneous external electric field, whereas the relaxation of surrounding dipoles is affected by the induced electric field, which becomes stronger as the dielectric constant ε_0 becomes larger.

Real systems deviate from the simplest physical picture mentioned earlier because of the specific donor–acceptor interaction, the effect produced by the electric field of the excited chromophore on the dynamics of the environmental molecules, and the intrinsic microheterogeneity of the sample (Demchenko 1986; Pavlovich et al. 1979a,b; Kotelnikov et al. 1983; Rubinov et al. 1982; Likhtenshtein & Kotelnikov 1983; Mazurenko 1989).

In propanol, for example, the kinetics of the relaxation shift of $v_{max}^f(\tau)$ for 4-amino-N-methylphthalimide is confirmed by three exponential decays. In glycerol the uniform distribution of relaxation times provides the best fit to the experimental data. The broadening of the fluorescent spectral width at low temperatures and the dependence of $v_{max}^f(t)$ and v_{max}^f on the wavelength of the incident light (λ_{inc}) appear to be other indicators of structural and dynamic heterogeneity. For the model systems mentioned earlier, the change in λ_{inc} leads to a drastic qualitative alteration of the kinetics of the relaxation shift: high frequencies of the incident light cause the "usual" long-wave shift of the fluorescent spectra, whereas low frequencies provide an "abnormal" short-wave shift (Rubinov et al. 1982).

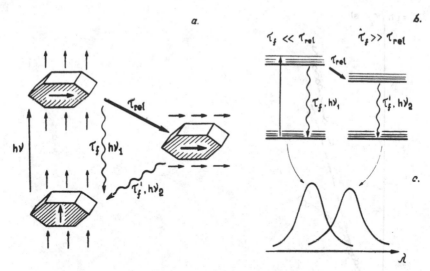

Figure 3.5. Schematic representation of physical phenomena that lead to relaxation shift in fluorescence spectra: τ_f and τ_{rel} are the excited-state lifetime and relaxation time, respectively; ν, ν_1, and ν_2 are frequencies of incident light and emission light at $\tau_f \ll \tau_{rel}$ and at $\tau_f \gg \tau_{rel}$, respectively: (a) directions of transition moment (large bold arrows) and electric dipoles of surrounding particles (short arrows) before and after excitation; (b) energy levels of the system under irradiation; (c) fluorescence spectra.

Microheterogeneity is manifested in the dependence of the relaxation shifts (ν_{max}) of fluorescent and phosphorescent spectra on temperature (Figure 3.6), while the characteristic times of phosphorescence (τ_{ph}^*) and fluorescence (τ_f^*) differ from each other by five to six orders to magnitude. Both types can be observed in a definite temperature region. The temperature dependence of the τ_r parameter can be seen (Figure 3.6) to be not greatly different from that for the dielectric relaxation time (τ_D) in the high-temperature region. However, the lower the temperature, the greater the divergence between the Arrhenius dependences of τ_r and τ_D. At low temperature, the deviation reaches five to six orders of magnitude.

The physical reasons responsible for such effects appear to be as follows: At low temperatures the relaxation processes induced by the dielectric dipole moment of the chromophore excited state make the main contribution to solvation. Spontaneous relaxation processes in the bulk of the sample are too slow to contribute significantly to the solvation of the excited molecules. When the temperature is increased, the contribution of spontaneous relaxation to the solvation increases. If $\tau_r \ll \tau_f^*$, the induced solvation is too fast to affect the relaxation shift. Yet another factor is

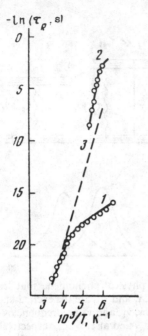

Figure 3.6. Arrhenius plots of the correlation times (τ_R) of dipole relaxation calculated from equation (3.4) for N-(1-oxy-2,2,6,6-tetramethyl-piperid-4-yl)-8-methylaminonaphthalenesulfonamide in a mixture of glycerol (75%), water (20%), and ethanol (1) and 2,7-dibromproflavin in water–glycerol solution (2). Dashed line denotes extrapolation of "high-temeperature" τ_R^v for the bulk relaxation to the low-temperature region (Likhtenshtein *et al.* 1990).

kinetic microheterogeneity caused by the cooperative reorientation of the whole ensemble of parameters.

If one assumes a Gaussian distribution over the free activation energies ($\Delta F^{\neq}$) of the reorientation of surrounding particles, it is possible to find an expression to relate the experimentally determined apparent activation energy (E_{max}) to the second moment of the distribution curve (ΔF_0^2) (Likhtenshtein & Kotelnikov 1983; Kotelnikov *et al.* 1983):

$$E_{app} = E_{max} - \Delta F_0^2/RT. \tag{3.10}$$

This expression is valid for cases in which the measurements are performed in a narrow range of relaxation times, τ_r, e.g., within the framework of one method (fluorescence or phosphorescence). In a joint analysis of data that differ in the characteristic frequency by six or more orders of magnitude, the value of E_{max} can be estimated from the equation

$$E_{max} \simeq (\ln \tau_f^* - \ln \tau_{ph}^*)(1/T_{ph} - 1/T_f), \tag{3.11}$$

where T_f and T_{ph} are the temperatures at the inflection points of the temperature dependences of v^f_{max} and v^{ph}_{max}, respectively. Experiments have shown that the value $E_{max} = 37\,kJ/mol$ for eosin in isobutanolic solution to be close to the activation energy for the dielectric relaxation, E_D. In the case of different specific interactions of the particles in the medium in the chromophore singlet and triplet states, the values of E_{max} and E_D are expected to be different.

Experimental values for the distribution parameter ΔF_0 for a number of simple compounds fall within the range of 6–14 kJ/mol. These values agree well with those determined by time-resolved techniques (Bystryak *et al.* 1990).

In mixed systems including a chromophore, a nonpolar solvent, and a polar ingredient, the kinetics of the relaxation shift can be limited by the translational and rotational diffusions of polar particles (Lumry & Hershberger 1978; Mazurenko 1989). Similar effects can be expected in biological objects, taking into account their chemical and structural heterogeneities. In membranes and proteins, for example, the water molecules or polar fragments can contribute significantly to the relaxation shift effect.

The role of specific interactions between the excited singlet state of the chromophore and surrounding molecules can be exemplified by the complexation of excited indole with hydroxy-containing solvents. This interaction is reflected in the anomalously large shift of the fluorescence spectra as compared with that observed when nonpolar solvents are used.

Considering the possibilities for including fluorescent and phosphorescent probes in various portions of proteins, membranes, nucleic acids, etc., one can conclude that the aforementioned approach holds great promise for solving a wide range of dynamic problems in molecular biology.

3.5. Study of local acidity and electrostatic and polar properties of biological objects

3.5.1. Measurement of pH

Many fluorescent probes can react with protons under physiological conditions. If the fluorescence intensity, I_f, is different for protonated and deprotonated forms, preliminary calibration is needed to estimate the acidity of the environment. A list of pK_a values for some fluorescent probes is as follows (Waggoner 1986; Dobretsov 1989):

quinoline	4.94
acridine	5.45
fluorescein	6.8
carboxyfluorescein	7.0
umbelliferone	8.0

    α-naphthol   8.6

    8-oxyquinoline   9.71

It is also of some practical use that some probes (such as acridine) associate in certain pH ranges, and the fluorescence of the associated molecules is quenched.

The value of pK_a is affected by local charges, ionic strength, and the polarity of the environment. This effect should be taken into account when investigating complex biological objects.

3.5.2. Measurements of electric charge density, transmembrane potential, and ion concentration

According to electrostatic theory (Waggoner 1986; Woolley *et al.* 1987; Dobretsov 1989; Kamino *et al.* 1989; Pratar *et al.* 1990), the electric potential $\varphi_{(\Delta R)}$ in solution at a distance ΔR from a flat surface is equal to

$$\varphi_{(\Delta R)} = 4\pi\sigma \exp(\ae\Delta R)/\varepsilon_0\ae, \tag{3.12}$$

where ε_0 is the dielectric constant, and $\ae$ is the effective ionic radius:

    $\ae = (8\pi\mu N_A^2 e^2/1{,}000\varepsilon_0 RT)^{\frac{1}{2}}$,

    $\mu = 0.5\Sigma C_i Z_i^2$,

    $\sigma = Ze/4\pi R_1^2$,

where μ is the ionic strength, C_i and Z_i are the concentration and charge of ions in solution, σ is the charge density, and R_1 is the radius of the given ion.

At an ionic strength $\mu = 0.1$ M, for example, the value of $\ae \approx 1\,\text{nm}^{-1}$, and the potential $\varphi_{(\Delta R)}$ drops by a factor of 2.7 at a distance of 1 nm from the surface.

We can calculate the σ value by measuring how the fluorescence intensity I_f of the absorbed probe depends on the ionic strength:

$$\sigma = \varepsilon_0 RT[\ln I_f'/I_f''/4\pi Ze N_A(1/\mu_1 - 1/\mu_2)], \tag{3.13}$$

where I_f' and I_f'' are the fluorescence intensities at the ionic strengths μ_1 and μ_2, respectively.

It should be noted that these deviations from the simple electrostatic formulas were observed in real biological objects. The reasons for such deviations are the specific interactions of probes with the complex mosaic of polar and hydrophobic portions of the objects.

The ratio of the concentration of a fluorescent probe inside (C_c) a membrane to that outside (C_m) is

$$C_m/C_c = \exp(-96.17\Delta\varphi_m/RT), \tag{3.14}$$

where $\Delta\varphi_m$ (in volts) is the transmembrane potential, and RT is given in kilojoules per mole. If this ratio is known, it can be used to calculate $\Delta\varphi_m$ in the object under investigation.

The cation forms of fluorescent probes listed below are commonly accumulated in the intramembrane space:

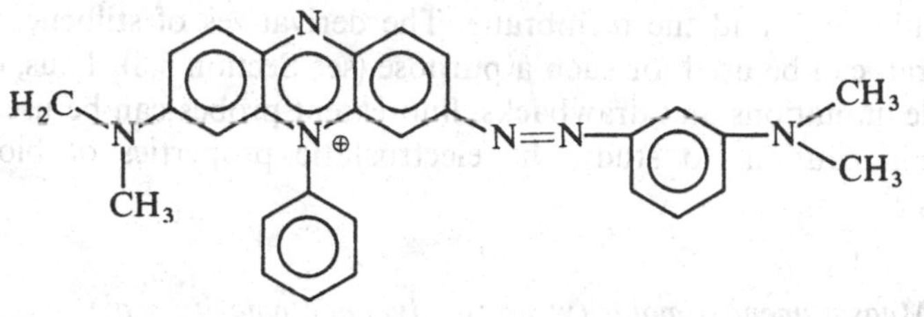

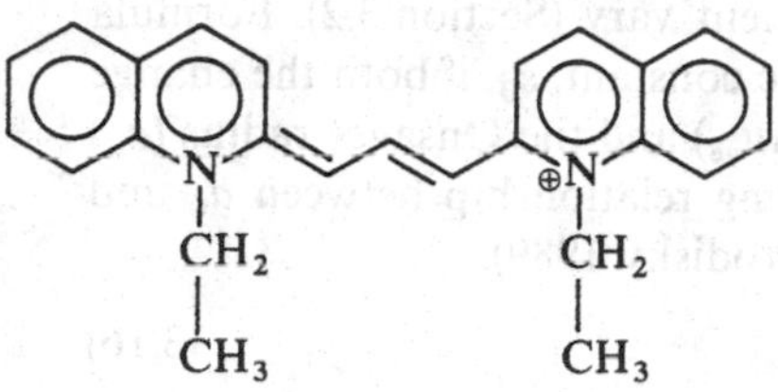

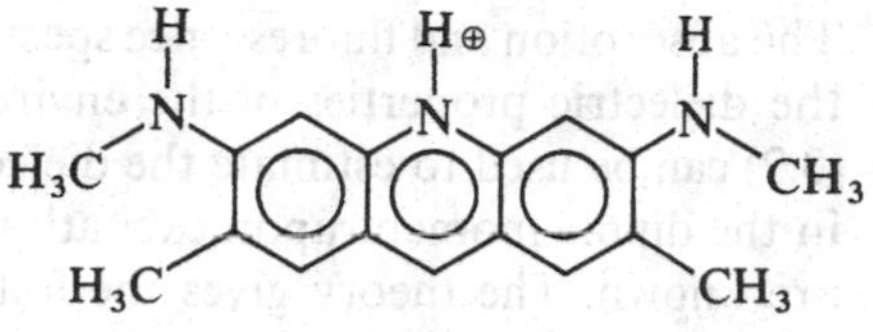

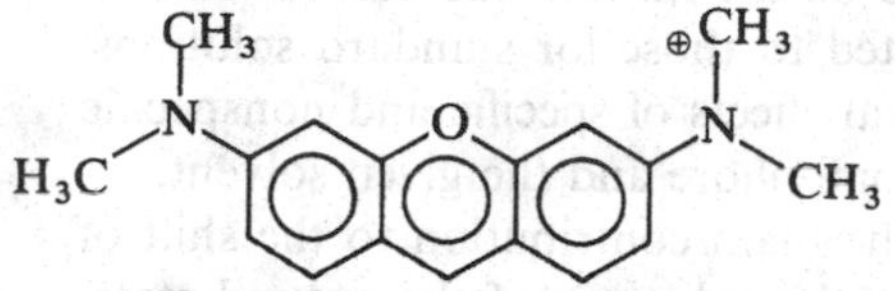

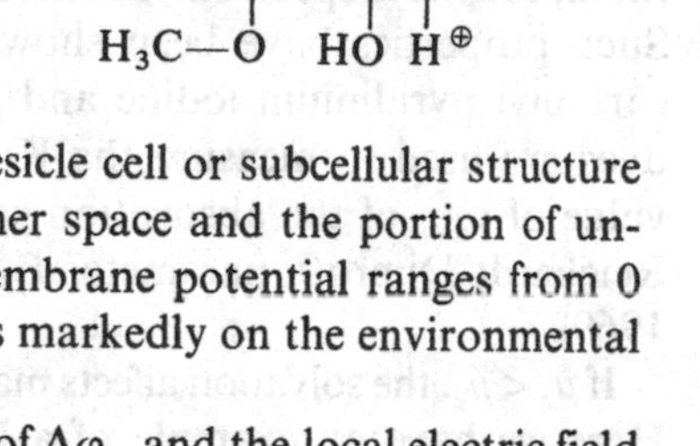

Determination of $\Delta\varphi_m$ in the closed vesicle cell or subcellular structure requires knowledge of the volume of inner space and the portion of unbound probes. The value of the transmembrane potential ranges from 0 to 0.25 V for various objects and depends markedly on the environmental and experimental conditions.

Two other methods for determination of $\Delta\varphi_m$ and the local electric field have been developed. The first is based on the formation of nonfluorescent associations of $\Delta\varphi_m$-dependent probes. In the second approach, one measures the effect of an electric field of intensity $\bar{E}$ on the shift of the absorption maximum or the fluorescence spectrum:

$$\Delta\nu_{max} = (\mu_g - \mu_e)\bar{E}/h + (\alpha_g - \alpha_e)E^2/2, \tag{3.15}$$

where μ_g and μ_e are the electric dipole moments of the chromophore in the ground and excited states, respectively, and α_g and α_e are the corresponding polarizabilities.

Certain dyes that form fluorescent complexes with ions are used for determination of the ion concentration in the intramembrane space. The antibiotic chlortetracycline, for example, forms a luminescent ternary complex with Ca^{2+} and the membrane. The derivatives of stilbene, indole, and furan can be used for such a purpose (see Section 3.8). Thus, despite possible limitations and drawbacks, fluorescent probes can be used, with a certain caution, to study the electrostatic properties of biological objects.

3.5.3. Measurement of polarity: on the dynamic polarity scale

The absorption and fluorescence spectra of chromophores are shifted when the dielectric properties of the environment vary (Section 3.2). Formula (3.2) can be used to estimate the dielectric constant, ε_0, if both the change in the dipole moment upon excitation ($\Delta\mu_{ge}$) and the Onsager radius (a_0) are known. The theory gives the following relationship between a_0 and the radius of the chromophore (r_{ch}) (Gorodisky 1989):

$$a^3 = a_0 r_{ch}^3 \cdot 10^{-24}\,cm^3, \tag{3.16}$$

where $a_0 = 0.97$ for nonassociative liquids, and $a_0 = 0.30$ for associative liquids, e.g., alcohol.

The empirical polarity scales based on the shifts of the absorption and fluorescence spectra ($\Delta\nu_{max}$) are related to those for standard solutions. The value of $\Delta\nu_{max}$ reflects the integral effects of specific and nonspecific interactions in the system of the chromophore and the given solvent.

For chromophores with $\mu_g > \mu_e$, the main contribution to the shift of the absorption spectrum comes from the solvation of the ground state. Such properties have been shown to be typical for 1-ethyl-4-methoxy-carbonyl pyridinium iodine and N-phenyl pyridinium biotin. The latter dyes are used to measure the Kosover polarity parameter Z_K (with the value of ν_{max} of the absorption spectrum expressed in kcal/mol) and the Reichardt-Dimroth parameter E_T, respectively (Kosover 1968; Reichardt 1969).

If $\mu_g < \mu_e$, the solvation affects mainly the shift in the fluorescence spectrum. Here we have an example of a highly sensitive fluorescent probe (Shin et al. 1988).

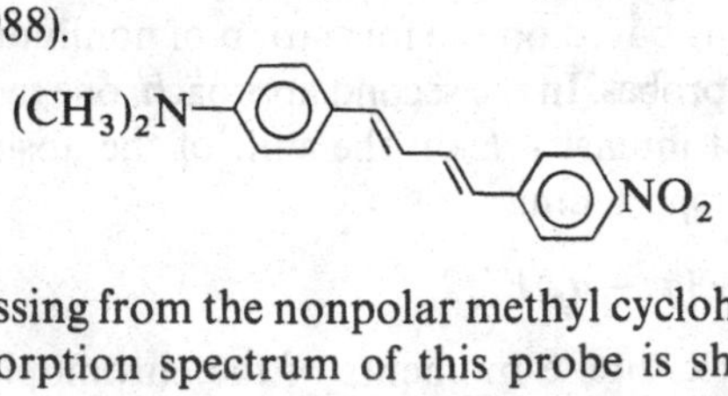

In passing from the nonpolar methyl cyclohexane to the polar methanol, the absorption spectrum of this probe is shifted by 16 nm, whereas the fluorescence spectrum is shifted by 111 nm.

In model mixed solutions, the polar components make the main contribution to the Δv_{max} value (Bakhshiev 1989). In the ternary system consisting of 3-aminophthalimide (or its analogues), a nonpolar solvent, and propanol, involvement of one molecule of the alcohol in the coordination sphere of the chromophore is accompanied by shifts of 500–600 and 1,000–1300 cm^{-1} in the absorption and fluorescence spectra, respectively.

It is noteworthy that all traditional spectral approaches to estimating polarity work well when the change in solvation between the ground and excited states of the chromophore is fast enough or small enough that the solvating layer is at equilibrium during the excited-state lifetime. However, a considerable number of biological and, in particular, cryobiological processes occur when the characteristic solvation time of the excited state (τ_r) is equal to or larger than the excited-state lifetime ($\tau^*_{f,ph}$). In such a case, the apparent polarity of the environment will depend not only on the static dielectric constant ε_0 and other parameters of solvation but also on the relaxation time τ_r.

In the framework of the classic Debye model of dielectric relaxation, the value of time-dependent $\varepsilon_0(\tau)$ is a function of the characteristic time of relaxation, τ_D:

$$\varepsilon_0(\tau) = \varepsilon_0(\infty) - [\varepsilon_0(\infty) - \varepsilon_0(0)]/(1 + \omega_D^2 \tau_D^2), \qquad (3.17)$$

where $\varepsilon_0(\infty)$ and $\varepsilon_0(0)$ are dielectric constants at $\tau_D \to \infty$ and $\tau_D \to 0$, respectively, and ω_D is the frequency of the electromagnetic radiation. The formula can be applied practically to only the simplest homogeneous polar liquids.

To estimate the local dynamic polarity of complex systems consisting of proteins, membranes, nucleic acids, etc., an approach was developed based on analysis of the temperature dependence of the relaxation shifts of the fluorescence and phosphorescence spectra (Likhtenshtein *et al.* 1990). As will be shown in Section 3.4, the Arrhenius plots of the relaxation times τ_r calculated from formula (3.4) encompass two regions. The high-temperature region is related to spontaneous relaxation in the bulk solvent. If the high-temperature Arrhenius plot is extrapolated to the low-temperature region, formula (3.4) can be used to calculate the theoretical values of the spectral maximum $v^v_{max}(T)$ caused by spontaneous relaxation, at a given temperature T. The value of the calculated relative shift in the overall temperature region is

$$P_v = [v^v_{max}(T) - v_{max}(0)]/[v_{max}(\infty) - v_{max}(0)], \qquad (3.18)$$

where indices ∞ and 0 correspond to $\tau_r \to \infty$ and $\tau_r \to 0$; the P_v value can be considered as the parameter that summarizes all the effects, including solvation energy and the characteristic time, τ^v_r, whose value is comparable to the lifetimes of the excited singlet (τ^*_f) and triplet (τ^*_{ph}) states. For each value of τ^*_f and τ^*_{ph} of a standard chromophore, the dependence of P_v on temperature can be plotted following analysis of the correlation between the kinetic parameters of the process of interest, e.g., electron transfer, and the parameter of bulk relaxation, P_v (Section 3.6).

At low temperatures, the induced dielectric relaxation can be quantitatively characterized by the parameter of the total shift:

$$P_i = [v^v_{max}(T) - v_{max}(0)]/[v_{max}(\infty) - v_{max}]. \tag{3.19}$$

A correlation can be expected between the P_i value and the kinetic parameters of the processes limited by the induced solvation of the intermediate products; this correlation will be characterized by a time τ_r that is comparable to τ_f^* and τ_{ph}^* for a standard chromophore.

3.6. Inductive resonance energy transfer as a method of investigating structures and dynamics of biological objects

3.6.1. Mechanism of inductive resonance energy transfer

The energy transfer from an electronically excited molecule to an unexcited one is known to occur via several different mechanisms:

1. radiative energy transfer, when a photon is emitted by the excited molecule (donor, D) and absorbed by the other molecule (acceptor, A);
2. exchange resonance energy transfer, when the donor and acceptor electron clouds overlap; radiation is not emitted by this transition;
3. inductive resonance energy transfer, due to the dipole–dipole interaction between the chromophores; no emission.

For each of these mechanisms it is useful to know how, in theory, the probability of energy transfer depends on the D–A distance.

The probability of radiative energy transfer without light scattering in the medium is independent of the D–A distance. The probability of exchange resonance energy transfer depends exponentially on the D–A distance and usually is zero for distances longer than 1.4 nm. Thus it is impossible to use these two processes of energy transfer to estimate long D–A distances. As will be shown later, the third mechanism, i.e., inductive resonance energy transfer, can be fruitfully used for quantitative estimation of the D–A distance in biological systems.

In the model of paired interactions between the excited donor molecule D^* and the unexcited acceptor A, the energy transfer process may be presented either as

$$D^* + A \rightarrow D + A^*$$

or as in the energy scheme given in Figure 3.7. The donor fluorescence and acceptor absorption spectra are shown in the same figure.

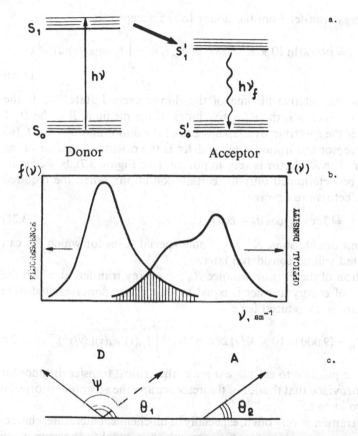

Figure 3.7. Scheme of the inductive resonance energy transfer process: (a) system of energy levels of the donor and the acceptor; (b) donor fluorescence and acceptor absorption spectra; (c) dipole model of the donor–acceptor $(D-A)$ interaction; θ_1 is the angle between the donor dipole (transition moment vector of the donor) and the direction of the vector connecting the donor and the acceptor; θ_2 is the angle between the acceptor dipole (transition moment vector of the acceptor) and the vector connecting the donor and the acceptor; ψ is the angle between the donor and acceptor dipoles.

According to theory, the interaction between the donor and acceptor can proceed nonradiatively, so that two intramolecular transitions (excited donor → unexcited donor, and unexcited acceptor → excited acceptor) occur simultaneously without fluorescence emission from the donor.

The energy transfer for the dipole–dipole interaction within the scope of quantum mechanics was first treated by Forster (1948). The rate constant for the energy transfer from the donor to the acceptor is

$$k_{DA} = (9{,}000\ln 10 \times K^2/128\pi^5 n^4 N_A \tau_f R^6)\cdot \int f_D(v)\cdot\varepsilon_A(v)(dv/v^4),$$

$$(3.20)$$

where τ_f is the relative lifetime of the donor excited state, N_A is the Avogadro number, n is the refractive index of the medium, R is the D–A distance, $\int$ is the spectrum overlap integral of the donor fluorescence $f_D(\varphi)$ and the acceptor absorption $\varepsilon_A(v)$, and K^2 is the orientation factor of the dipole pair. The K^2 factor is very important (see Figure 3.7); however, its value can be determined only for certain conditions, when the required geometric factors are known:

$$K^2 = |3\cos\theta_1\cdot\cos\theta_2 - \cos\theta_3|^2.$$

$$(3.21)$$

For random orientations, K^2 is $\frac{2}{3}$. Some special cases for which K^2 can be calculated will be considered later.

The notion of the critical distance R_0 for energy transfer, at which the probability of energy transfer is equal to that of the donor excited-state deactivation, is introduced here:

$$R_0^6 = (9{,}000\ln 10 \times K^2/128\pi^5 n^4 N_A)\cdot \int f_D(v)\cdot\varepsilon_A(v)(dv/v^4).$$

$$(3.22)$$

Thus, it is possible to calculate *a priori* the critical transfer distance for each pair provided that the donor fluorescence and the acceptor absorption spectra are known.

Energy transfer is very often, especially in biochemical literature, characterized by the energy transfer efficiency, which is equal to the proportion of the energy transferred from the donor to the acceptor:

$$E_{eff} = (\varphi_0 - \varphi)/\varphi_0,$$

$$(3.23)$$

where φ and φ_0 are the quantum yields of the donor fluorescence with and without the acceptor. The E_{eff} value expressed in terms of the R and R_0 distances is

$$E_{eff} = 1/[1 + (R/R_0)^6].$$

$$(3.24)$$

3.6.2. Estimation of the distance between donor and acceptor groups

From simple considerations, the D–A distance (when one donor interacts with only one acceptor) is obtained by the donor fluorescence quenching ratio:

$$R = R_0/(\varphi_0/\varphi - 1)^{\frac{1}{6}}.$$

$$(3.25)$$

Several other cases are described next (Weber & Teal 1959; Stryer 1978; Alfimova & Likhtenshtein 1979):

1. one donor and n acceptors:

$$\varphi_0/\varphi = 1 + \sum_{i}^{n} (R_0/R_i)^6; \tag{3.26}$$

if $R_i = R$,

$$R = R_0 \cdot [n \cdot (\varphi_0/\varphi - 1)]^{\frac{1}{6}}. \tag{3.27}$$

[Two different acceptor arrangements in a sphere (or a circle) with a donor (or acceptor) in the center correspond to the latter case.]

2. m donors and one acceptor:

$$\varphi/\varphi_0 = 1/m \sum_{j}^{m} [1 + (R_0/R_j)^6]^{-1}; \tag{3.28}$$

if $R_j = R$,

$$\varphi/\varphi_0 = [1 + (R_0/R)^6]^{-1}. \tag{3.29}$$

3. m donors and n acceptors:

$$\varphi/\varphi_0 = 1/m \sum_{j}^{m} \left[1 + \sum_{i}^{n} (R_0/R_j)^6 \right]^{-1}. \tag{3.30}$$

It is of interest to estimate the contributions to the quenching ratio of two acceptors spaced at R_1 and R_2. The main contribution to quenching is made by the nearest neighbor. The contribution of the other acceptor is substantial (20%) if $R_1 < R_2 < 1.05R_1$ and is quite negligible at $R_2 > 1.05R_1$. In complicated cases, when there is a whole set of distances, the fluorescence decay will not obey an exponential law. The distribution effect of the distance between the donors and acceptors on the kinetics of the fluorescence decay has also been analyzed in detail (Grienvald, Haas, & Steinberg 1972).

3.6.3. Orientation factor

The expressions for estimation of the donor–acceptor distance involve the value of the orientation factor K^2, which may range from 0 to 4 according to equation (3.21). This uncertainty is an obstacle for precise determination of the distance R in the general case. Nevertheless, in many instances the experimental data and theoretical approaches permit reasonable values of K^2 to be found and hence the value of R to be determined with sufficient accuracy. As shown in recent papers on inductive resonance energy transfer, K^2 is equal to $\frac{2}{3}$ whenever time of random reorientation of transition vectors in chromophores is much shorter than the lifetime of the donor excited state. The use of this value may introduce noticeable errors to the

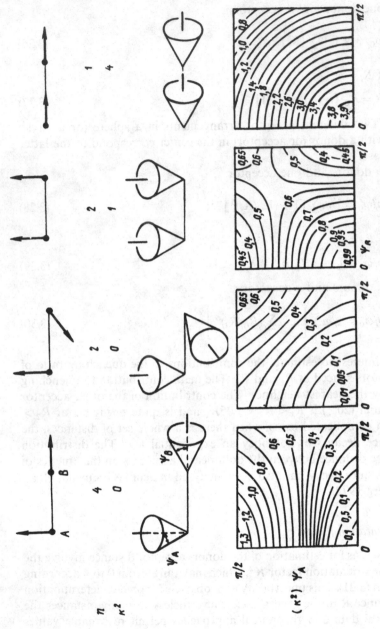

Figure 3.8. Eisinger diagrams for calculating the orientation factor K^2 [equation (3.21)] at different orientations of the donor and acceptor groups with regard to chromophore rotation; n is the weight of the state; ψ_A and ψ_B denote the angles limiting the rotation cones of transition dipoles.

distance estimation in cases where the real value of K^2 is different. Consequently, the problem of the chromophores' orientation requires careful analysis for each particular system.

Average values of K^2 have been calculated for a number of mutual orientations of corresponding vectors for donor and acceptor groups (Haas *et al.* 1974; Dale & Eisinger 1976; Grinvald *et al.* 1972) (Figure 3.8). To see how these calculations can be used practically, let us denote the distance calculated from the formula with $\langle K \rangle^2 = \frac{2}{3}$ by the symbol $\bar{R}$. As can be seen from the figure, the real distance $R \leqslant 1.35\bar{R}$ if appreciable energy migration is detected in the experiment. Thus, the method allows estimation of an upper limit for the distance between the chromophores, even without knowing their mutual orientation. Analysis of the Eisinger diagrams leads to the following conclusions:

1. If a chromophore (donor or acceptor) rotates isotropically because of some structural peculiarities (which is the case for transition metals), the value of K^2 ranges from 0.35 to 1.3, and $0.9\bar{R} \leqslant R \leqslant 1.1\bar{R}$.

2. Rotation of chromophores at an angle of 30° gives the value of $0.8\bar{R} \leqslant R \leqslant 1.2\bar{R}$, and $0.9\bar{R} \leqslant R \leqslant 1.1R$ at 45°, for practically all possible mutual orientations, and in the case of parallel transition axes, $0.95\bar{R} \leqslant R \leqslant 1.05\bar{R}$ at any rotation angle.

3. If chromophores reorient themselves with a correlation time greater than the lifetime of the donor excited state (static distribution), the foregoing calculations are not suitable. Nevertheless, calculating R when $E_{eff} \leqslant 0.5$ leads to only minor discrepancies between the static and dynamic models. An extra analysis of the Eisinger diagrams results in the conclusion that the indeterminacy in R (estimated with $K^2 = \frac{2}{3}$) does not exceed 20% at $0.1 > E_{eff} \geqslant 0.0$.

As shown elsewhere (Chapter 6), spin and luminescence labels attached to protein macromolecules feature rotational mobility relative to protein globules, which is realized by rotation about several chemical bonds.

Thus, the aforementioned features of certain systems may restrict the accuracy of distance estimation to 10–20% in some cases. However, even at this accuracy the inductive resonance energy transfer may provide information on the locations of spaced sites in allosteric or multienzyme systems. The data on the chromophores' depolarization may increase the accuracy of the results of the D–A distance estimation. The method can be successfully applied to comparative investigations (e.g., to the study of conformational transitions) because of the high sensitivity of the experimental parameters to the distance between the chromophores.

Membrane structure can also be studied by this approach based on measurements of the efficiency of the inductive resonance energy transfer

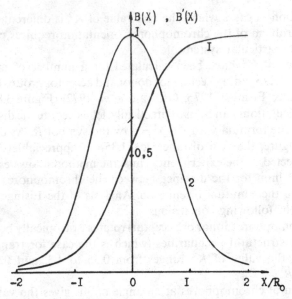

Figure 3.9. Dependences of the $B(X)$ and $B'(X)$ coefficients [equations (3.31) and (3.32)] on the thickness of the membrane X. R_0 is the critical radius of the inductive resonance energy transfer (Dobretsov 1989).

between donor and acceptor molecules located in different portions of the membrane system (Dobretsov *et al.* 1982; Dobretsov 1989). If the donor molecules are distributed uniformly in the space at one side of a flat membrane of thickness x, and the acceptor molecules are on the other side, the theory proposes the following expressions:

$$\ln(I_f/I_f^0) = -1.6 \cdot \alpha \cdot B(X) N_0 \pi R_0^3 \tag{3.31}$$

and

$$\ln(I_f/I_f^0) = -1.14\alpha' \cdot B'(X) N_0' R_0^2, \tag{3.32}$$

where I_f and I_f^0 are the intensities of fluorescence with and without an acceptor, respectively; N_0 and N_0' are the spatial and surface concentrations of the acceptor; $B(X)$ and $B'(X)$ are functions of the membrane thickness (Figure 3.9); and α is a coefficient that depends on the mutual orientation of the transition dipole moments of the donor and acceptor chromophores, where $\alpha = 1$ for rapid rotation of the chromophores. Equations (3.31) and (3.32) allow determination of the membrane thickness if the R_0, N_0, and N_0' values are known.

Table 3.1 lists values of the critical radius (R_0) for a number of donor–acceptor pairs that are used in biophysical investigations.

Table 3.1

Donor	Acceptor	R_0, nm
Tb^{3+}	Rodamine B	6.0
Anthracene	Perylene	3.35
Pyrene	Perylene	3.8
Fluorescein	Heme	5.1
Rodamine	Heme	5.5
Tyrosine	Tryptophan	1.9
Tyrosine	Heme	3.1
Indole	Pyrene	4.0
Indole	ANS	3.7
Tryptophan	Pyrene	2.7
Tryptophan	Anthracene	2.7
Tryptophan	Heme	3.7

By keeping a clear view of all the *pros* and *cons* of the energy transfer method, utilizing all the recent achievements in protein, membrane, and nucleic acid chemistry, and making use of independent physicochemical approaches, the investigator will have efficient tools for studying a wide range of complex biological systems.

3.7. Dynamic quenching of fluorescence as an approach to the study of molecular dynamics

The phenomenon of quenching of excited states has been widely used in physicochemical and biophysical investigations (Burshtein 1976; Blackwell *et al.* 1987; Dobretsov 1989; Chen & Blumenthal 1990). For a number of systems the experimental dependences of relative intensities (I_f) and quantum yields (φ_f) of fluorescence on the concentration of a quencher [Q] are described by the Stern-Volmer equation:

$$\varphi_f^0/\varphi_f = I_f^0/I_f = 1 + K_Q[Q] = 1 + \tau_f^* k_q[Q], \qquad (3.33)$$

where K_Q is the Stern-Volmer constant, and k_q is the quenching rate constant.

The direct proportionality of φ_f^0/φ_f and I_f^0/I_f constitutes evidence for dynamic quenching. When nonfluorescent complexes are formed between an excited chromophore and a quencher, quenching is static. Quenching of the excited state of a chromophore upon its collision with the quencher may occur by any of the following mechanisms: (1) inductive resonance or exchange resonance transfer of energy, if the energy levels of donors and acceptors overlap fairly well; (2) catalysis of the interconversion transition by heavy atoms; (3) quenching by the mechanism of charge transfer, which may be either reversible or irreversible.

The last mechanism, which is widespread and deserves to be considered in more detail, generally includes a number of stages: complexation, electron transfer, and recombination:

$$X + Q \xrightarrow{h\nu} X^* + Q \underset{k_{-1}}{\overset{k_1}{\rightleftharpoons}} [X^* \cdots Q] \underset{k_{-e}}{\overset{k_e}{\rightleftharpoons}} [X^{\pm} \cdots Q^{\mp}] \rightarrow \text{product}.$$

In polar media the component of the ion pair $X^{\pm}$ and $Q^{\mp}$ separate from each other; in nonpolar media pair recombination predominates. While the electron transfer and recombination steps are fast, the rate-determining step is the complexation, and $k_q \cong k_1 \approx k_D$ (where k_D is the rate constant for particle collision). Limitation of the process by electron transfer leads to the equation $k_q = k_1 k_e / k_{-1}$. The k_q value increases with higher environmental polarity and higher driving energy of electron transfer.

Molecules with polar groups (—COOH, COO⁻, —NH$_3^+$, imidazoline, etc.), ions (I⁻, NO$_3^-$, CNS⁻, etc.), molecules with heavy atoms, and many aromatic chromophores quench excited chromophores, with rate constants $k_q = (3-6) \cdot 10^9 \, M^{-1} s^{-1}$, which are close to the diffusion-controlled rate k_D.

According to experimental data, quenching of a number of chromophores by aromatic molecules appears to be even more efficient; this efficiency is usually explained in the framework of the simple Stock–Einstein theory of collisions (Kapinus & Kucherova 1984). In the author's opinion, the efficiency may be due to harpoon electron transfer before the actual collision of the chromophores.

The phenomenon of dynamic quenching is widely used to study the microviscosity of biological systems and the accessibility of chromophores and quenchers.

3.8. Charge transfer complexes, excimers, and exciplexes as luminescent probes

Aromatic compounds such as Durene, naphthalene, N-phenylcarbazole, anthracene, pyrene, chrysene, and phenanthrene form fluorescent charge transfer complexes with the acceptor molecules Durene, chloranyl, 1,3,5-trinitrobenzene, 1,2,4,5-tetracyanobenzene, tetrachlorophthalanhydride, etc. The complexes are characterized by high quantum yields, large changes in the transition dipole moment upon excitation into the first excited singlet state (up to 7 D), and relatively large relaxation shifts (up to 1,600 cm⁻¹). The unique properties of these complexes allow them to be used as probes for studying the micropolarity and local dynamics of the objects under investigation (Leonhardt & Weller 1963; Ware 1975; Waggoner 1986; Dobretsov 1989; Kapinus 1988).

High chemical reactivity of molecules in the singlet excited state causes the formation of complexes with other molecules in the ground state. According to the current terminology, complexes between identical molecules (monomers) are named excimers; those between different molecules are referred as exciplexes. The differences in the fluorescence parameters of monomers (M), dimers (M_2), and exciplexes ($M \cdot X$) allow the fluorescence spectra of these compounds to be detected in a single experiment.

The processes of formation of exciplexes, which involve association, dissociation of the complex, electron transfer, and luminescence of M^* and MX^*, are commonly accepted to proceed according to the following scheme:

$$M \xrightarrow{h\nu} M^* + X \underset{k_{-1}}{\overset{k_1}{\rightleftharpoons}} M^*X \xrightarrow{k_f^e} MX$$

$$\begin{array}{c} k_f \\ -h\nu_f \\ k_{da} \end{array} \qquad \begin{array}{c} -h\nu_f^e \\ k_{da}^e \end{array}$$

where k_f and k_f^e are the emission constants of the monomer and the exciplex, respectively, and k_{da} and k_{da}^e are the respective constants of radiationless deactivation.

According to this scheme, the relative values of the quantum yields and fluorescence intensities of the excimers and monomers can be expressed as

$$\varphi_f^e/\varphi_f = I_f^e/I_f = (k_f^e/k_f)[k_1/(k_f^e + k_{da}^e + k_{-1})]. \tag{3.34}$$

This equation allows the value of the collision constant k_1 to be determined if $k_{-1} \ll (k_f^e + k_{da}^e)$ and k_f, k_f^e, and k_{da}^e are known. At $k_{-1} \gg (k_f^e + k_{da}^e)$, the value of the equilibrium constant $K_{MX} = k_1/k_{-1}$ can be estimated. In low-viscosity media, typical values of these constants are $k_1 \simeq 5 \cdot 10^9 \, \text{M}^{-1} \text{s}^{-1}$, $k_{-1} = 5 \cdot 10^7 \, \text{s}^{-1}$, $k_f \approx k_f^e \approx 5 \cdot 10^9 \, \text{s}^{-1}$.

The value of k_1 is directly proportional to the diffusion coefficient and can be used to study microviscosity in biological objects.

The characteristic feature of exciplexes is the large value of the dipole moments, amounting to up to 16 D (e.g., the exciplex of excited naphthalene and triethylamine). When passing from nonpolar ($\varepsilon_0 = 2$) to polar ($\varepsilon_0 > 10$) solvents, the long-wave length relaxation shifts of the exciplex fluorescence spectra lie in the range of 3,500–4,000 cm^{-1}. Thus, exciplexes can serve as sensitive indicators of micropolarity and of the relaxation dynamics of the environment. The values of the enthalpy ($\Delta H = 4$–65 kJ/mol) and entropy ($\Delta S = 22$–130 J/mol/degree) of exciplex formation change symbiotically with variations in the chemical structures of components and solvents (compensation phenomena) (Kapinus 1988).

Molecules with donor and acceptor groups connected by flexible bridges seem to be the most promising in the study of biological objects (Anderson *et al.* 1974; Filipescu & Chang 1972; Davidson & Whelan 1977). Such molecules are exemplified by the formulas

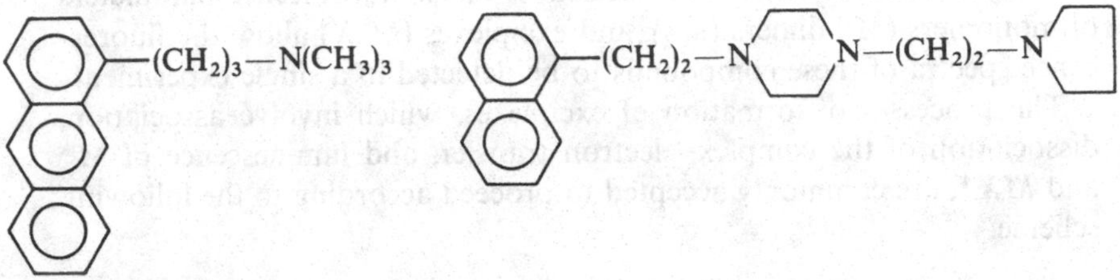

The "triple" intermolecular exciplexes have large dipole moments and relaxation shifts in a polar solvent.

The fluorescent bifunctional chromophores that feature complete charge separation after irradiation are of some special interest. For example, the bianthryl molecule forms an ion-radical pair in acetonitrile solution (Kang *et al.* 1988; Bystryak *et al.* 1992)

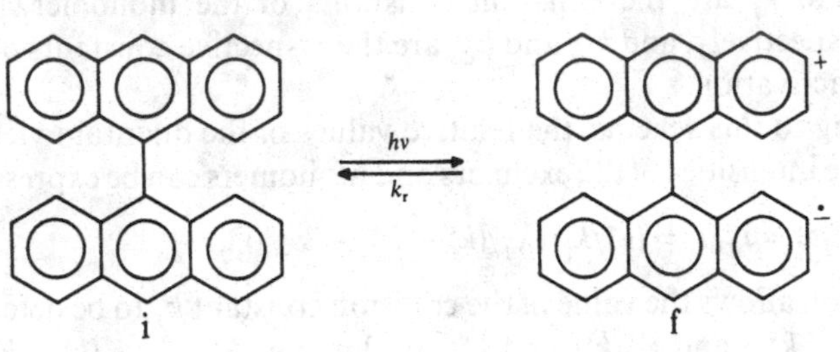

that is characterized by $v_{max}^f = 21,000 \, cm^{-4}$ and $\mu_{eg} = 5.5 \, D$; these parameterers are extremely sensitive to the polarity and relaxation dynamics of the environment. The value of the recombination constant k_r is even more sensitive to the aforementioned parameters of the medium.

3.9. Study of slow translational diffusion: photobleaching and fluctuation techniques

Relatively slow diffusion of fluorescent probes and labeled macromolecules is studied by photobleaching techniques, which are based on photobleaching of small portions of the object under study by laser pulses, followed by monitoring of fluorescence recovery as a result of diffusion from adjacent portions of the system (FRAP techniques) (Cherry 1979; Lakowics 1983; Beddard 1986; Edidin 1987; Dobretsov 1989).

When the area of photobleaching is small enough, the kinetics of the recovery of fluorescence intensity follow a simple exponential law (Yguerabide & Foster 1981):

$$[I_f(\infty) - I_f(t)]/[I_f(\infty) - I_f(0)] \approx \exp(-t/\tau'_D),$$

where the indices 0, t, and ∞ correspond to the observation time; the value of $\tau_D \approx 0.32R^2/D_{\tau_r}$, where D_{τ_r} is the translational diffusion coefficient equal to 10^{-11}–$10^{-15}\,\mathrm{m^2\,s^{-1}}$ for biological molecules in biomembranes.

Recently, some improved versions of the photobleaching method have been developed (Peters *et al.* 1981; Koppel & Sheetz 1983). These are invasive techniques of photobleaching using stationary light of relatively low intensity to simultaneously photobleach several spots. The first approach may be exemplified by the work of Ferrieres *et al.* (1989), in which an algorithm was proposed to determine the value of the diffusion coefficient for the chromophore-monomer and that of the association constant for dimers. In the second approach, developed by Smith and McConnell (1981), diffusion coefficients can be measured based on the phenomenon of recovery of polarization of the chromophore fluorescence.

Microscopic diffusion processes can be studied by analyzing the auto-correlation function $I_f(t) \cdot I_f(t + \Delta t)$, which is measured experimentally (Magde *et al.* 1974; Geertes 1983). Spontaneous diffusion is accompanied by fluctuations in the fluorescence intensities of small portions of the labeled surface ($\sim 0.2\,\mu\mathrm{m}$). The fluctuation frequency is a function of the diffusion coefficient.

A method has also been developed to determine the diffusion coefficients of macromolecules by analyzing the broadening of the light-scattering spectra and the Doppler shift in the luminescence spectra.

In conclusion, let us summarize the main possibilities for the methods of fluorescent labels and probes. These methods allow (1) measurement of the distance between chromophores up to 0.6 nm, (2) determination of the values of the coefficients of rotational ($D_R = 10^{-8}$–$10^{-10}\,\mathrm{s^{-1}}$) and translational ($D_{tr} = 10^{-11}$–$10^{-15}\,\mathrm{ms^{-1}}$) diffusion and the collision rate constant [$k_d = 5 \cdot (10^{-8}$–$10^{-10})\,\mathrm{M^{-1}\,s^{-1}}$], (3) study of relaxation processes with correlation times of $\tau_c = 10^{-8}$–$10^{-10}\,\mathrm{s^{-1}}$, (4) estimation of the polarity and acidity of the objects of interest, and (5) determination of potentials and other electrostatic characteristics of biological systems.

4

Triplet labeling methods

In the past few years, experimental techniques have so improved that it has become possible to study the triplet excited state of a chromophore by its phosphorescence or delayed fluorescence emission both in frozen solution and in liquids at room temperature. A number of the properties of phosphorescence, such as radiative and radiationless deactivation of the excited states, polarization, relaxation shifts, dynamic quenching, formation of exciplexes, etc., are similar to those of fluorescence described in Chapter 3. On the other hand, there are some essential features of triplet probes that distinguish them from fluorescent tables. In particular, paramagnetism and the relatively long lifetimes of excited triplet states allow us to extend the boundaries of investigation of biological systems as compared with the method of fluorescence (Kotelnikov *et al.* 1979; Maniara *et al.* 1988; Likhtenshtein *et al.* 1982; Shin *et al.* 1988; Mekler *et al.* 1983; Sudha *et al.* 1984; Mekler & Likhtenshtein 1986; Van der Kooi *et al.* 1990).

4.1. Peculiarities of triplet excited states

Direct absorption of a light quantum by the ground singlet state (S_0), with the formation of triplet states (T_1, T_2), accompanied by a change in the spin of the system, is forbidden according to the law of conservation of momentum (spin conservation). Because of spin–orbital and spin–spin electron–nucleus interactions, the probability of the $S_0 \rightarrow T_i$ transition is not equal to zero, though it is very small. However, in general the triplet state is reached via an $S_0 \rightarrow S_1$ transition, followed by $S_1 \rightarrow T_1$ interconversion. The value of the internal conversion rate constant, $k_{ST} = 10^4 - 10^{12}\,\mathrm{s}^{-1}$, depends on the chromophore structure and the presence of other molecules, e.g., paramagnetics or heavy atoms.

The value of the rate constant for radiationless transition $k_{T_1 S_0}$ ranges from 10^{10} to $10^0\,\mathrm{s}^{-1}$; the rate constant for triplet-state deactivation by radiation is $k_{ph} = 10^6 - 10^0\,\mathrm{s}^{-1}$. The competition between the radiationless transitions, $S_1 \rightarrow S_0$ and $S_1 \rightarrow T_1 \rightarrow S_0$, is governed by the value of the appropriate energy gap ($\Delta E_{S_0 S_1} = E_{S_1} - E_{S_0}$). If $\Delta E_{S_0 S_1} > 2 \cdot 10^4\,\mathrm{cm}^{-1}$, the

116

second mechanism, internal conversion, prevails (Ermolaev 1963; Parker 1968; Ermolaev *et al.* 1977; Kapinus 1988).

The theory also predicts the reverse $T_1 \rightarrow S_1$ transition, followed by the $S_1 \rightarrow S_0$ transition with emission of a quantum (the delayed fluorescence). This type of fluorescence is temperature-dependent (following the Boltzmann law) and is referred as E-type delayed fluorscence. Another type of delayed fluorscence, P-type, is caused by triplet–triplet annihilation upon collision between two excited chromophores (Figure 4.1). The temperature dependence of the latter results from the dependence of collision frequency on the viscosity of the medium.

The low transition rates of excited triplet states make it possible to study very slow rotational and translational diffusion and relaxation processes with characteristic times of 10^0–10^{-6} s.

One more principal difference between the phosphorescence and fluorescence phenomena is that the excited triplet state has two unpaired electrons and can be considered as a specific biradical label. The processes of inter-conversion and deactivation of the triplet state can be accelerated owing to exchange spin–spin interactions with paramagnetics, including other chromophores in the triplet state, radicals, and transition ions (Figure 4.1). The efficiency of these processes will depend on the structure of the reagents and the medium, their dynamics, distances, mutual reorientation, etc. Thus, the aforementioned properties of the object of interest can be studied by the triplet labeling method.

The triplet and singlet excited molecules differ also in a number of physicochemical properties. In the long-lived triplet state the probability of absorption of a second quantum and that of a chemical reaction are increased. However, the chemical reactivity of the triplet state is, as a rule, less than that of the high-energy singlet state.

Three mechanisms of energy transfer with participation of excited triplet states can be distinguished, e.g., triplet–singlet, singlet–triplet, and triplet–triplet (Erolaev *et al.* 1977). The two former mechanisms, of electrostatic inductive resonance nature, are forbidden by the law of spin conservation. For a triplet–singlet transition, the low rate of energy transfer can be compensated for by the long decay time of the donor triplet state if the overlap integral is large enough. At very short distances, in particular at the direct contact of chromophores during the act of collision and in the case of a strong spin orbital interaction, the efficiency of these formally forbiden mechanisms can be fairly high. For example, the rate constant for the forbidden energy transfer between the donor–acceptor pairs 9,10-dibromanthracene-acetophenone and 9,1-dibromanthracene-fluorene in liquid solution is $k = 5 \cdot 10^8$–$1.5 \cdot 10^9 \, M^{-1} s^{-1}$.

The exchange mechanisms in which excited triplet states participate will be considered in subsequent sections.

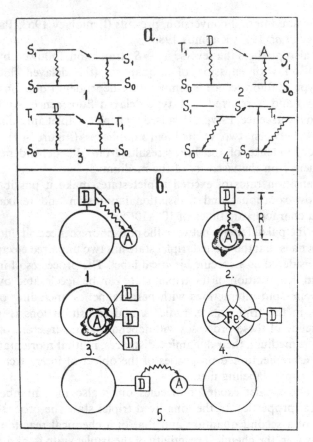

Figure 4.1. Fluorescent and triplet labeling techniques: (a) The processes of energy migration and exchange used in the method. Solid arrows indicate paths of energy migration from donor to acceptor. A vertical wavy arrow indicates the direction of energy transfer inside the chromophore; 1, $S_1 \rightarrow S_1$ transition; 2, $T \rightarrow S$ transition; 3, $T \rightarrow T$ annihilation. (b) Types of structural and dynamic problems solvable by the method (D, donor group; A, acceptor group); 1, distance between labeled portions of the biological matrix and its variation; 2, distance between the active center and the group at the surface; 3, depth of the center; 4, delocalization of the electron density over the protein globule; 5, the collision between molecules, including macromolecules (Likhtenshtein 1988b).

4.2. Structures and chemical properties of triplet probes

Five types of phosphorescent chromophores are known (Parker 1968; Ermolaev *et al.* 1977; Kapinus 1988):

    1. aromatic compounds with condensed rings,
    2. aromatic compounds with heavy atoms (Br, I, Pt, etc.),

3. compounds with diketone groups (diacetyl, dibenzyl),
4. porphyrins and their metal complexes (Zn, Co, Cd, etc.), and
5. lanthanides (Tb^{3+}, Eu^{3+}, and Tl^{3+}).

A list of the chemical formulas of a number of phosphorescent molecules that are used or may be used as triplet probes is as follows:

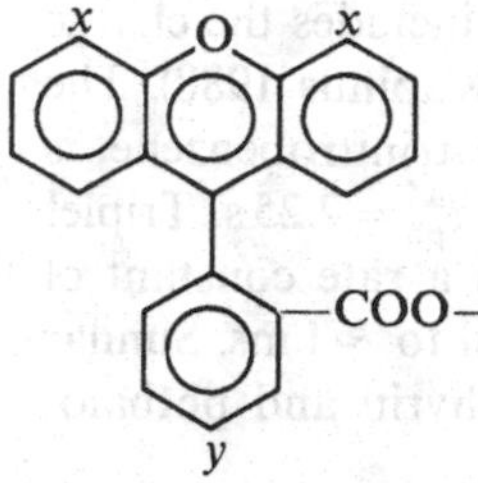

$x = I$, $y = H$; $x = Br$, $y = H$;
$x = I$, $y = SCN$

TPI

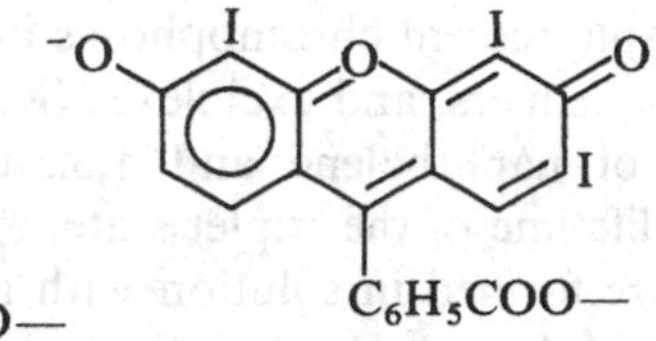

TPII

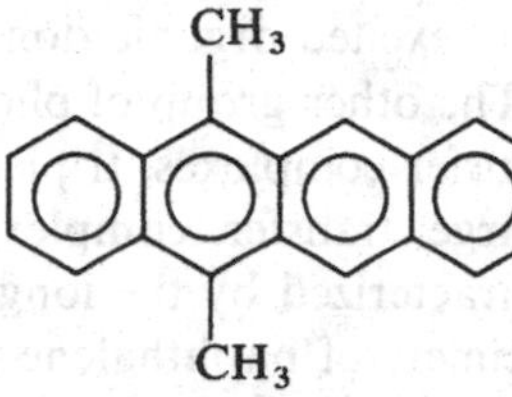

TPIII

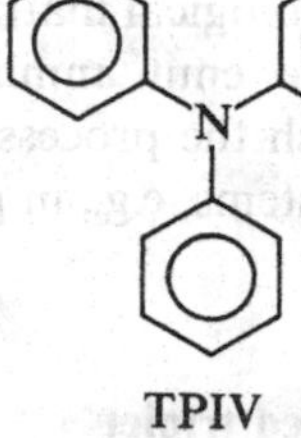

TPIV

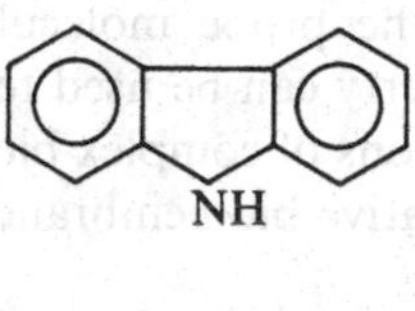

TPV

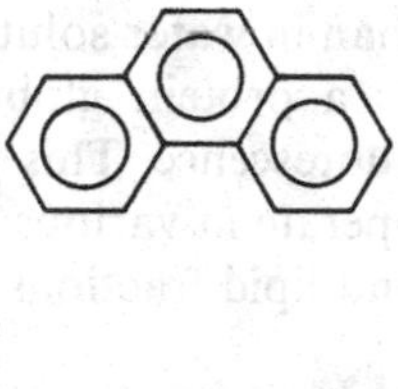

TPVI

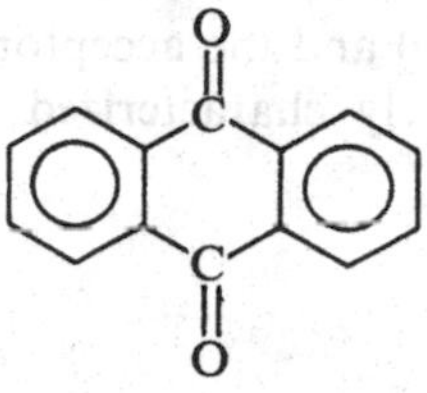

TPVII

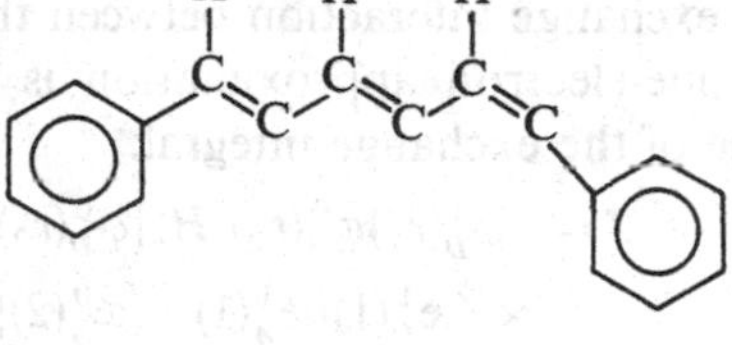

TPVIII

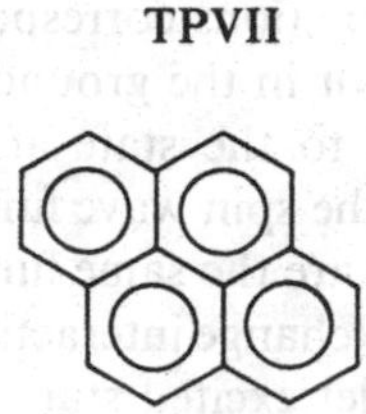

TPIX

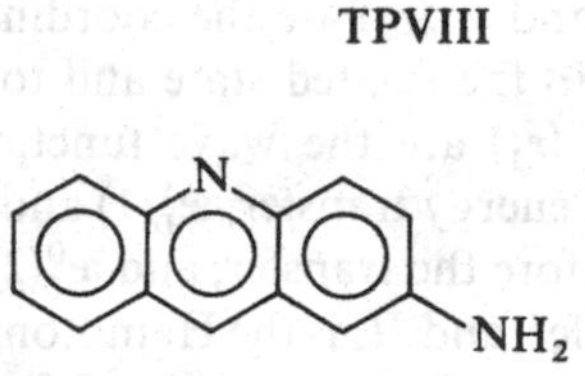

TPX

Among "classical" fluorescent and phosphorescent chromophores such as anthracene, pyrene, phenanthrene, etc., those with predominantly phosphorescent properties (eosin, Erythrosin) may be found listed as well. The aqua and β-diketonate complexes of Eu^{3+} and Sm^{3+} have low extinction coefficients ($\varepsilon = 5\text{--}15\,M^{-1}\,s^{-1}$) in the region of the $S_0 \rightarrow T_1$ transition, but they can accept energy via the mechanism of inductive energy transfer from excited organic donors with $\varepsilon \simeq 10^4\,M^{-1}\,s^{-1}$ (Figure 4.1).

The other group of phosphorescent chromophores includes the charge transfer complexes, triplet excimers, and exciplexes (Kapinus 1988). The charge transfer complex of naphthalene and 1,3,5-trinitrobenzene is characterized by the long lifetime of the triplet state, $\tau^*_{\text{ph}} = 2.25\,s$. Triplet excimers of naphthalene are formed in solution with a rate constant of $2\cdot10^7\,M^{-1}\,s^{-1}$, the lifetime of the complex being equal to $\sim 1\,ms$. Similar kinetic parameters have been observed for the porphyrin and phthalocyanine excimers.

Aromatic molecules such as pyrene, anthracene, or 1,2-benzopyrene emit intensive annihilation fluorescence, whereas triplet–triplet annihilation is weaker for caronene, perylene, and ANS. The essential advantage of the triplet probes under consideration is that when included into biological or model membranes they emit longer and more intensive fluorescence than in water solutions. When inserted into a rigid biological matrix such as a protein globule, the probe molecule ceases to emit annihilation fluorescence. This property can be used to distinguish the processes that operate in various portions of complex biological systems, e.g., in protein and lipid fractions of native biomembranes.

4.3. Exchange interactions with participation of excited triplet states: elements of theory

The exchange interaction between the donor (D) and the acceptor (A) in the one-electron approximation is quantitatively characterized by the value of the exchange integral:

$$J = \langle \varphi^1_D(r_1)\varphi^0_A(r_2)|H_e|\varphi^0_D(r_2)\varphi^1_A(r_1)\rangle,$$
$$\times \langle \mathbf{æ}^1_D(1)|\mathbf{æ}^1_A(1)\rangle\langle\mathbf{æ}^0_A(2)|\mathbf{æ}^0_D(2)\rangle,$$

where $\varphi_D(r_1)$ and $\varphi_D(r_2)$ are the coordinate wave functions corresponding to the donor in the excited state and to the acceptor in the ground state, $\varphi_D(r_2)$ and $\varphi_A(r_1)$ are the wave functions related to the state after the triplet–triplet energy transfer, $\mathbf{æ}^1_D(1)$ and $\mathbf{æ}^1_A(1)$ are the spin wave functions in the state before the transfer, and $\mathbf{æ}^0_D(2)$ and $\mathbf{æ}^0_A(2)$ are the same functions after the transfer, and H_e is the Hamiltonian of the exchange interaction.

The relaxation of a chromophore from the triplet excited state occurs via the following routes of the exchange mechanism (Figure 4.2): (1)

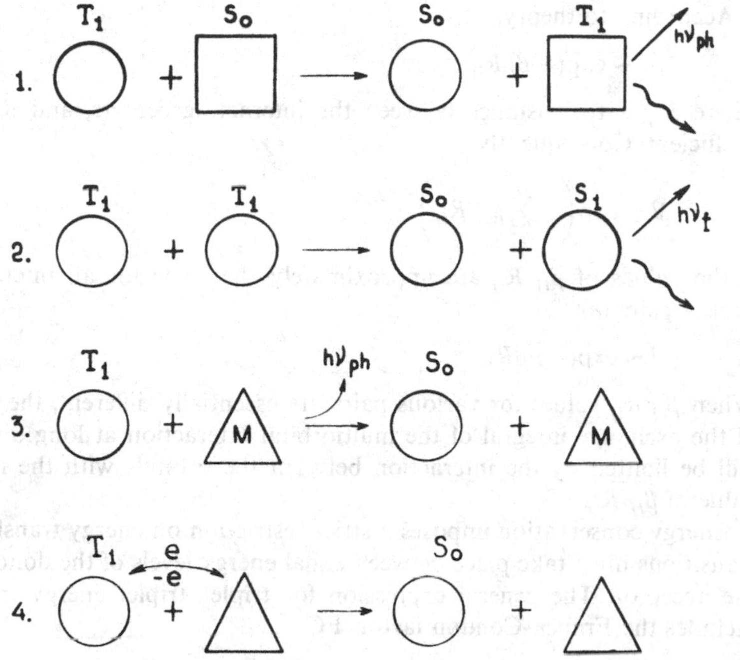

Figure 4.2. Schematic representation of the mechanism of quenching in the excited triplet state (T_1): (1) triplet–singlet energy transfer; (2) triplet–triplet energy transfer; (3) $T_1 \to S_0$ interconversion; (4) electron transfer.

exchange resonance transfer of energy from the triplet level of the donor to the triplet (multiplet) level of the acceptor; (2) catalysis of $T_1 \to S_0$ interconversion under the influence of external paramagnetic species; (3) triplet–triplet annihilation; (4) quenching by the mechanism of reversible or irreversible charge transfer (Parker 1968; Tatikolov & Kuzmin 1975; Ermolaev 1977; Mekler *et al.* 1983; Kapinus 1988). The expression for the rate constant of the exchange process contains the value of the exchange integral J, which depends, in turn, on the value of the overlap integral:

$$S_{ij} = \int \varphi_i \varphi_j, \tag{4.1}$$

where φ_i and φ_j are the one-electron wave functions for the ith and jth orbitals being overlapped.

In the first approximation,

$$J \sim \prod^{ij} S_{ij}. \tag{4.2}$$

Equation (4.2) takes into account all pairwise interactions of the orbitals.

According to theory,

$$S_{ij} \sim \exp(-\beta_{ij}R_{ij}), \tag{4.3}$$

where R_{ij} is the distance between the interacting centers, and β_{ij} is a coefficient. Consequently,

$$J \sim \exp\left(-\sum^{ij} \beta_{ij} \cdot R_{ij}\right). \tag{4.4}$$

If the values of $\beta_{ij} \cdot R_{ij}$ are approximately the same for all interacting orbital pairs (n);

$$J \sim \exp(-n\beta R).$$

When $\beta_{ij} \cdot R_{ij}$ values for various pairs are essentially different, the decay of the exchange integral of the multiorbital interaction at long distance will be limited by the interaction between the orbitals with the largest value of $\beta_{ij} \cdot R_{ij}$.

Energy conservation imposes a strict restriction on energy transfer: all transitions must take place between equal energy levels of the donor and the acceptor. The general expression for triplet–triplet energy transfer includes the Franck-Condon factor, FC:

$$K_{TT} = 2\pi J_{TT} FC/h, \tag{4.5}$$

where $FC = \int I_{ph}^D(v)\varepsilon_A(v)dv$, and $I_{ph}^D(v)$ and $\varepsilon_A(v)$ are, respectively, the intensity of the donor fluorescence and the extinction coefficient of the acceptor.

Triplet–triplet energy transfer, like electron transfer, is accompanied by a reorganization of the chromophores and the environment. If the contribution of the reorganization is significant, the approximate FC term should be included in equation (4.5).

Figure 4.3 shows the theoretical dependence of the overlap integral S/S_0 on distance for various pairs of atoms (Berdnikov & Bogdanchikov 1979; Larson 1982; Newton 1982).

Particular attention should be paid to the following regularities: (1) the interaction of the d–d-orbitals is weakened even more drastically than that of the s–s-orbitals when the distance between the centers is increased; (2) the involvement of molecular orbitals from the medium in the interaction contributes to an increase in the effective overlap, and the S/S_0 ratio diminishes more slowly with distance than under vacuum; (3) conjugated bridges between interacting centers strongly promote the exchange processes.

The probability of long-distance exchange interactions is affected not only by the distance between the donor and acceptor centers but also by their relative orientations (Sider *et al.* 1984). According to the theoretical

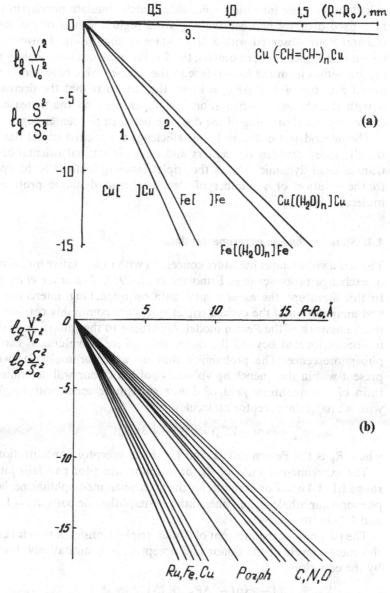

Figure 4.3. Dependence of theoretically calculated exchange (S) and resonance (V) integrals on the distance between two atoms plotted on the basis of data reported elsewhere (Berdnikov & Bogdanchikov 1979; Larsson 1982; Newton 1982): (a) the effect of the medium between the interacting atoms (1, vacuum; 2, water molecules; 3, conductive bridge); (b) the effect of the nature of the atoms; Porph, porphyrin (Sider *et al.* 1984).

calculations made for oblate spheroids, which simulate porphyrin rings, at a fixed distance between the edges the highest values of the overlap integral S are observed with a face-to-face configuration. However, at a constant distance between centers, the S value increases only slightly upon the transition from the face-to-face to the edge-to-edge configuration (by about a factor of 3 at $\Delta R_{DA} = 1\,\text{nm}$); the reason is that the decrease in S with the change in orientation is compensated by the increase in S caused by the shortening of the distance between the center edges.

The dependence of the exchange efficiency for excited triplet states on the distances between the centers and on their mutual orientation and translational dynamics allows the triplet labeling method to be applied to the solution of a number of structural and dynamic problems of molecular biology.

4.4. Static exchange: experimental data

There is a voluminous literature concerned with quantitative investigation of exchange processes (e.g., Ermolaev *et al.* 1977; Zamaraev *et al.* 1981). In this literature, the experimental data on quenching, interconversion, and annihilation of the excited triplet states are commonly considered in the framework of the Perren model. According to that model, only donor molecules located beyond the quenching sphere of volume V can emit phosphorescence. The probability that no acceptor molecules will be present within the quenching volume V of the donor will determine the ratio of the quantum yield of donor phosphorescence with (φ_{ph}) and without (φ_{ph}^0) the acceptor molecules:

$$\varphi_{ph}/\varphi_{ph}^0 = \exp(-VC_A) = \exp(-4/3\pi R_P^3 C_A), \qquad (4.6)$$

where R_P is the Perren radius, and C_A is the acceptor concentration.

The experimental value of R_P for a donor–acceptor pair falls into the range 1.1–1.4 nm. For example, for the acetophenone-naphthalene, benzophenone-naphthalene, and phenanthrene-naphthalene pairs, $R_P = 1.1$, 1.3, and 1.4 nm, respectively.

The paramount requirement of triplet–triplet transfer, a match between the energy levels of the donor and acceptor, is quantitatively described by the equation

$$K_{TT} = k_d[1 - \exp(-^3\Delta E_{DA}/RT)]^{-1}, \qquad (4.7)$$

where k_d is the rate constant of diffusional collisions, K_{TT} is the rate constant of the triplet–triplet transition, and $^3\Delta E_{DA} = E_D^3 - {}^3E_A$ is the gap between the triplet levels of the donor and acceptor.

As seen from Figure 2.8, the quantitative exchange parameters, the rate constant K_{TT}, and the exchange integral of the spin–spin interaction depend on the distance R between the centers. When plotted, the experimental data lie on two approximately exponential curves described by the following equation:

$$J_1, K_{TT} \approx 10^{16} \exp(-LR). \tag{4.8}$$

For systems in which the exchanging centers are separated by a non-conducting medium (molecules or groups with saturated chemical bonds, $L = 25\,\text{nm}^{-1}$. For systems in which the radical centers are linked by conducting conjugated bonds, $L = 3\,\text{nm}^{-1}$. In nonconducting systems, the exchange intensity is damped by about a factor of 10 when the centers move apart by 0.1 nm; in conducting systems, the distance must be increased by 0.7 nm to obtain damping by a factor of 10.

It should be emphasized that the values of the Perren radius appear not to depend on the gap $\Delta E_{S_0 T_1}$ between the ground singlet state (S_0) and the triplet state (T_1) for the molecules of the medium if the value of the gap is much higher than the $\Delta E_{S_0 T_1}$ value for the acceptor (Siegel & Judeikis 1964). Thus, for the donor–acceptor pair phenanthrene-naphthalene, the values of K_{TT} in solid solutions of aromatic compounds (benzene, diphenyl, xylene) do not differ significantly from those in solutions of nonsaturated compounds (ethanol, esters, etc.), despite the significant difference in the $\Delta E_{S_0 T_1}$ values.

Experimental data have confirmed the theoretical prediction that the triplet–triplet transition efficiency depends on the orientations of the donor and acceptor groups. However, attempts to observe such a transition failed in hybrid molecules containing donor and acceptor fragments with perpendicularly oriented aromatic rings, despite the thermodynamic preference of the transition and the short distance between the centers (less than 1.0 nm) (Filipescu & Chang 1972).

The empirical equation (4.8) offers interesting possibilities for using exchange data to solve the following structural problems:

1. Characterization of the conductivity of the medium if the values of J_1, K_{TT} and R are known.
2. Determination of the distance between the centers if the values of J_1 and K_{TT} and the nature of the conductivity are known.
3. Estimation of J_1 and K_{TT} if R and the nature of the conductivity are known.

The detailed analogy between electron transfer and spin exchange processes has been described repeatedly in the literature. The latter process may be considered as an idealized electron transfer without reorganization. Therefore, analysis of the data on the exchange interaction permits assessment of the role of purely electronic factors in the complex electron transfer process.

4.5. Dynamic exchange processes

4.5.1. Elements of theory

As mentioned earlier, all types of exchange interactions are efficient at short distances, and the probability of interaction is high upon collisions between the exchanging centers in solution.

The exchange processes between excited donor molecules (D^T) and an acceptor in solution are generally analyzed in the framework of the following scheme: formation of a collision complex, exchange within the complex, and dissociation of the complex:

$$D^T + A \underset{k_{-1}}{\overset{k_1}{\rightleftharpoons}} D^T A \overset{k_1}{\longrightarrow} DA^T \overset{k_1}{\longrightarrow} D + A^T.$$

In the case of fast dissociation of the complex, e.g., $k_3 \gg k_2$, the apparent value of the rate constant is

$$k_{ex} = k_1 k_2 / (k_{-1} + k_2). \qquad (4.9)$$

If $k_2 \gg k_{-1}$, the exchange process is referred as a strong exchange, and

$$k_{ex} = k_1 \approx k_d \sim T/\eta. \qquad (4.10)$$

In the case of weak exchange, $k_{-1} \gg k_2$, and

$$k_{ex} = k_1 \cdot k_2 / k_{-1}. \qquad (4.11)$$

Taking into account that $k_1 \simeq k_d$ and $k_{-1} = \tau_1^{-1}$, where τ_1 is the lifetime of the $D^T A$ complex, and using equation (4.8), the following expression can be derived:

$$k_{ex} \approx k_d \cdot \tau_1 \exp(-L R_{DA}) \, M^{-1} s^{-1}. \qquad (4.12)$$

Formulas (4.10)–(4.12) can be used to determine the microviscosity of the medium, the steric hindrances, and the depth of the triplet or of quenching centers immersed in the biological matrix. The interaction of charged chromophores can be studied by the method developed for estimation of the electrostatic factor in the spin–spin interaction (see Section 2.4.3).

The long lifetimes of triplet labels in the excited state ($\tau_{ph}^* = 10^{-6}$–10^0 s) offer unique possibilities for measuring the rates of slow dynamic processes (collision of proteins, diffusion in rigid membranes, concentration of dioxygen in biological systems, etc.). Another promising trend in the application of triplet labels is the determination of the steric factor f in pseudodiffusion processes (Kotelnikov *et al.* 1986; Likhtenshtein *et al.* 1986a; Vogel *et al.* 1986). Although they are controlled by diffusion, these slow dynamic processes nevertheless feature small values of the rate constant (k'_d), according to theoretical considerations (Temkin & Jacobson 1984):

$$k'_d \approx \sqrt{f_D f_A} \cdot k_d, \tag{4.13}$$

where f_D and f_A are the geometric steric factors for the donor and the acceptor, respectively.

In biological systems, many enzymatic and biochemical processes proceed in the membrane structures. In membranes, diffusion actually takes place in two-dimenstional space (lateral diffusion). In this case, the dependence of the rate constant of the collision between two pairs on the lateral diffusion coefficient D_L is described by the following formula:

$$k'_d = (8/\pi)R^2 \int_0^\infty \exp(-D_L t u^2/R^2)\{U[I_0^2(u) + Y_0^2(u)]\}^{-1} du, \tag{4.14}$$

where I_0 and Y_0 are the zero-order Bessel functions of the first and second kind, respectively, U is an auxiliary variable, R is the particle radius, and t is the reaction time (Razi-Nagvi 1974).

It is the long-range character of the exchange processes combined with the long lifetime of the excited triplet state that can prove the validity of the "harpoon-like" mechanism of long-distance exchange in viscous media. In this case, the value of the apparent radius is (Burshtein 1976)

$$R_{app} = R_{DA} + [\ln(k_0/a_B^2 D_{tr}) + 1.15]/a_B, \tag{4.15}$$

where a_B is the Bohr radius of the exchange interaction between the A and D particles, R_{AD} is the Van der Waals distance between the same particles, and k_0 is the rate constant at a distance of R_{AD}.

4.5.2. Experimental data

As can be expected from theoretical considerations, the values of the rate constants for the triplet–triplet energy transfer, interconversion, and annihilation, if they are energy-allowed, are close to the value for the diffusion constant (k_d) (Parker 1968; Ermolaev *et al.* 1977; Kapinus 1988). Thus, the quenching rate constants for the excited triplet–triplet states of the donor–acceptor pairs diacetyl-pyrene, acetophenol-naphthalene, triphenylene-*trans*-stilbene, and Erythrosin-hemin range within $(7-10)\cdot 10^9$ $M^{-1} s^{-1}$. Triplet–triplet energy transfer at Van der Waals distances tends to be fairly fast (Siegel & Judeikis 1964; Filipescu & Chang 1972; Anderson *et al.* 1974). The rate constant for the intramolecular transition in 4-(1-naphthylmethyl)benzophenone in benzene solution is equal to $10^{11} s^{-1}$. The time of intermolecular energy transfer from the triplet excited state of benzophenone to liquid 1-methylnaphthalene was measured to be $2 \cdot 10^{-11}$ s.

The values of the rate constants for quenching of excited molecules by electron acceptors are fairly large as well. Quenching of the phosphorescene of Mg phthalocyanine by benzophenone in solution occurs with a rate constant of $k_q = 1.4 \cdot 10^9 \, M^{-1} \, s^{-1}$. In low-viscosity nonpolar solutions, triplet excimers of naphthalene form with a rate constant of $2 \cdot 10^7 \, M^{-1} \, s^{-1}$, the lifetime of the excimer being approximately equal to 10^{-3} s. Exciplexes of porphyrins have similar lifetimes.

The experimental rate constants for phosphorescence quenching by the triplet–triplet annihilation mechanism are reported to be less by a factor of 9 than those of fluorescence quenching. This fact can be explained by taking into account the static spin factor $\frac{1}{9}$ that is required for the $TT \rightarrow S_0 S_1$ transition (Kapinus 1988).

An approach has been advanced based on the cascade scheme of the exchange process (Mekler *et al.* 1983). The proposed scheme

$$D^T + A \longrightarrow D + A^T,$$
$$A^T + A^T \longrightarrow A^{S_0} + A^{S_1},$$

features triplet–triplet annihilation upon collision of the excited triplet molecules of an acceptor (A^T) with each other, and quenching of D^T by the acceptor molecule A. The main advantage of such an approach is the possibility of monitoring the dynamic interaction of biologically active acceptors (quinones, porphyrins, dyes, etc.) with the triplet probes using routine fluorscence techniques.

Lanthanides and transition metal complexes, including active centers of metal enzymes, ferrocene, ferricyanide, hemin, acetylacetonate, and NRs, are efficient quenchers that are widely used in spin–labeling methods (Chapters 1 and 2) (Figure 4.4) (Tatikolov & Kuzmin 1975; Likhtenshtein *et al.* 1986; Likhtenshtein 1990a). The method of dynamic quenching of excited triplet states has been proved to be applicable to phosphorescent biological objects such as chlorophyll, tryptophan- and tyrosine-containing proteins, and heme proteins, with the Fe atom being substituted by Mg, Zn, etc.

Among quenchers in biological systems, dioxygen plays a peculiar role (Van der Kooi *et al.* 1990). The exchange interaction with O_2 significantly decreases the lifetimes of excited triplet states of chromophores and hinders the recording of phosphorescence. On the other hand, the phenomenon of quenching can be used to study important biological processes connected with the formation and consumption of dioxygen.

The energy of the triplet ground state of O_2 ($^2\Sigma_g^-$) differs from that of the first excited state $^1\Delta_g$ by only 7,882 cm^{-1}. Thus, dioxygen should be a good acceptor in the spin-allowed triplet–triplet energy transfer from

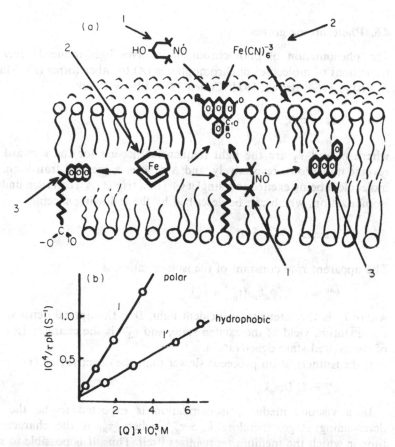

Figure 4.4. Study of location and translational diffusion of triplet and spin probes in the membrane: (a) Diagram illustrating the principle of the method: 1, nitroxide probes; 2, Fe-containing probe; 3, triplet probe. (b) Dependence of the eosin fluorescence quenching rate on spin probes 1 and 1' in lecithin liposomes (Mekler *et al.* 1982; Likhtenshtein 1990a).

excited organic donors. This prediction from theory has been experimentally confirmed. Thus, the values of the rate constants ($M^{-1}s^{-1}$) for triplet–triplet energy transfer from donors to dioxygen are $2.7 \cdot 10^9$ (anthraquinone), $3.2 \cdot 10^9$ (pyrene), $3.3 \cdot 10^9$ (anthracene), $1.1 \cdot 10^9$ (eosin), and $1.1 \cdot 10^9$ (chlorophyll) (Ermolaev *et al.* 1977). Dioxygen is a strong quencher of the phosphorescence of metal complexes. The excited triplet states of Ru(II) with aromatic ligands (1,1-phenanthroline, 2,2-dipyridyl) and Cr(III)-acetylacetonate are quenched by O_2 with $k_{TT} \simeq 5 \cdot 10^9 \, M^{-1}s^{-1}$ (Demas *et al.* 1973).

Thus, the method of dynamic quenching of the excited triplet state has a reliable theoretical and experimental basis.

4.6. Photochrome probes

The phenomenon of photochromism involves light-induced reversible transitions of molecules of chromophores (*A*) to other forms (*B*) (Malkin & Fisher 1964; Brown 1971; Cohen 1975; Eltsov 1982):

$$A \underset{h\nu_2, \Delta}{\overset{h\nu_1}{\rightleftarrows}} B,$$

where ν_1 and ν_2 are the light frequencies needed for the forward and reverse transitions, respectively, and Δ denotes a thermal transition. The molecule *A* being excited to a singlet (A^S) or a triplet (A^T) can then undergo isomerization, which can be described by the following shceme:

$$A \underset{\tau_d^{-1}}{\overset{h\nu_1}{\rightleftarrows}} A^{S,T} \xrightarrow{k_{is}} B.$$

The apparent rate constant of the isomerization is

$$k_{is}^{ap} = D \cdot I_{in} \varphi_d k_{is} / (\tau_d^{-1} + k_{is}), \tag{4.16}$$

where I_{in} is the intensity of incident light, D is the optical density, φ_d is the quantum yield of the excited state, and τ_d^{-1} is the characteristic time of the excited-state deactivation.

If the isomerization proceeds slower than the deactivation ($\tau_d^{-1} \gg k_{is}$),

$$k_{is}^{ap} = I_{in} D \varphi_d k_{is}. \tag{4.17}$$

In a viscous medium, isomerization is expected to be the rate-determining stage; therefore $k_{is} \simeq \tau_{RI}^{-1}$, where τ_{RI} is the characteristic time in which the medium reorganizes itself. Thus, it is possible to study the dynamics of the medium by measuring the value of k_{is} if the experimental values of I_{in}, τ_d, and φ_d are known. The sensitivity of the approach to various ranges of the relaxation time is governed by the τ_d value. If a photochrome molecule is photochemically stable and the $A \rightarrow B$ transition is irreversible, a long exposure to light allows one to determine the parameters of very slow relaxation processes with $\tau_{RI} \geq 1$ s.

The following photochrome systems appear to be the most promising for studying biological system dynamics (Cohen 1975; Eisenback 1979; Eltsov 1982):

1. Tioindigoid dyes undergoing photoisomerization:

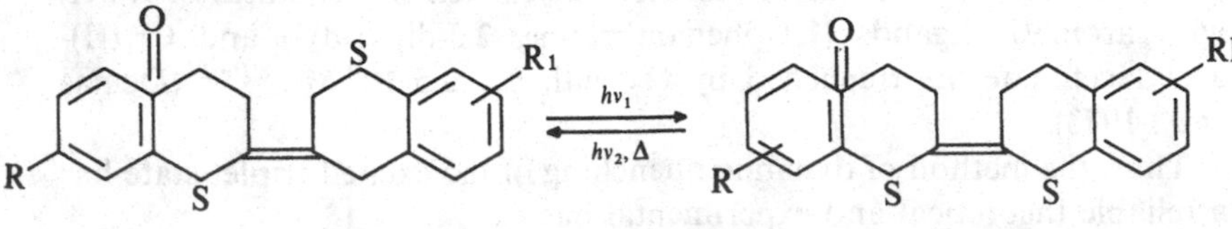

E-isomer   Z-isomer

The values of the maxima of the absorption spectra are equal to 543 and 484 nm for the E- and Z-isomers of tioindigo (R, $R_1 = H$), respectively. Therefore these isomers can be excited separately by irradiation at different frequencies. This effect can be enhanced by variations in the chemical structures of R and R_1. The fluorescence intensity of the E-isomer is much higher than that of the Z-isomer.

2. Azo and cyano dyes:

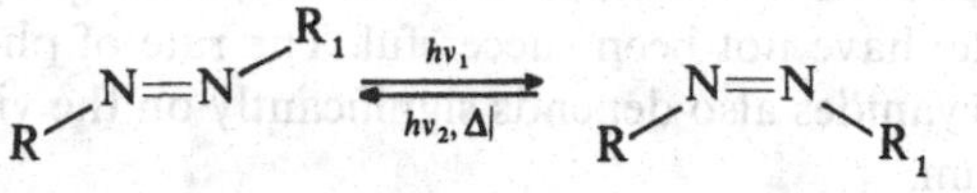

The difference in the maxima of the adsorption spectra for the Z- and E-isomers can amount to as much as 80 nm.

3. Stilbenes $R_1 R_2 - C = C - R_3 R_4$, such as

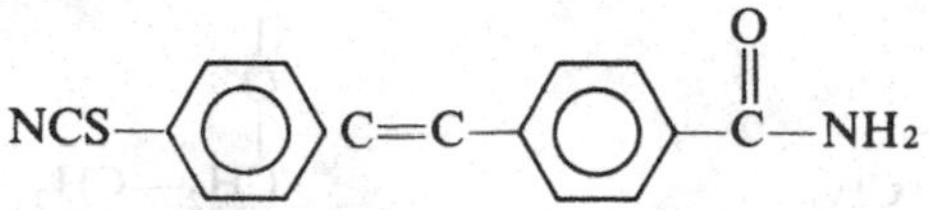

which is characterized by high values of the quantum yield for the E-isomer.

4. Photochrome tautomeric systems, which may undergo photo-reactions:

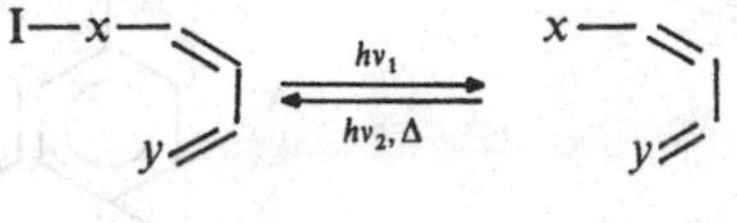

or

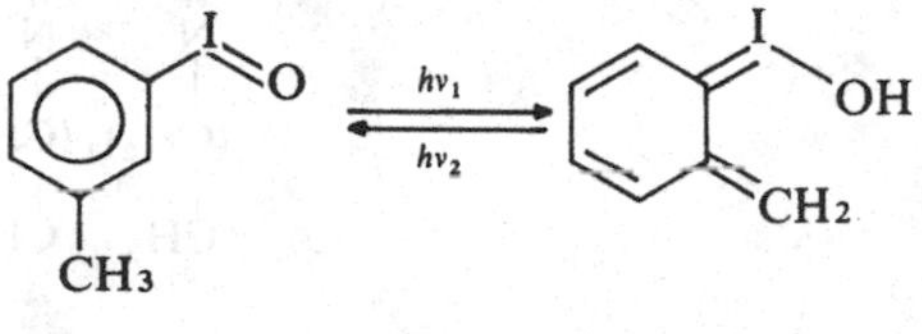

5. Systems involving photodimerization:

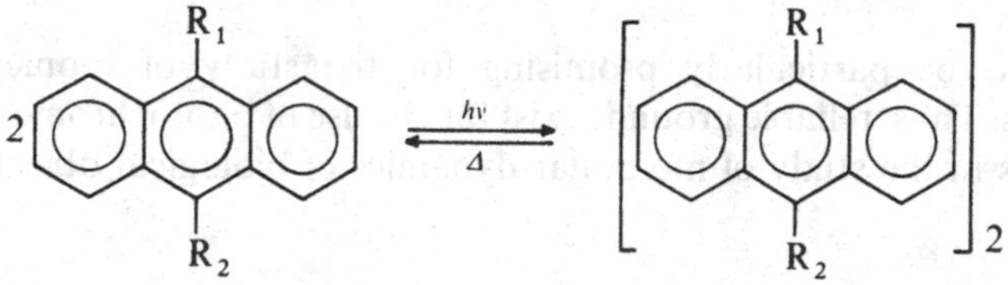

which can be used to study translational diffusion.

Thus, synthetic chemistry currently offers a wide collection of photo-chromic compounds with useful ranges of polarity, photochemical activity, size and form of substituents, and optical properties.

The available experimental data confirm the significant effects of microviscosity and molecular dynamics of the medium on the rates of photochromic processes predicted theoretically (Eltsov 1982; Görner & Schulte-Frohlinde, 1979). For example, $E \rightarrow Z$ phototransformation of tioindigo does not occur at a viscosity of $\eta > 5 \cdot 10^8$ pascals (Pa). Attempts to observe the $E \rightarrow Z$ photoisomerization in solid polymers (polysterol, polymethylmethacrylate) have not been successful. The rate of photoiso-merization of azomerocyanides also depends significantly on the viscosity of the polymeric medium.

Lipid derivatives of photochromic azo compounds (Morgan *et al.* 1987),

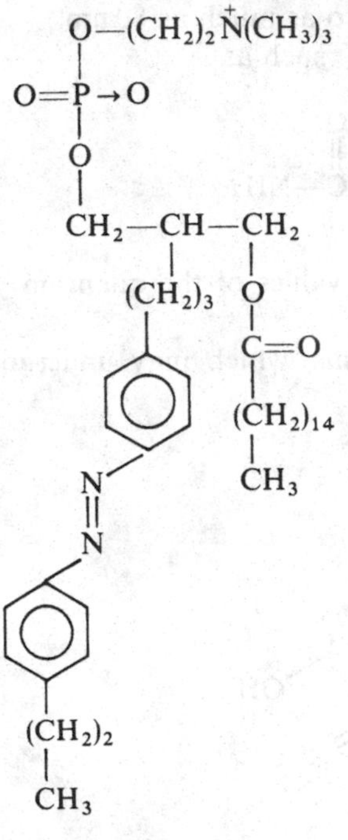

$$O-(CH_2)_2 \overset{+}{N}(CH_3)_3$$
$$O=P \rightarrow O$$
$$O$$
$$CH_2-CH-CH_2$$
$$(CH_2)_3 \quad O$$
$$C=O$$
$$(CH_2)_{14}$$
$$CH_3$$
$$N$$
$$\parallel$$
$$N$$
$$(CH_2)_2$$
$$CH_3$$

Palmityl azo-PC

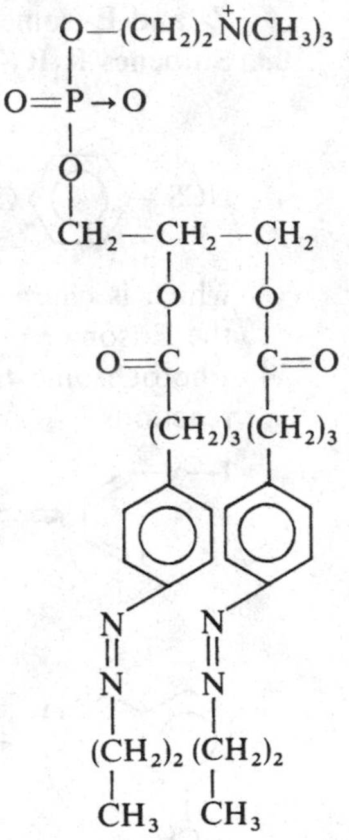

$$O-(CH_2)_2 \overset{+}{N}(CH_3)_3$$
$$O=P \rightarrow O$$
$$O$$
$$CH_2-CH_2-CH_2$$
$$O \quad O$$
$$O=C \quad C=O$$
$$(CH_2)_3 (CH_2)_3$$
$$N \quad N$$
$$\parallel \quad \parallel$$
$$N \quad N$$
$$(CH_2)_2 (CH_2)_2$$
$$CH_3 \quad CH_3$$

Bisazo-PC

appear to be particularly promising for the study of biomembrane dynamics. Thus, reliable grounds exist for the use of photochrome systems as probes in the study of molecular dynamics of biological objects.

4.7. The triplet probe–photochrome labeling method

In biological systems, rare diffusional collisions play a significant role in helping physiologically active compounds to achieve active contact with enzymes and membrane proteins, especially in such rigid locales as a membrane whose temperature is below the phase transition point.

The traditional ESR and fluorescence methods for recording collision frequencies do not allow the study of slow translational diffusion because of the short characteristic times of these methods ($\tau_{ch} = 10^{-7}$–10^{-8} s). Great progress has been made in this field because of the availability of triplet labels, whose characteristic times are as long as $\tau_{ph}^* = 10^{-3}$–10^{-5} s (Section 4.5). Nevertheless, the sensitivity of the latter is not high enough to study diffusion of proteins in membranes.

An approach was developed to determine the rate constants of very rare collisions, including those between macromolecules in rigid media, by Mekler and Likhtenshtein (1986). The approach is based on the cascade scheme involving a photochrome reaction sensitized by the triplet–triplet energy transfer between a triplet label D^T and a photochrome molecule A_{ph}. The energy levels of the sensitizer and the photochrome should involve the cascade shown in Figure 4.5. The triplet state of the sensitizer can be excited by light of lower energy than that of singlet-state excitation of the photochrome. In this scheme, the concentration of the A-isomer, $[A]$, approaches the photostationary level exponentially:

$$[A] = \exp - [(k_T^A Z_T^A + k_T^B Z_T^B)C_T t],$$

where k_T^A and k_T^B are the rate constants for the triplet–triplet transfer from the sensitizer (D) to the photochrome isomers A and B, respectively, Z_T^A and Z_T^B are the probabilities of isomerization after the triplet energy transfer for the A and B isomers, C_T is the concentration of the sensitizer in the excited triplet state, and t is the time of light exposure.

At low optical density,

$$C_T = I_{in} D \tau_d \varphi_d \tag{4.18}$$

[symbols as in equation (4.17)]; hence,

$$[A] = \exp - [(k_T^A Z_T^A + k_T^B Z_T^B)I_{in} D \tau_d \varphi_d t]. \tag{4.19}$$

The values of k_T^A and k_T^B can be determined by measuring the sensitizer fluorescence quenching in the presence of the photochrome by the method of flash photolysis. Z_T^A and Z_T^B can be obtained by monitoring the sensitized direct and reverse isomerization.

This approach has been used to study rare collisions in a model system containing Erythrosin as a sensitizer and 4-acetoamido-4,4-iso-*trans*-thiocyanate stilbenedisulfuric acid as a photochrome (Mekler &

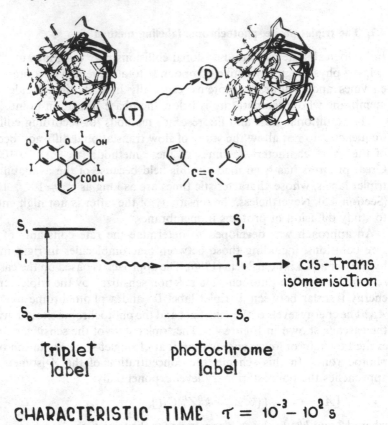

CHARACTERISTIC TIME $\tau = 10^{-3} - 10^{2}$ s

Figure 4.5. Schematic representation, chemical formulas, and energy levels in double triplet (T)–photochrome (P) labeling techniques.

Likhtenshtein 1986) (Figure 4.5). Both types of molecules were covalently bound to the chymotrypsin macromolecule. The kinetics of isomerization was monitored by the decrease in fluorescence of the photochrome label transisomer. The values of the isomerization probability, Z_T^A and Z_T^B, were found to be equal to 0.2 and 0.6, respectively. At pH 7 and 20°C the rate constants for triplet–triplet energy transfer between the sensitizer and the photochrome molecule were $k_T^A = 2 \cdot 10^7 \, M^{-1} s^{-1}$ and $k_T^B = 10^7 \, M^{-1} s^{-1}$.

It should be emphasized that in these experiments the concentration of the triplet sensitizer did not exceed $10^{-7} \, M$, and the collision frequency was close to $1 \, s^{-1}$, which is several orders of magnitude less than that determined according to routine luminescence or ESR techniques.

Figure 4.6 illustrates the progress made in the study of collisions between labeled molecules, which began with characteristic times τ_{ch} of $10^{-7} - 10^{-9}$ s

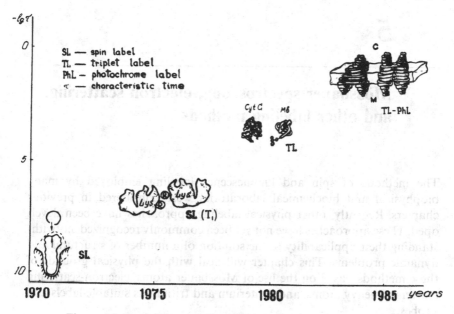

Figure 4.6. Logarithms of characteristic times (τ, s) plotted against the approximate periods of development of the methods of estimating the frequency of collisions between labels.

in the routine techniques of spin labeling and fluorescence labeling, and then progressed to $\tau_{ch} = 10^{-5} - 10^{-6}$ s in the spin relaxation techniques, $\tau_{ch} = 10^{-3} - 10^{-4}$ s in the triplet labeling method, and finally $\tau_{ch} = 10 - 10^{-1}$ s in the more sophisticated double triplet–photochrome labeling approach. The composition of a new cascade scheme including quencher-sensitizer-photochrome appears to make it possible to monitor very rare collisions between the quencher and the sensitizer molecules by recording photochrome isomerization.

There is no doubt that triplet probe–photochrome labeling will allow us to extend the range of investigation of very slow diffusion processes in biological membranes and other rigid systems.

Thus, the triplet labeling method is distinguished by higher sensitivity to long-distance exchange interactions (up to 1.6 nm) and the possibility of studying slow relaxation and diffusion processes. Novel approaches based on the cascade principle (quenching of delayed annihilation fluorescence and the triplet probe–photochrome labeling method) offer additional advantages. The use of the excited triplet state enables us to reduce the concentration of labels and probes by several orders of magnitude.

5

Mössbauer spectroscopy, electron scattering, and other labeling methods

The methods of spin and luminescence labeling employed by many biophysical and biochemical laboratories were considered in previous chapters. Recently, other physical labeling approaches have been developed. These approaches have not yet been commonly recognized, notwithstanding their applicability to the solution of a number of structural and dynamic problems. This chapter will deal with the physical grounds of these methods based on the use of Mössbauer atoms, electron-scattering clusters of heavy atoms, and deuterium and tritium as suitable labels and probes.

5.1. Mössbauer labels

Mössbauer or gamma-resonance (GR) spectroscopy is widely used in physics, chemistry, and biology (Suzdalev 979, 1988; Dickson 1984). As a rule, the main objects studied are iron-containing proteins with the natural isotope replaced by ^{57}Fe. The Mössbauer labeling techniques considerably extend the application of GR spectroscopy (Frolov et al. 1973, 1977; Likhtenshtein 1976a; Belonogova et al. 1978). The introduction of Mössbauer atoms into various portions of biological structures via chemical reactions, absorption, or biosynthesis, followed by analysis of the GR spectra, allows one to study the microstructure and dynamics of the object of interest.

5.1.1. Physical principles

The phenomenon of gamma resonance (γ-resonance) is based on the transition of a nucleus from the ground state to the excited state upon absorption of a γ-quantum (Josephson 1960; Blume 1967; Afanasiev et al. 1978; Thosar 1983; Suzdalev 1988). Nuclear γ-resonance can be observed only when the emitting (^{57}Co) and absorbing (^{57}Fe) nuclei are bound fairly tightly to the matrix and the energy given off is assimilated by the crystal lattice as a whole. In these experiments, the difference in the resonance energy is compensated for by the relative movement of the emitting and absorbing samples. The dependence of the intensity of the resonance absorption (f') on the movement rate is termed the γ-resonance or Mössbauer spectrum.

The f' value is related to the amplitude of the mean displacement of the nuclei in the recoil direction ($\langle x^2 \rangle$) during the lifetime of the excited state (τ_n^*):

$$f' = \exp(-4\pi\langle x^2\rangle/\lambda^2), \tag{5.1}$$

where λ is the γ-quantum wavelength and x is in nanometers.

In numerical form, calculated for ^{57}Fe,

$$f' = 10^{-2,300\langle x^2\rangle}. \tag{5.2}$$

In the case of the ^{57}Fe nucleus participating in a Brownian diffusion process with the diffusion coefficient D, if the Stokes-Einstein law is fulfilled, then the magnitude of the relative broadening of the spectrum is given as

$$\Delta\Gamma/\Gamma = 2D\tau_n^*/\lambda^2 = 10\,T/\eta R, \tag{5.3}$$

where η (poise) is the viscosity, and R (nm) is the radius of the particle that is rigidly linked to the nucleus.

In the case of a jump-wise motion with a jump time τ_{tr}, we have

$$\Delta\Gamma/\Gamma = 2\tau_n^*/\tau_{tr}. \tag{5.4}$$

The foregoing formulas give an excellent representation of the scope of the effects of the dynamics of the medium on the GR spectrum. For example, when the amplitude of the ^{57}Fe vibration increases by 0.05 nm, the f' value decreases by a factor of 10^5. Such properties of GR spectra can be used effectively in studying the dynamics of biological objects.

The other parameters of the GR spectrum are measured experimentally: the chemical (isomer) shift (δ), the line width (Γ), quadrupole splitting (ΔE), the line-shape anisotropy ($\Delta\varepsilon$), and the magnetic hyperfine structure (MHFS) (Figure 5.1).

The value of δ is related to the s-electron density on the Mössbauer atom and depends strongly on the atom valence and the nature of the ligand. The quadrupole splitting arises from the interaction between the nuclear quadrupole moment and the asymmetric electric field of the ligand environment of ^{57}Fe.

If the ^{57}Fe nucleus is in the external or internal magnetic field (i.e., the Mössbauer atom is paramagnetic), the γ-resonance spectrum may show a magnetic structure. In the ground state, the ^{57}Fe nucleus has a magnetic moment of $I_n = \frac{1}{2}$, which can be oriented along the magnetic field or against it. The magnetic moment of the excited ^{57}Fe nucleus $I_n = \frac{3}{2}$ can correspond to any of four orientations. Differently oriented magnetic moments will make different contributions to the interaction energy; therefore the

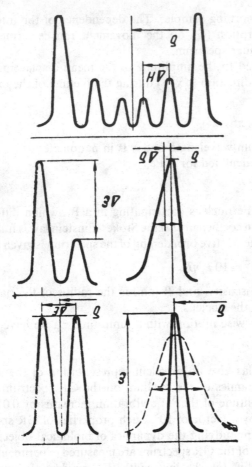

Figure 5.1. Types of Mössbauer spectra and spectrum parameters.

spectrum can consist of up to six components (Figure 5.1.). The value of the magnetic field strength H_z can be calculated from the distance between adjacent components that are proportional to H_z.

The magnitude of the internal magnetic field in paramagnetic compounds can reach 1–$10\,T$, and therefore in a number of cases splitting is observed in the absence of a magnetic field. In other cases, application of a high-strength magnetic field can reveal the hyperfine structure of the GR spectrum. It is evident that in high magnetic fields, the electrons in the Fe orbitals are paired to form a low-spin complex, since in a high-spin complex the fast electron relaxation $\tau_e \ll \tau_n^*$ and antiferromagnetic interaction between the atoms would average the magnetic fields, and the magnetic hyperfine structure would disappear. At slow relaxation (Afanasiev et al. 1978),

$$\tau_e = h/\Delta\Gamma. \tag{5.5}$$

The effect of the external paramagnetic atoms on τ_e by spin dipole–dipole or exchange interactions can be used to study the relative positions of ^{57}Fe and other paramagnetics. According to formulas (2.9) and (2.10), effects can be detected at distances of up to 2 nm for dipole–dipole interactions and up to 10 nm for exchange interactions. The theory predicts that the various MHFS components will differ in their sensitivity to these effects. Thus, the interactions leading to $\tau_e \ll 10^{-7}$ and to complete disappearance of the $S = \pm\frac{1}{2}$ and $\mp\frac{3}{2}$ components will only slightly affect the $S = \pm\frac{5}{2}$ component (Suzdalev 1988). Unfortunately, in structural research in molecular biology, Mössbauer spectroscopy is not currently exploited as fully as it could be.

5.1.2. Dynamic effects in Mössbauer spectroscopy

Theoretical considerations based on formulas (5.1)–(5.4) suggest that the dynamic state of the Mössbauer atoms is reflected in the parameters of the GR spectrum. The atom may oscillate without diffusion with a characteristic time $\tau_c \leqslant 10^{-7}$ s. Such oscillation over the lifetime of the nuclear excited state will drastically reduce the f' parameter if the amplitude of the motion exceeds 0.01 nm. This kind of motion does not affect the line width (Γ). On the other hand, diffusion without oscillation leads to line-width broadening without a change in the f' value.

Solid dynamics features, as a rule, nuclear oscillations of high frequency ($\nu_n = 10^8 - 10^{12}s^{-1}$) and low amplitude (0.0001–0.0001 nm). According to the Debye model, at $T \gg Q_D$,

$$\langle x \rangle^2 \sim E_R T/k_B Q_D^2,$$

where E_R is the recoilless energy of the γ-quantum, and Q_D is the Debye temperature. The line width in this system is not temperature-dependent.

The fact that particles with tightly bound Mössbauer atoms undergo rotational and translational diffusion characterized by diffusion coefficient of $D_R > 10^{-7}s^{-1}$ and $D_{tr} > 10^{-9}cm^2 s^{-1}$ leads to a sharp increase in the line width and a decrease in the intensity of their GR spectra. Mössbauer atoms included in biological macromolecules, e.g., proteins, are involved in at least three modes of dynamics: (1) the "usual" solid-state heat oscillation, (2) local displacement together with the site of the atom's incorporation taking part in conformational movement, and (3) diffusion with the macromolecule as a whole. Accordingly, the apparent value of the probability of the recoilless resonance absorption is

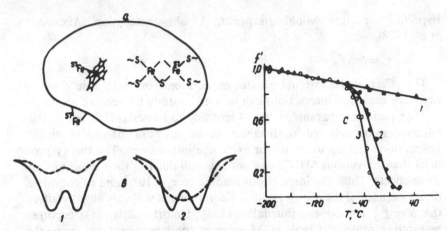

Figure 5.2. The Mössbauer labeling method (Frolov *et al.* 1973; Likhten-shtein 1976a, 1988b): (a) Possible methods of introducing labels (^{57}Fe) into proteins. (b) Expected changes in Mössbauer spectral parameters for label mobilities with $\tau_c \leqslant 10^{-7}$ s and $\langle x \rangle > 0.01$–0.02 nm; the solid lines show an absence of mobility, and the broken lines show its presence; 1, vibrational motion (f' decreases, $\Delta\Gamma$ does not change); 2, diffusional motion ($\Delta\Gamma$ increases, f' does not change). (c) Dependence of the relative value of the parameter f' for labels on proteins: 1, iron–sulfur protein, in the dry state; 2, iron–sulfur protein in aqueous solution; 3, human serum albumin, with labels at the surface.

$$f'_{app} = \exp\{(-4\pi/\lambda)[\langle x^2\rangle_S + \langle x^2\rangle_C + \langle x^2\rangle_D]\}, \tag{5.6}$$

where the indices S, C, and D are related to the appropriate dynamic modes. Figure 5.2 shows an example of the temprature dependence of the Mössbauer atom ^{57}Fe tightly bound to the serum albumin globule at various degrees of humidity.

The experimentally measured dependence of the decrease in f' on the temperature and moisture content may be described in terms of three groups of factors: (1) a monotonic increase in $\langle x^2\rangle$; (2) oscillations of the Mössbauer atom between several states of different f', with all these states characterized by a measurable value of f'; (3) practically complete disappearance of a portion of the spectrum when there are significant fractions characterized by a critical value of $\langle x^2\rangle_{cr}$, whose contribution to the measured f' is negligibly small.

It should also be noted that even apparently insignificant dynamic microheterogeneity of the medium must affect the experimental dependence of f' on T. Indeed, if it is assumed that there is a Gaussian distribution among the vibrational energies of Mössbauer atoms in the sample, $E_v \sim \langle x^2\rangle$, the experimental value of f' will be equal to

$$f'_{app} = \exp[-\beta_M(1 - \beta_M\langle\Delta x\rangle_0^2/2\langle\Delta x\rangle_{max}^2)], \tag{5.7}$$

where $\beta_M = 4\pi^2 \langle \Delta x \rangle^2_{max}/\lambda^2$, and $\langle \Delta x \rangle^2_{max}$ and $\langle \Delta x \rangle^2_0$ are the most probable amplitude and the second moment of the amplitude distribution, respectively (Likhtenshtein & Kotelnikov 1983; Likhtenshtein 1988a,b).

The dynamic state of the Mössbauer atoms is reflected in other parameters of the GR spectrum as well. Thus, when atoms move with an average velocity of $\langle V \rangle$, the second-order Doppler effect leads to an additional isomer shift (Josephson 1960; Suzdalev 1988):

$$\delta_D = - E_v \langle V^2 \rangle / 2c^2, \tag{5.8}$$

where c is the velocity of light.

The intensity and the mechanism of the solid-state phonon process affect the temperature dependence of the MHFS, with the line width ($\Delta\Gamma$) affected in its turn by the value of the spin–lattice relaxation time (τ_{1e}). For example, the line-width broadening of the component of the spectrum corresponding to the $S_z = \pm\frac{1}{2}$ transition (Suzdalev 1988) can be expressed as

$$\Delta\Gamma_s = 112P_1 + 32P_2, \tag{5.9}$$

where $P_{1,2} \sim \exp(\omega_E/\Gamma)$ and ω^7 for the Einstein and Debye models of oscillation, respectively.

The aforementioned ideas provide a reliable foundation for application of Mössbauer spectroscopy to studies of the dynamic problems of molecular biology. Examples of such applications will be given in Chapters 6 and 7.

5.2. NMR probes

NMR techniques provide a powerful method for studying the structures and dynamics of biological systems. Some possibilities for the method in paramagnetic systems with NRs and transition ions were described in Section 2.5. This section deals with probes modified specifically with ^2H (D), ^{13}C, ^{15}N, ^{17}O, ^{28}Na, ^{23}Al, ^{44}Ca, ^{63}Cu, and ^{67}Zn characterized by distinct and easily interpreted NMR spectra (Seeling 1977; Jost & Griffith 1980; Marsh 1989; Sixl et al. 1984; Watts 1985; Chimelka et al. 1989). Such modifications can be exemplified by the introduction of ^{13}C into the $C{=}O$ groups of lysine, ^{15}N into the amide group of glycine, and ^2H (D) into specific positions of lipid probes:

$$R{-}O{-}CD_2{-}CD_2{-}N(CH_3)_3$$
$$\text{DMPC-}d_4$$

$$R{-}O{-}CH_2{-}CH_2{-}N(CD_3)_3$$
$$\text{DMPC-}d_9$$

$$R{-}O{-}CD_2{-}CDOH{-}CH_2OH$$
$$\text{DMPG-}d_3$$

$$R = CH_3(CH_2)_{12}COOCH_2$$
$$|$$
$$CH_3(CH_2)_{12}COOCH$$
$$|$$
$$CH_2{-}O{-}\overset{\displaystyle O}{\underset{\displaystyle O^-}{\overset{\|}{P}}}{-}O^-$$

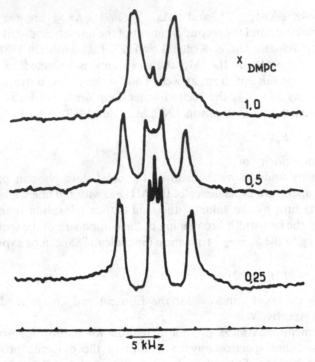

Figure 5.3. ^{3}H NMR spectra (46.1 MHz) of DMPC-d_4/DMPC mixtures of different compositions at 28°C. The smaller quadrupole splitting is due to the β-CD$_2$ segment; the outer signal arises from the α-CD group (Sixl & Watts 1982).

The use of labeled probes makes interpretation of NMR spectra much easier. For example, replacement of an adjacent proton by deuterium reduces the magnetic moment by a factor of 3. Two-dimensional NMR spectroscopy based on the dipole–dipole interaction between the nuclear spins is particularly effective in labeled systems (Wütrich 1986).

The most widely used NMR probes that are commonly combined with spin labeling are the deuterated probes (Watts 1985).

The characteristic feature of the ^{2}H NMR spectrum is the quadrupole splitting due to the electric quadrupole moment of the ^{2}H nucleus. The quadrupole moment (Q) and related magnetic moment (μ_Q) precess in the electric field gradient of the environment. An external magnetic field splits the energy of a system with a nuclear spin $I = 1$ into three levels:

$$E = -m\mu_Q H_0 + eQ[3m^2 - 2]/\partial^2 V/\partial Z^2, \tag{5.10}$$

where $m = 1, 0, -1$ is the magnetic quantum number, and $\partial^2 V/\partial Z^2$ is the gradient of the electric field.

If a ^{2}H NMR probe rotates at frequencies comparable to the quadrupole splitting ($\Delta v_Q^0 = 1.3\,\text{kHz}$), the value of Δv_Q decreases. Therefore, the value of Δv_Q can be used to characterize the dynamic state of the given ^{2}H probe. Figure 5.3 illustrates the effect produced by the medium on the dynamics of the ^{2}H DMPC-d_4 probe in the membranes of 1,2-dimyristoyl-*sn*-glycero-3-phosphatidylcholine. The μ-C^{2}H segment is characterized by a lower value of Δv_Q and consequently by faster mobility of the α-C^{2}H segment.

Another parameter of the ^{2}H NMR signal, spin-lattice relaxation rate ($1/T_{1n}$), is also related to the probe correlation time of the rotational diffusion (τ_c). At $\tau_c \ll v_L$, where v_L is the frequency of Larmor precession;

$$1/T_{1n} = 3e^4(\partial^2 V/\partial Z^2)Q^2 \tau_c/10h^2. \tag{5.11}$$

The values of T_{1n} for the deuteron and proton are 0.5 s and 3 s, respectively, in a 50% mixture of ^{2}H$_2$ and ^{1}H$_2$O. Thus, measurements of T_{1n}, T_{2n}, and the Δv_Q parameter of the deuterium probe can be valuable sources of information on the dynamic state of the system under investigation.

Lipid NMR probe techniques are effective tools for studying the precise mechanisms of the effects of the properties of physiologically active reagents and membrane ingredients, including proteins, on the dynamics of various parts of the biological membrane.

5.3. Total tritium labeling technique

The total tritium labeling technique (TTLT) has been proposed for studying the structural elements and appropriate conformational transitions of biological objects of interest (Shishkov *et al.* 1976; Mitsova *et al.* 1981; Skrypkin *et al.* 1985; Volynskaya *et al.* 1985). In TTLT, samples (proteins, nucleic acids, membranes, etc.) are treated with tritium atoms, which replace protons in the surface layers. One can then observe the tritium content in various components (e.g., amino acids, nucleotides) of the system under study to obtain information on the distribution of these components relative to the surface and on conformational changes.

The tritium atoms are incorporated into organic compounds either by a direct high-energy ($E \geqslant 1.5\,\text{eV}$) replacement reaction,

$$^3\text{H}^{\cdot} + \text{RH} \longrightarrow \text{R}^3\text{H} + \text{H}^{\cdot}, \tag{5.12}$$

or by a reaction with thermolized atoms ($E = 0.3\text{--}0.4\,\text{eV}$),

$$^3\text{H}^{\cdot} + \text{RH} \longrightarrow \text{R}^{\cdot} + \text{H}^3\text{H},$$
$$^3\text{H}^{\cdot} + \text{R}^{\cdot} \longrightarrow \text{R}^3\text{H}. \tag{5.13}$$

Scheme (5.12) leads, as shown *a posteriori*, to less damage to the biopolymer than does scheme (5.13). High-energy tritium ions are obtained by means of a high-frequency (40-MHz) electric discharge at a low pressure of $^3H_2$. The ions are accelerated by an electric field with a strength of $20-100 \, V \, cm^{-1}$ and, after being recharged by dihydrogen, attack the object under investigation. Thermolized tritium atoms are obtained by the dissociation of $^3H_2$ on tungsten wire at 2,000 K.

The tritium atoms are introduced onto lyophilized dry films or frozen aqueous solutions. It has been shown in special experiments that the tritium does not penetrate beyond 0.3 nm below the surface. Other experiments indicate a definite specificity in labeling amino acids. This specificity was corrected for in an analysis of data on two model proteins: whale myoglobin and egg-white lysozyme. The portions of the protein that were observed to be accessible to labeling agreed well with theoretical expectations of accessibility.

Experiments with ribonuclease, α-chymotrypsin, subtilysine, thermolysine, lysozyme, and carboxypeptidase have indicated that the value of relative specific radioactivity is proportional to the $\frac{2}{3}$ power of the molecular mass of the protein. The theoretical value of the "sliding" area of a water molecule over a protein globule is

$$S_t = 11 M_r^{2/3}.$$

Thus, surfaces of biological objects can be covered with a specific radioactive "tan," followed by analysis of the degree of "tanning" for individual elements of the structure under investigation. Such an approach may provide data on the distribution of specific elements over the surface and on the change in distribution caused by conformational transitions.

5.4. Electron-scattering labels

5.4.1. General

The advantages of contemporary electron microscopy are derived from the development of high-resolution techniques, improvements in methods of sample preparation, the use of a "noiseless" matrix, and advances in quantitative physical and mathematical methods of analysis. Another recent trend in this field is the use of heavy metals as markers.

The intensity of electron scattering or X-ray scattering increases, other conditions being equal, with the atomic number of the component atoms (Figure 5.4). Biological objects consist almost entirely of light elements and must be stained by layers of heavy metals (Mo, Os, Pb, V, etc.) to produce clear electron microscopy images. However, such added contrast leads to loss of many essential details in the object images, e.g., the active

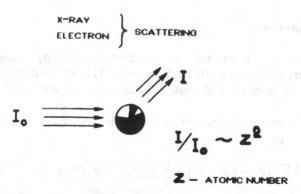

Figure 5.4. Schematic representation of the main principle of the X-ray- and electron-scattering methods.

centers of enzymes, individual subunits of protein complexes, protein and lipid components in the membrane, etc.

Similar problems are encountered in X-ray-scattering techniques. Although X-ray scattering has been successfully used to determine the sizes and shapes of macromolecules and membranes, it faces difficulties in resolving fine structural details.

One can solve a number of structural problems – determination of distances between chosen portions of an enzyme protein macromolecule, localization of these portions on the macromolecule, and the distribution of the enzyme macromolecules or ligands in a complex biological system, e.g., in membranes – by means of the electron-scattering labeling techniques proposed by a number of investigators.

Starting with Beer's seminal paper (1964), a procedure marked by the introduction of single heavy atoms into DNA and proteins, followed by study of the modified samples by electron microscopy, has been reported in a number of works (Beer & Moudrianakis 1962; Ulanov *et al.* 1967, 1979; Crewe *et al.* 1970; Strothkamp & Lippard 1976; Beer *et al.* 1978). Unfortunately, significant problems have been encountered in the acquisition and unambiguous interpretation of micrographs for such large objects.

A second trend in this field involves the application of ferritin as an electron-scattering marker. Ferritin is a metal protein 16 nm in size, with $M_r = 600$ kD, having about 2,000 iron atoms per protein molecule. Its large size limits the use of this marker in solving structural problems.

This section has been concerned with the physical and chemical bases of various versions of labeling by recently developed electron- and X-ray-scattering techniques. These labels include assemblies of heavy atoms (clusters) that are large enough to provide a significant scattering effect but small enough to be comparable to the dimensions of many biological objects of interest.

5.4.2. Physical grounds

Electron microscopy. To produce an electron microscopy image, a microobject must efficiently scatter fast electrons. This ability depends, in particular, on the number and mass of atoms in the scattering particle.

The relative intensity of electron scattering by molecules in the gas phase is given by the formula

$$I/I_0 = a \sum [(Z_i - f_i)^2 + S_i] + \sum_i \sum_j (Z_i - f_i)F_{ij}(0,r), \qquad (5.14)$$

where Z_i is the atomic number, f_i and S_i are the factors specific for the atom, and F_{ij} is a function of the scattering angle and distribution of nuclei in the molecule. To simplify equation (5.14), it can be suggested that $I/I_0 \sim Z^2$ (Figure 5.4). In such a case, one atom of mercury of $Z = 80$ is characterized by a scattering ability that is almost 100 times as high as the average scattering ability of protein atoms. It is estimated that a cluster of four to six atoms of mercury can scatter fast electrons about 10 times more efficiently than portions of a protein globule of equal size.

Small-angle X-ray scattering. When a solution of macromolecules is irradiated with X-rays, the scattering intensity is dependent on the angle between the incident and reflected rays, because of the difference in reflection of rays from different portions of the molecule. This dependence is variable for macromolecules of different sizes, densities, and shapes. The experimental measurement of this dependence and the elucidation of its relationship to the molecular parameters of microobjects are the concerns of a special method whose efficiency has already been demonstrated (Kratky 1963).

Here we shall consider only the problem of determining the distance between heavy atoms residing on a macromolecule. According to the theoretical treatment (Vainshtein *et al.* 1970; Feigin *et al.* 1978), for spherical macromolecules randomly distributed in solution, where each molecule is modified by two heavy atoms separated by a distance r, the function of the dependence of the scattering intensity on the angle θ contains three terms:

$$I(\theta) = k[N^2\varphi^2(S) + 2f_1(S)\{1 + \sin Sr/2Sr\} + 4f_1(S)\varphi(S)(\sin Sr/Sr)], \qquad (5.15)$$

where $S = 4\pi \sin \theta/\lambda$, $\varphi(S) = 3[(\sin Sr - Sr \cos Sr)/(Sr)^3]$, λ is the irradiation line width, $f_1(S)$ is the atomic factor of a heavy atom, and N is the number of electrons in the molecule.

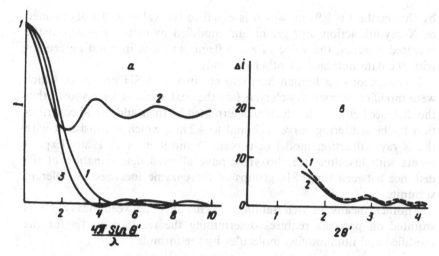

Figure 5.5. Curves of small-angle X-ray scattering by protein solutions with electron-dense labels (Feigin *et al.* 1978; Vainshtein *et al.* 1980): (a) Normalized components of the scattering intensity for a homogeneous sphere of radius R (a macromolecule) with "point" scatters (labels) added along the diameter; 1, geometric characteristics of the macromolecule; 2, determined by the label–label interaction; 3, depends on the label–macromolecule interaction. (b) Experimental dependence of the scattering-intensity difference (ΔI) on the scattering angle θ for solutions of histidine carboxylase with three labels and without them; 1, experimental curve; 2, calculated curve.

The first term of the equation describes the scattering of the macromolecules, which decreases sharply with increasing θ. An important property of the second term, which reflects the participation of heavy atoms in a scattering event, is that it is described by a function with a number of clear-cut maxima (Figure 5.5), whose positions and intensities are associated with the distance between the heavy atoms. Starting from a certain value of θ, this term becomes dominating. For the second term to be experimentally separated, the inequality $Z_e > N_a/100$ must satisfied, where Z_e is the number of electrons in the heavy atom, and N_a is the number of light metals in the protein. For a protein with a molecular mass of 50,000, the X-ray-scattering label must contain at least six or seven mercury atoms. This number should be somewhat increased in proportion to increasing molecular mass of the macromolecule.

The foregoing relationships have been tested on molecules of known structure, and the approach has been found to be highly informative (Vainshtein *et al.* 1970, 1980; Lanzin *et al.* 1985). For example, the distance between the iodine atoms in a solution of diiodoanthracene as determined by this method is 0.9 nm, which is equal to the value of 0.9 determined by X-ray diffraction. For gramicidin modified by mercury atoms at the oriented residues, the value of r is 1.0 nm, which is in good agreement with the data obtained by other methods.

Investigations on human hemoglobin, two β-93 SH groups of which were modified by mercury clusters (see the next section), have shown that the distance between the clusters determined by their additional contribution to the scattering curves is equal to 4.2 nm, which is consistent with the X-ray diffraction model of protein (Vainshtein *et al.* 1980). Experiments with histidine decarboxylase have allowed determination of the distance between three SH groups of the enzyme localized at different subunits.

Another means of determining the distance between heavy labels situated on proteins requires determining the radii of inertia for the modified and nonmodified molecules by the formula

$$(M + 2m)R_M^2 = MR_0^2 + 2r^2,$$

where m and M are the numbers of electrons in the label and the protein molecule, respectively, $2r$ is the distance between the labels, and R_0 is the radius of the molecule (Lanzin *et al.* 1985).

It may be expected that the scattering labeling technique will find wide application in small-angle X-ray scattering for determining changes in the distances between certain portions of protein molecules during their functions, as well as in studies of membrane systems modified at different positions with electron-scattering labels.

5.4.3. *Modification of biological objects by electron-scattering labels*

A number of polynuclear complexes and clusters of heavy metals are used nowadays as electron-scattering labels: (1) polymercury mercarbides; (2) colloid particles and clusters of gold; and (3) compounds of palladium and iodine (Likhtenshtein *et al.* 1973; Schwab & Thoenen 1978; Ottensmeyer 1982; Eskelinen & Peura 1986; Sombyo 1986; Levi-Setti 1988).

Mercarbides appear to be the most convenient reagents to use as electron-scattering labels (Levchenko *et al.* 1973, 1979; Strothkamp *et al.* 1978; Lipka *et al.* 1979; Marakushev *et al.* 1986). Since electron microscopy requires highly scattering centers in separate parts of the biological matrix, different versions of modification by mercarbide electron-scattering labels have been developed. This approach allows (1) revelation of several closely spaced SH groups in compounds containing single heavy atoms [in particular, *p*-chloromercuribenzoate (PCMB)]; (2) modification of a single SH group by addition of compounds containing several closely spaced atoms of heavy metals [e.g., mercarbide electron-scattering labels (MESL)], which are water-soluble parts of the Hoffman bases; (3) study of the parts that do not contain SH groups by thiolization, with subsequent blocking of the SH groups inserted by MESL; (4) modification of hydrophobic portions of proteins and membranes by hydrophobic mercaptans, followed by mercurization.

Polymercarbides are organometallic compounds with the H atoms replaced by mercury. It was Hoffman who first described the synthesis and properties of compounds that are obtained by mercurization of alcohols, aldehydes, and their organic derivatives with mercury oxides. According to physicochemical and X-ray data, the crystallized fraction of the Hoffman base is a derivative of methanetetramercarbide (Grdenic *et al.* 1974):

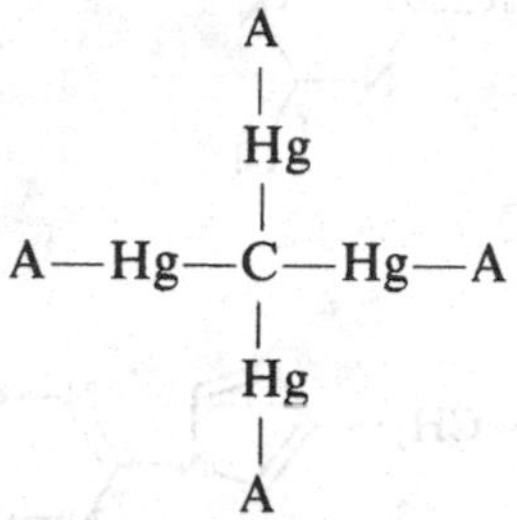

$$
\begin{array}{c}
A \\
| \\
Hg \\
| \\
A-Hg-C-Hg-A \qquad (M) \\
| \\
Hg \\
| \\
A
\end{array}
$$

where —A is an anion

$$
\begin{array}{c}
O \\
\parallel \\
CH_3C-O-,\ I^-
\end{array}
$$

The physical and chemical properties of compound M are identical with those of a product obtained by Matesson (Matesson *et al.* 1970) in the mercurization of tetrakis(dimethoxyboric)methane with mercury acetate. The soluble fraction of the Hoffman base can be repeatedly used as MESLs, which are seen on the electron micrographs as spots of a size corresponding to an ensemble of four or five MESL molecules. Particles of similar size were obtained for assemblies of $C(HgAC)_4$ molecules (Levchenko *et al.* 1979).

Chemical modification of functional groups of proteins and lipids can be performed by the procedures described next.

The basic reaction for attaching a MESL (M) to a biological object with SH groups is mercurization:

$$R-SH + M \longrightarrow R-S-M \tag{5.16}$$

Here, 2,4,6-trichloro-1,3,5-triazine (TCTA) appears to be the most convenient reagent, for a variety of reasons. Alkylation of deprotonated primary and secondary amines, mercaptans, and imidazole can be performed via a nucleophilic reaction with this strongly temperature- and pH-dependent reagent. When performed at 0°C, this reaction leads to replacement of only one chlorine atom in the TCTA, with the formation of a very strong covalent C—N bond. The other two Cl atoms can be easily replaced by an SH group at room temperature, and the procedure can then be repeated. Being pH-dependent, the reaction of alkylation by TCTA can be used for specific modification of lysine and histidine groups of proteins and NH_2 groups of lipids. Modification of protein groups is readily realized by the following schemes (M, MESL; X, TCTA):

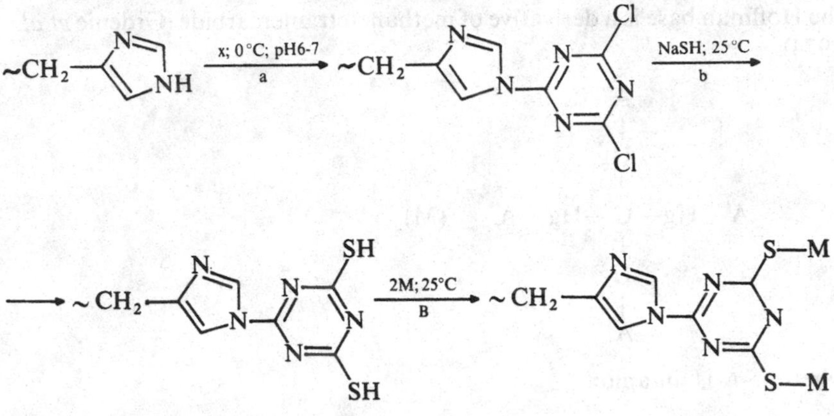

$$(5.17)$$

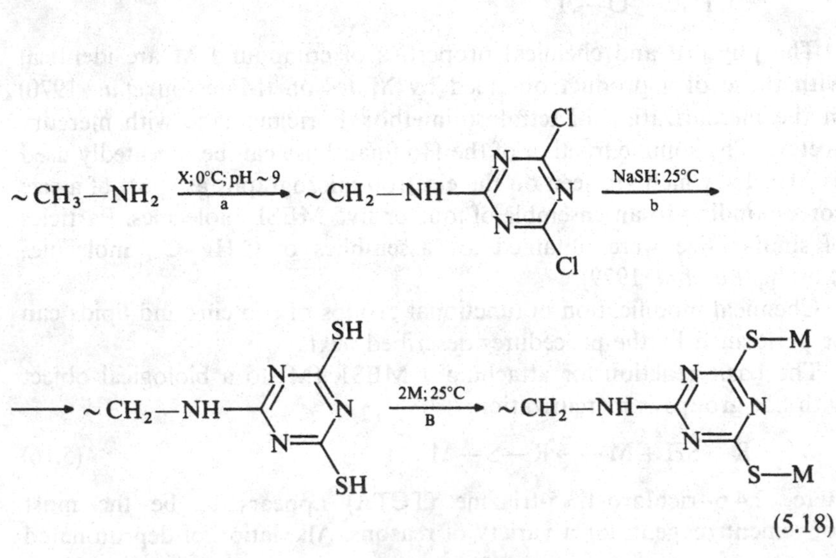

$$(5.18)$$

(5.19)

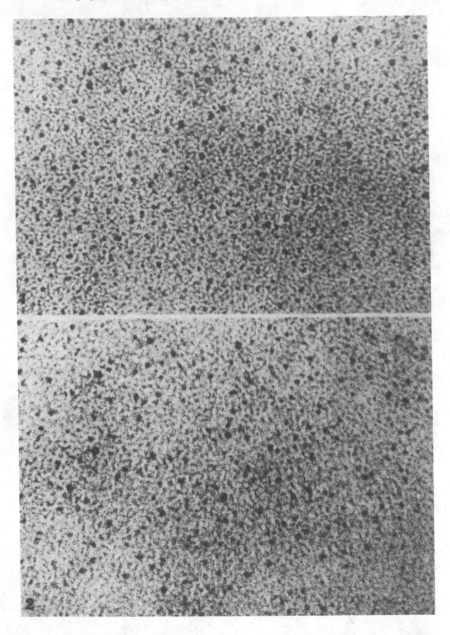

Figure 5.6. The electron-scattering (electron-dense) labeling method. ▶
Electron micrographs of biological objects labeled with electron-scattering
labels (ESL). In parentheses are the numbers of ESL per molecule (magni-

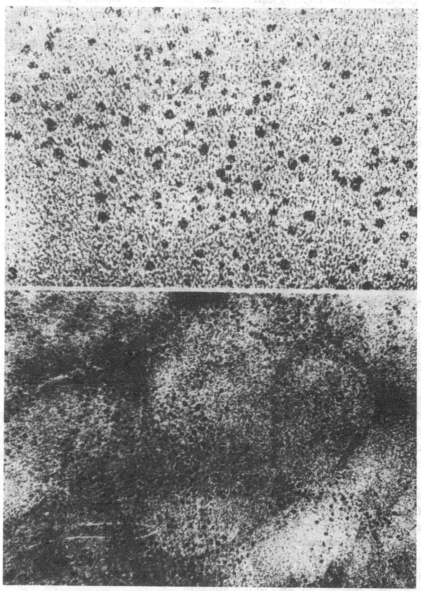

fication is indicated in square brackets): 1, trichlorotriazine (2–3) [700,000]; 2, lysozyme (2) [700,000]; 3, human serum albumin (4–5) [700,000]; 4, bacteriorhodopsin in membrane in the light (6) [300,000] (Levchenko *et al.* 1979).

Reactions (5.17)–(5.19) can be used for "multiplication" of the SH groups. Thus it becomes possible to increase the scattering power of the center. These reactions can also be applied to modification of bio-membranes by the introduction of MESL into the hydrophobic heads of phosphatidylethanolamine or phosphatidylserin. Figure 5.6 shows microphotographs of the modified trichlorotriazine and the single imidazole residue of lysozyme. Biomembranes can be MESL-labeled with the help of three types of SH-containing reagents (Levchenko *et al.* 1979; Marakushev *et al.* 1986): (1) hydrophobic molecules (e.g., dodecylmercaptan); (2) lipids that are modified via schemes similar to (5.17)–(5.19); (3) mercaptoundecanoic acid. The two former compounds become localized in the superficial portion of the membrane. The third type of reagent is believed to be situated in the inner portion of the membrane.

5.4.4. *Electron microscopy determination of shape and size of electron-scattering particles*

Analysis of electron micrographs has been fairly well developed for negatively stained objects with a regular structure (Klug 1978). To analyze micrographs of objects labeled by electron-scattering markers, some special approaches are needed.

The problem of reproducing the shapes of electron-scattering particles using a set of projections is dealt with in the work of Serzhanov *et al.* (1992a). The figure of interest is envisaged to be describable mathematically with a set of geometric parameters. The choice among alternative models can be made by analyzing the projection distribution function $f(S)$. Figure 5.7 shows distribution functions for electron-scattering particles of different shapes. The figures are clearly seen to be qualitatively distinguishable from each other.

Another approach to this problem is based on comparing the scattering area of the object of interest with that of the single MESL and plotting the distribution function $f(S)$. This approach offers the possibility of estimating the object's homogeneity and the number of MESLs in the spot being analyzed. The area of the single MESL is taken from the molecular model of $C(HgAc)_4$ (Serzhanov *et al.* 1992b). This approach has been used for the molecules of MESL and for MESL-labeled TCTA, lysozyme, myoglobin, and serum albumin.

According to the electron micrographs, MESL molecules form closely arranged ensembles with four to six MESLs in each. The labeled TCTA holds about three labels. The areas occupied by the electron-scattering portions of labeled lysozyme, myoglobin, and serum albumin correspond approximately to the areas expected from the chemical analysis.

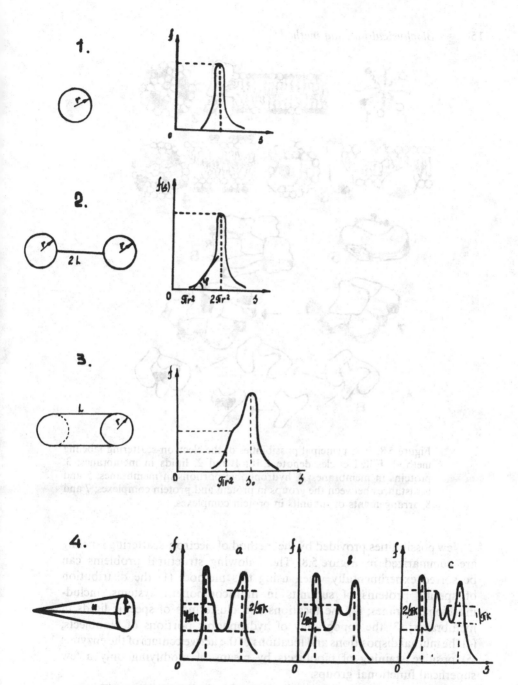

Figure 5.7. Distribution functions $f(S)$ (to the right) of the projection areas (S) of different figures (to the left) upon random orientations: 1, ball; 2, gantel $(tg\varphi = 4rL^{-2})$; 3, cylinder $[S_0 = (4L^2r^2 + \pi^2r^4)^{\frac{1}{2}}]$; 4, cone $(K = r/H$; a, $K < 0.319$; b, $K = 0.319$; c, $0.319 < K < 0.337)$ (Serzhanov et al. 1992a).

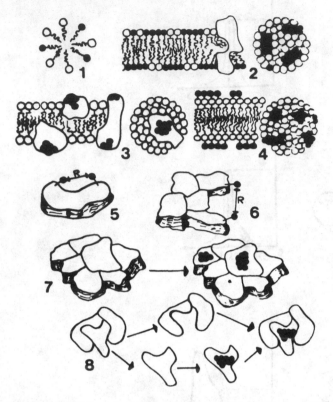

Figure 5.8. The principal possibilities of the electron-scattering labeling method. Filled circles denote ESL: 1 and 2, lipids in membranes; 3, proteins in membranes; 4, hydrophobic portions in membranes; 5 and 6, distances between the groups in protein and protein complexes; 7 and 8, arrangements of subunits in protein complexes.

New possibilities provided by the method of electron-scattering labeling are summarized in Figure 5.8. The following structural problems can be solved experimentally when using this method: (1) the distribution of specific proteins of subunits in multicomponent systems, including biomembranes; (2) the locations and clusterings of specific lipids in membranes; (3) the topography of hydrophobic portions of bioobjects; (4) the mutual dispositions and locations of the active centers of the enzyme; (5) positive staining of bioobjects by means of modifying only a few superficial functional groups.

The prospects for further application of techniques that use electron-scattering clusters of heavy metals appear to be fairly promising. Based on the observation of electron-scattering images in micrographs, a super-sensitive assay can be developed for very small amounts of labeled physiologically active compounds, (drugs, prostaglandins, hormones, effectors of memory, etc.). The distributions of these compounds in complex bio-objects like membranes, including the ligands of receptors, can be studied by such means.

Clusters of heavy metals (e.g., gold, platinum, palladium, lead) containing various number of atoms with a variety of configurations (rods, triangles, squares, circles, etc.) can be obtained synthetically.

The electron-scattering labeling method, combined with new electron microscopy techniques such as scanning microscopy, microanalysis of micrographs, and devices for visualization of stereoscopic images, is certain to be widely used in molecular biology in the very near future.

The use of Mössbauer spectroscopy, NMR, and total tritium and electron-scattering labels like the spin and luminescent labels considered here can significantly expand the arsenal of tools for studying complex biological structures, and these undoubtedly are valuable additions to the conventional physicochemical techniques employed by biochemists.

6

Studies of proteins and enzymes: structure, dynamics, and mechanism of action

As follows from Chapters 1–5, problems such as the structures of active centers of enzymes and receptors, localization of specific zones on macromolecules, and mutual disposition of centers of complex protein and membrane systems can be studied by means of spin, fluorescence, triplet, tritium, and electron-scattering labeling. The first three methods are used to determine the parameters of rotational and translational diffusion and to monitor conformational and phase transitions. The intramolecular dynamics of protein globules and the surrounding water shell are monitored by the whole collection of spin, fluorescent, triplet, and Mössbauer labels. Modification of proteins by spin and triplet labels offers the opportunity to characterize the conducting ability of protein globules and to determine the values of steric factors. Nitroxide spin labels can be used to probe the local redox properties of the electron transport system.

In this chapter we shall consider the main results of applying biophysical labeling to proteins and enzymes. Since there has been a great variety of publications in this field, we must limit the examples cited to only the most typical.

6.1. Active centers of enzymes

6.1.1. Sterine proteases

Serine proteases are uses as a "proving ground" for various versions of the spin-labeling method. The active centers of these enzymes consist of two portions: (1) the catalytic hydrolase center, with an active serine group, and (2) an additional center providing specific binding to a substrate and precise orientation of the reacting groups of the substrate and the catalyst (so-called aryl- and amide-binding segments).

The first attempt to study the active center of an enzyme by means of spin labeling was undertaken by Berliner and McConnell (1966), who modified the Ser-195 group of α-chymotrypsin at pH 3. The ESR spectrum

158

indicated that there was fairly strong immobilization of the nitroxide fragment due to inclusion of the label into a relatively narrow cleft of the active center. Later there were proposals for a whole series of spin-labeled reagents for the active serine group (phosphate, sulfonyl, and nitrobenzene derivatives), as well as for the histidine and methionine-192 groups and for probes for the hydrophobic packet of the enzyme (Hsia *et al.* 1969; Kosman 1972; Berliner 1990).

The first investigation proved that the ESR spectra of the spin-labeled preparations were highly sensitive to fine changes in the structures of the active centers at their interactions with the substrate analogues. It was found, in particular, that the involvement of indole in the hydrophobic packet of α-chymotrypsin significantly affected the rotation of the labels attached to Ser-195 and Met-192 groups and the rate of deacylation of spin-labeled acyl enzymes. Monitoring of the rotation of a spin label on Met-192 at various pH values showed that the ionization of isoleucine-16 groups allosterically controlled the transition of an active conformation to a nonactive conformation of the enzyme.

On the basis of the experimental results it was stated that enzymatic hydrolysis occurred in the following sequence: (1) substrate and enzyme interaction at the aryl- and amide-binding centers, (2) binding-induced conformational changes in the enzyme that orient the amide-binding segment and the catalytic center, and (3) bond cleavage in the aryl segment.

The preparation of α-chymotrypsin with the Met-192 group modified by a spin label was used for monitoring the conformational state of the enzyme active center in inverse micelles (Belonogova *et al.* 1983). The latter are formed in a nonpolar medium [e.g., octane in the presence of a detergent (aerosol OT)] with the addition of a water solution of the enzyme. The increase in the ratio $a = [H_2O]/[detergent]$ led to an increase in polarity in the region of the active center, traced by the value of superfine splitting A_z of the ESR spectrum of the spin label. The mobility of the spin label increased as the a parameter decreased down to the value of 1:15–20. This effect was accompanied by acceleration of the deacylation of the *trans*-cinnamoyl of α-chymotrypsin. A further increase in hydration caused the label's mobility to increase and limited the deacylation rate. Thus, the catalytic ability of the enzyme required the optimum conformation of the active center and optimum polarity of the medium as well.

The use of spin labels enabled these researchers to compare the active centers of various proteins with the active center of α-chymotrypsin, whose structure was well known. The investigated enzymes can be arranged in the following sequence according to the degree of immobilization of the biradical spin label:

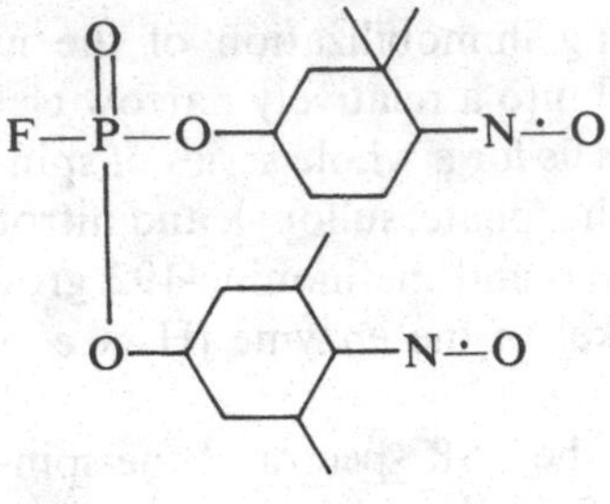

SLXXIV

α-chymotrypsin > trypsin > elastase > thrombin > subtilin BPN ~ bovine erythrocyte cholinesterase > eel cholinesterase > horse serum butyryl cholinesterase (Hsia *et al.* 1969). Similar differences in conformation of the serine protease active center were observed with the aid of the spin-labeled derivatives of ortho-, meta-, and parabenzoylsulfonyl fluorides. The meta-derivatives were shown to be the most sensitive to the structure of the enzyme and the labels.

Active-site-directed fluorosulfonyl nitroxide inhibitors were used for the blood-clotting enzyme thrombin (Berliner 1990). The natural coagulant, α-thrombin, consists of two polypeptide A- and B-chains, the latter being a homologue to serine protease and including His-15, Asp-102, and Ser-195 catalytic residues. With controlled tryptic treatment, α-thrombin converts to γ-thrombin, which being catalytically active does not exhibit fibrinogenous clotting activity. The meta-substituted derivative in the active center of γ-thrombin is more nearly immobilized than that of the α form. The difference in the conformations of the active centers is estimated to be about 0.28 nm. Interactions between the regulatory protein thrombomodulin and spin-labeled α-thrombin lead to propagation of subtle conformational changes to the catalytic site in the thrombin molecule upon complexation.

The spin-labeled oligopeptides, substrate analogues, were used to study subtle differences in the active sites of a number of enzymes (leucine aminopeptidase, α-chymotrypsin, etc.) (Damerau 1988):

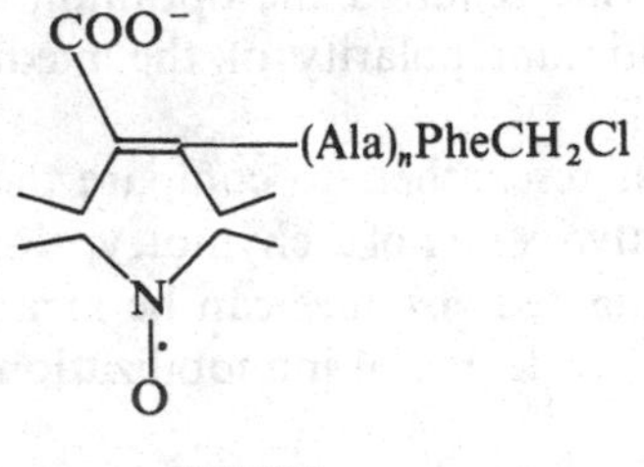

SLXXV

The active sites of cholinesterases have been studied with the whole set of spin probes that are analogues to the substrate.

Methylcholinesterase is an enzyme that catalyzes very efficiently the hydrolysis of acetylcholine. The enzyme consists of the serine catalytic center, anionic centers (α and β), and ahydrophobic allosteric site. Introduction of a nitroxide fragment to the substrate molecules and analogous derivatives of tetraallylammonium and to hydrophobic compounds allowed monitoring of the conformational state of the enzyme active center. In a number of cases, including a nitroxide fragment enhanced the inhibiting ability of the substrate analogues (Bulger *et al.* 1978).

The use of bifunctional and biradical probes of the type (Sinha & Chignel 1975)

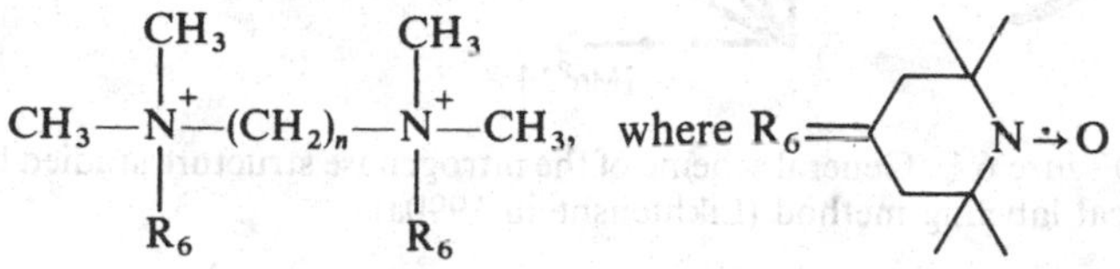

$$CH_3-\overset{\overset{\displaystyle CH_3}{|}}{\underset{\underset{\displaystyle R_6}{|}}{N^+}}-(CH_2)_n-\overset{\overset{\displaystyle CH_3}{|}}{\underset{\underset{\displaystyle R_6}{|}}{N^+}}-CH_3, \quad \text{where } R_6=$$

SLXXVI

allowed estimation of the distances between the esterase sites and the anion sites and between the α and β anionic sites (1–1.2 nm). Nitroxide derivatives of physiologically active compounds (e.g., novocaine, metacine, and γ-carboline) have been employed as spin probes to study the binding of inhibitors and the paramagnetic ions Co^{2+} and Ni^{2+} (Dorokhov *et al.* 1988).

Fluorescent chromophores (naphthalene, acridine, dansyl chloride, etc.) localized in the hydrophobic sites of proteases are used as probes for the determination of a number of binding centers and for the recording of subtle conformational changes upon functioning of the enzymes (Narayanan & Balaram 1981). Thus, an additional relaxation shift has been observed in the fluorescence spectrum of densyl fluoride attached covalently to the active serine groups upon benzamidine binding.

The modification procedures and methods of investigation for proteases may be extensively applied to other enzymes.

6.1.2. Nitrogenase

The central enzyme of biological fixation of molecular nitrogen catalyzes their coupled reactions (Lowe & Smith 1985; Likhtenshtein 1988a, b): (1) N_2 reduction to NH_3 under the action of natural electron donors (ferredoxin, flavodoxin) or artificial electron donors (dithionite) in the

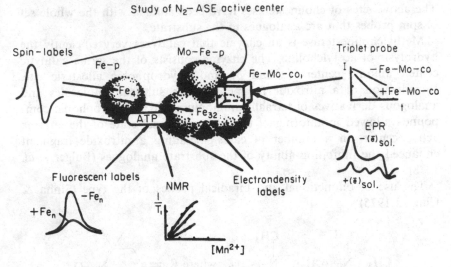

Figure 6.1. General scheme of the nitrogenase structure studied by physical labeling method (Likhtenshtein 1990a).

presence of adenosine triphosphate (ATP); (2) ATP hydrolysis to adenosine diphosphate (ADP) and inorganic phosphate in the presence of an electron donor; (3) ATP-dependent evolution of H_2. The active form of the enzyme is formed under the combined action of two compounds, an iron-containing protein (FeP) with a Fe_4S_4 cluster per molecular mass of ~ 60,000 and an iron-molybdenum protein (FeMoP) with 32 atoms of nonheme iron, 24–26 sulfur atoms, and 2 molybdenum atoms per molecular mass of 220,000.

The methods of spin, fluorescence, and electron-scattering labeling were applied to determine the mutual arrangement of metal atoms and the ATP center and the localization of these centers on the enzyme macromolecules (Likhtenshtein *et al.* 1986a) (Figure 6.1). First, nitroxide fragments R˙ were inserted into the active centers of FeP and FeMoP by replacing iron atoms by a *p*-chloromercuribenzoate derivative (PCMB-R˙). The PCMB-R˙ molecules were bound covalently to cysteine groups, which bind iron atoms in the active centers of native proteins. Spin-labeled preparations of Fe- and FeMo-proteins and other complexes give an ESR spectrum consisting of a single line at 77K. This fact is the evidence for a strong spin exchange interaction between closely spaced nitroxide fragments aggregated in clusters with no less than three or four centers.

In order to allow study of the mutual arrangement of clusters, the iron of the active centers was substituted by PCMB. Electron microphotographs of modified samples showed all the mercury atoms to be located in a

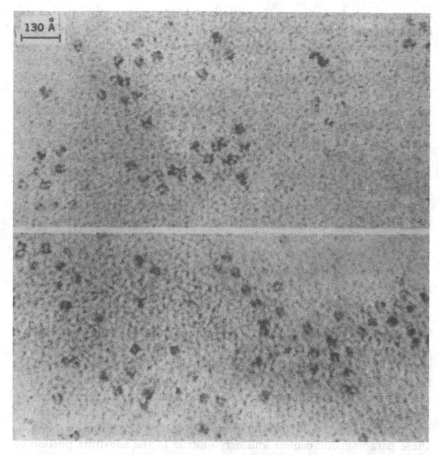

Figure 6.2. Electron micrographs of the nitrogenase preparation modified by 100 equivalents of *p*-chloromercuribenzoate, without contrasting (Levchenko *et al.* 1979).

particular part of the enzyme macromolecule. When no contrast agent was used, the electron photographs revealed that elementary cells of clusters contained no less than four mercury atoms, and these cells themselves were assembled into groups (Figure 6.2). Quite recently, on the basis of ESR, Mössbauer, EXAFS, and X-ray studies, it has been concluded that the active center of the nitrogenase FeMoP component involves Fe_4S_4 clusters (P-clusters) and Fe_6MoS_6 clusters (FeMo cofactor) (Sosfenov *et al.* 1986; Bolin *et al.* 1990).

The phenomenon of fluorescence quenching was used to estimate the distance between the fluorescing chromophore of the labels bound by mercury and the groups alkylated to the SH residues of the nitrogenase or incorporated as ethylene derivatives of ATP. The incorporation of the label into the active site was accompanied by an abrupt quenching of fluorescence; when the iron was removed, the initial fluorescence was regained. On the basis of the pertinent theory (Section 3.6), the distance between the chromophore fragment of the label and the nearest iron cluster was found not to exceed 1.6 nm (Alfimova & Likhtenshtein 1979). Taking into account the finite size of these groups, it is possible to assume that the ATPase portion is rather close to the neighboring nonheme group. An analogous conclusion has been drawn from analysis of the spin–spin interaction of the NR introduced into the ATPase site and the cluster by the method of electron-scattering labeling (Levchenko *et al.* 1979; Likhtenshtein *et al.* 1982).

An NR has also been used as a redox reagent, bound to triphosphate and then introduced into the ATP site (Likhtenshtein *et al.* 1986a). An ESR study showed that the reduction rate of this radical by nitrogenase was much higher than that for a free radical. Addition of ATP sharply decreased the rate of the triphosphate derivative reduction. This experiment showed that an electron was transferred from a Fe_nS_n cluster that located in the immediate vicinity of the adenosine fragment of ATP being replaced by the nitroxide.

There are arguments in favor of the viewpoint that the nitrogen-binding site is remote from the ATPase site. For instance, no dipole–dipole broadening of the spectrum of acrylonitrile (which is an inhibitor of nitrogenase and is likely to interact with the nitrogen-binding site) could be detected. By introducing paramagnetic species into the ATPase site [i.e., a spin label on the SH group or Mn^{2+} in the composition of ternary complexes (Mn^{2+}-ATP or -ADP nitrogenase)], the distance between these paramagnetic centers and acrylonitrile in the substrate portion of the active site was found to be greater than 0.6 and 1.0 nm in the cases of Mn^{2+} and NR, respectively. Introduction of Mn^{2+} into a ternary nitrogenase complex produced no noticeable effect on the line width of the ESR signal of the FeMoCo cluster.

Probing of FeMoCo and P-clusters of FeMoP by a triplet probe (eosin linked covalently to the insulin macromolecule) showed that FeMoCo is located in the vicinity of the protein globule surface, with the P-cluster immersed to a depth exceeding 0.7 nm (Druzhinin *et al.* 1988). Blocking of histidine and lysine groups of the nitrogenase preparation (genetically devoid of FeMoCo) did not prevent complexation with the FeMoCo in solution.

A hypothetical model of the active-site system of nitrogenase, taking into account the data discussed earlier, is shown in Figure 6.3. The suggestion of rather close spacings among the P-clusters, FeMoCo, and ATP center was recently confirmed by X-ray data. However, an

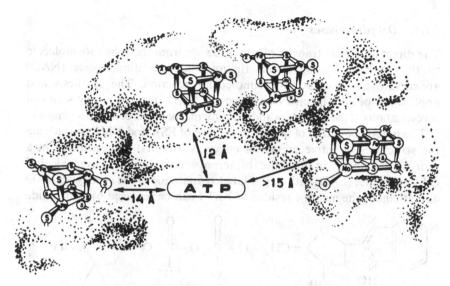

Figure 6.3. Scheme of mutual disposition of clusters and ATP centers of nitrogenase.

earlier suggestion that the two molybdenum atoms were closely spaced (Likhtenshtein 1976a), which then seemed to be confirmed (Sosfenov *et al.* 1986), did not agree with this model (Bolin *et al.* 1990). The reason for such a discrepancy is not yet clear.

An analysis of the available data allows one to presume the following scheme of nitrogenase reactions: The electrons from a one-electron reducing agent are transferred first to the iron-containing component of the enzyme. In the ATPase site of the enzyme, which is located at the boundary between the components and is adjacent to the assembly of iron-containing clusters at FeMoP, the high-energy phosphate bond of ATP is completed with the electron transfer, with the formation of super-redox states of FeMoP. The key step of coupling is the compulsory protonation of the cluster ligand by a portion released upon ATP hydrolysis in the active-site hydrophobic cavity. Protonation in its turn promotes the transfer of an additional electron from the FeP to the FeMoP, with the formation of a super-reduced state after the reversible release of a proton.

Electrons in the super-reduced form are transferred one by one to the FeMoCo. The decisive reduction step of molecular nitrogen complexes at the active sites occurs as a synchronous four-electron attack, with electron transfer either from the two molybdenum atoms of the FeMoCo clusters or from the MoFe atoms within the same cluster.

6.1.3. Dehydrogenases

The direct reversible transfer of a hydride ion from the substrate molecule to the nicotine residue in nicotinamide-adenine dinucleotide (NAD) appears to be an important bimolecular reaction. The synthesis and application of spin-labeled derivatives of NAD inspired a new series of investigations of the structures and reaction mechanisms of dehydrogenases.

The unique potential of the combined use of ESR and NMR in studying the structures of active centers of enzymes was first demonstrated in a thorough manner in the case of ternary complexes of liver alcohol dehydrogenase (LAD) with a spin-labeled analogue of the coenzyme NAD and ethanol. The nicotine residue of XXVI can be replaced by a nitroxide

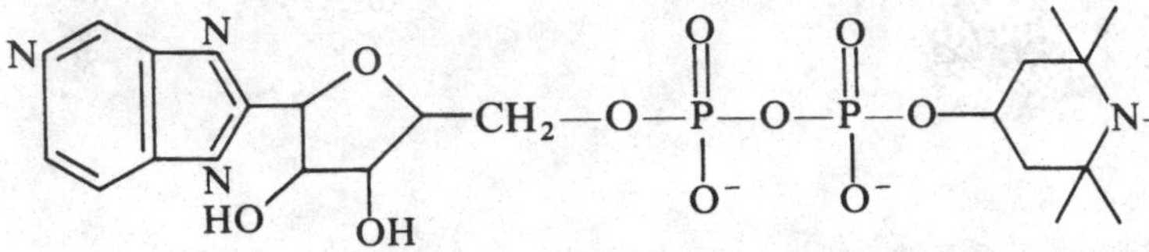

SLXXVII

fragment that competes with NAD and, according to the ESR spectrum, is tightly wrapped in the enzyme molecule. The entry of the label is accompanied by a sharp increase in the relaxation time of the water protons and broadening of the NMR spectrum of ethanol. Another interesting result was a nonuniform broadening of various CH groups. The application of a computational method makes it possible to determine the distance between the ethanol protons and the nitroxide, which is dependent on the foregoing reaction mechanism. The kinetic parameters of the ethanol exchange in a ternary complex can be determined from measurements of T_{2n} at different temperatures ($E_a = 23 \, \text{kJ} \, \text{mol}^{-1}$, $\Delta S_a = 85 \, \text{e.u.}$).

A number of probes analogous to NAD and ATP containing ^{2}H and ^{15}N isotopes in nitroxide fragments were used in another interesting series of experiments (Park & Trommer 1989; Rudy *et al.* 1989):

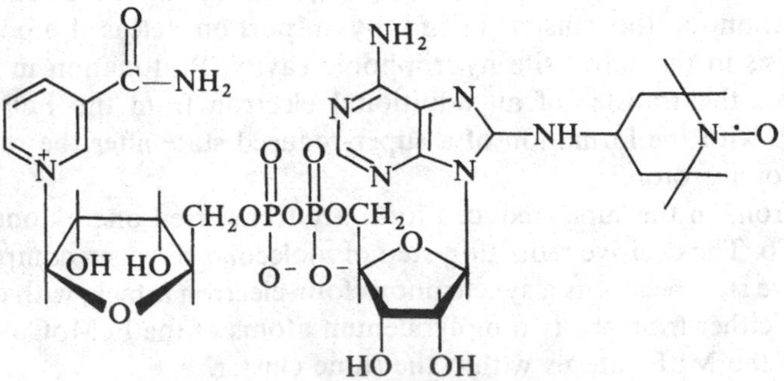

SLXXVIII

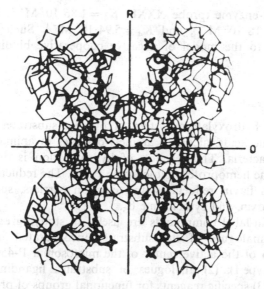

Figure 6.4. Spatial arrangement of four monomers of crystalline GAPDH with four N^6-SL-NAD$^+$ bound in the catalytic centers. The α-carbon backbone of the tetramer is viewed along the R axis. The four spin labels in the anti-anti-binding geometry are shown in dark outline (Park & Trommer 1989).

The high resolution of ESR techniques (Section 1.4) allows one to distinguish between the three individual groups of components related to (1) the free probe in solution, (2) the isolated probe in the active center, and (3) a pair of interacting molecules, each consisting of the probe on the active center, in a solution of glyceraldehyde-3-phosphatase (GADPH) tetramer labeled with the spin probe. The spin–spin dipole–dipole splitting of the ESR spectrum of the latter corresponds to the 1.1–1.3-nm distance between the radicals (Figure 6.4). According to the X-ray analysis, the distance between nitroxide fragments located on adjacent subunits of the enzyme is 1.28 nm. Thus, the conformations of the enzyme in crystal and solution are practically identical.

The interaction of the substrate (glyceraldehyde-3-phosphatase) in the active center is accompanied by an increase in the dipole–dipole splitting of the spin-labeled enzyme from 9.36 to 9.6 mT. This fact is evidence for a noticeable contraction in the enzyme-substrate complex. Conformational transitions were also observed in subunits of the tetramer, induced by the introduction of the NAD molecule to the other subunits and by the interaction with regulatory proteins.

The high accuracy of recording of the ESR spectrum of the spin label provides high accuracy in determining the association constant (K) for the complex ADP-R-enzyme (probe XXXI): $K_1 = 1.86 \cdot 10^7 \, \text{M}^{-1}$, $K_2 >$ $3.9 \cdot 10^6 \, \text{M}^{-1}$, $K_3 = 1.18 \cdot 10^6 \, \text{M}^{-1}$, and $K_4 = 5.94 \cdot 10^4 \, \text{M}^{-1}$. Such a sequence corresponds to the model of negative cooperative binding of coenzymes.

6.1.4. Cytochrome P-450

A powerful enzyme hydroxylation system of organic substrates (e.g., hydrocarbons, organic acids, alcohols, amines) operates in animal and plant tissues and bacteria. The reaction of hydroxylation is directly affected by the enzyme hemoprotein cytochrome P-450. The reduction of the enzyme involves flavin reductases and electron carriers, such as adrenodoxin, rubredoxin, and cytochrome b_5.

Three types of spin-label compounds are useful in studying enzyme hydroxylation: (1) analogues of the substrates incorporated in the hydrophobic portion of the active center of the microsomal P-450 and cytochrome P420 (type I); (2) analogues of substrates liganding the heme iron (type II); (3) specific reagents for functional groups of proteins (Mock *et al.* 1982; Weiner 1986). In one study, spin-labeled metyrapone (SLM) and other substrate analogues were bound to the active center of cytochrome P-450 from *pseudomonas putida* (P-450 camphor) with a high afinity, $K = 10^{-4} - 10^{-5}$ M. From the value of the dipole-dipole splitting, the distance between Fe^{3+} of the heme and the $>N\!-\!O^\bullet$ fragment of the spin label,

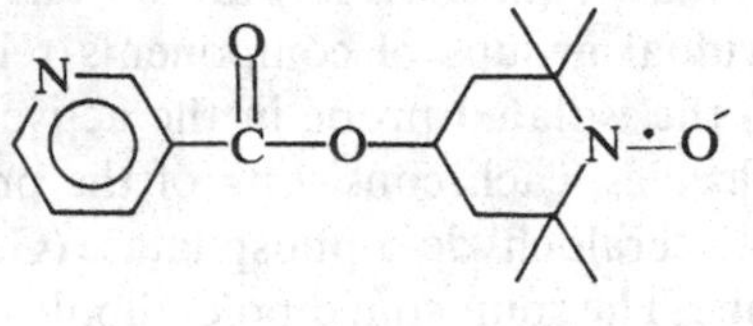

SLXXIX

is 0.57 nm. This distance and those for the modified microsomal P-450 (P-450 m) agree well with the model of binding of the second-group substrates. The nitroxide fragments of the probes in the active centers were not accessible for spin–spin interactions with the spin probe ferricyanide, freely diffused in solution.

A collection of nitroxide spin labels containing ligand groups (cyanide, amine, pyridine residues) was used to study the active center of cytochrome P-450 from *P. putida* (Subkhankulova *et al.* 1986; Lyakhovich *et al.* 1988). The distances between the nitroxide fragments of the labels and heme iron (0.6–0.9 nm) corresponded to those expected from the chemical formula of the label. A study of the spin-label binding

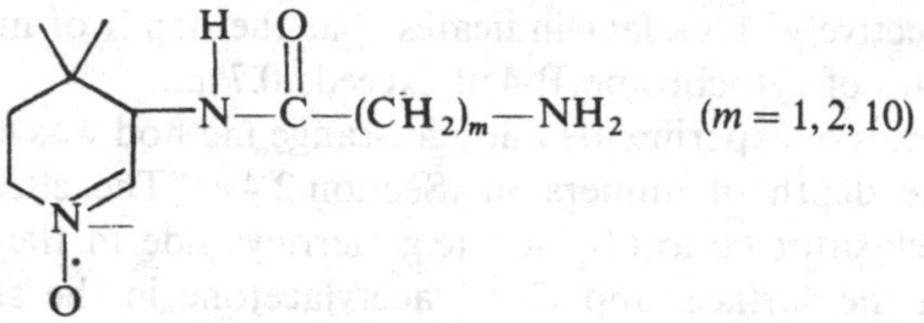

$$\text{—N—C—(CH}_2)_m\text{—NH}_2 \quad (m = 1, 2, 10)$$

SLXXX

of cytochrome P-450 indicated that the label with $m = 10$ possessed the highest affinity for the heme group (Weiner 1986).

In the absence of molecular oxygen, the electron donors of cytochrome P-450 reduced nitroxide analogues of the substrates of both types I and II. Upon aeration, oxidation of the corresponding hydroxylamines was observed (Weiner 1986).

The nitroxide hydrophobic probes, analogues of the substrates of the first groups, such as

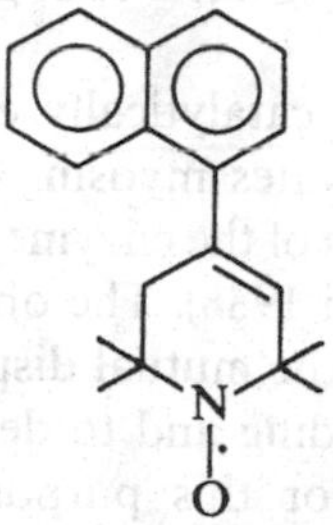

SLXXXI

were involved in the active center as well. However, attempts to observe the effect of the spin–spin interaction between the $>N—O$ fragment and the heme group on the ESR spectra of the probes did not succeed. To estimate the distance between these centers one needs a more sensitive technique based on measurements of the spin–lattice relaxation parameters of the spin probes (Section 2.3).

In some studies an approach was used that was based on analyzing the interaction of the cytochrome P-450 heme group with the whole collection of spin and triplet probes located in the various portions of the microsome membrane and in the aqueous phase (Yudanova *et al.* 1986; Cherepanova *et al.* 1990; Likhtenshtein, 1990a). In the first series of these experiments it was shown that the paramagnetic heme group did not interact with the spin-labeled lysozyme molecules and that triplet anthracene derivatives diffused freely in the aqueous phase and in the lipid phase of the microsome membrane, respectively. This fact indicates that the depth of immersion of the heme group of cytochrome P-450 exceeds 0.7 nm.

In a second series of experiments a longer-range method was employed to determine the depth of immersion (Section 2.4.4). The effect of the increasing ion-relaxator concentration (e.g., ferricyanide in the aqueous phase, Co^{2+} on the surface, and Co^{2+} acetylacetone in the superficial portions of the membrane) on the saturation curve of the ESR signal of the heme group of cytochrome P-450 was analyzed. The latter is located in the hydrophobic part of the enzyme-protein globule and is immersed by 1.2–1.4 nm in the case of the aqueous phase and is ~ 1.2 nm from the surface of the lipid protein of the microsomal membrane (Cherepanova *et al.* 1990).

6.1.5. Myosin and actin

Myosin is a contracting protein that catalyzes the coupled reaction of ATP hydrolysis. The protein consists of six polypeptide chains, which form a superhelical elongated conformation ("tail") turning into two globular "heads."

Earlier biochemical investigations revealed two catalytically essential SH groups: SHI, which upon being blocked activates myosin, and the SHII group, whose modification causes deactivation of the enzyme (Burley *et al.* 1972; Grebenshchikov *et al.* 1972; Charkviani 1988). The objectives of investigations by spin labeling were to estimate the mutual disposition of the SH groups and centers of the ion-metal binding and to detect the feasible conformational transitions in myosin. For this purpose, spin reagents on SH groups and nitroxide derivatives of ATP were used.

The selective blocking of the SHI group allows one to follow allosteric softening of the conformational state in the vicinity of the SH group, which appears to be necessary for the catalytic activity of the enzyme. The allosteric and catalytic centers seem to be remote from each other; this conclusion is based on an analysis of data from the spin–spin interaction between the spin label and Mn^{2+} bound with the enzyme and from the excited energy migration between the fluorescent label of the SHI group and the chromophore analogue of ATP in the myosin active center.

The strong spin–spin interaction between the nitroxide fragment of the probe (adenine derivative with the NR fragment attached to the ribose hydroxyl group) and the Mn^{2+} ion in the complex disappears in the presence of myosin. The distances between the bound Mn^{2+} ion and the probe range from 40 to 45 nm. The latter conclusion seems to contradict some of the data implying that the ATP-Me^{2+} complex is commonly a substrate of ATPases. In our opinion, the problem needs to be treated with due care.

The investigations of spin-labeled preparations of actin have led to the following conclusions: (1) a nitroxide analogue containing the adenine group is bound tightly to the active center of actin; (2) spin label XI attached to the SH group of the enzyme is located at a distance of about 2.0 nm from the center of binding of the Mn^{2+} ions; at such a distance the actin polymerization does not produce any significant changes; (3) the conformational changes of spin-labeled actin in the actin-tropomin-tropomyosin complex and of spin-labeled tropomyosin in the presence of tropomin are observed under the action of Ca^{2+} ions (Burley *et al.* 1972).

The data on the effects of Mg^{2+} on the rotational motion of myosin heads modified by the eosin triplet label provided direct evidence that physiological concentrations of cations influence the rotational dynamics of myosin heads in thick synthetic filaments. The quantitative results indicated that even in the cross-bridge compact conformation the label underwent large-amplitude microsecond rotations (Ludescher *et al.* 1988). The addition of ATP to the spin-labeled oriented myosin filaments induced ESR spectra corresponding to randomly reoriented labels (D. Thomas *personal communication* 1990). This finding is not consistent with the current concepts suggesting directed concerted reorientation of myosin heads in the filaments under the action of ATP.

6.1.6. Other enzymes and proteins

The method of a "spin ruler" was used to estimate the size of a narrow hydrophobic cleft in a number of enzymes and proteins. The most detailed results were obtained for carboanhydrase (Chignell 1979). When nitroxide derivatives of sulfamides with various lengths of bridges between the aromatic groups were included into the active center of carboanhydrase from erythrocytes and nitroxide fragments, the mobility of such a fragment was shown to be drastically enhanced if the length of the probe exceeded 1.45 nm. Similarly, increasing the chain lengths of radicals introduced by a dinitrophenyl fragment into the active centers of antibodies affects the ESR spectrum at a certain critical value (1.0–1.2 nm) (Hsia & Piette 1969).

Dipole–dipole interactions between the nitroxide head of a spin-labeled dinitrophenyl derivative (haptene) and aromatic groups of the Fab fragment of immunoglobulin have been studied thoroughly by Frey *et al.* (1988). They also investigated the influence of spin labels on the widths and positions of the NMR spectra of aromatic groups of the antibody. In these experiments, parts of the native aromatic groups were specifically replaced by deuterated groups in biosynthesis. Those authors placed a total of three tryptophane, six tyrosine, one phenylalanine, and one histidine groups in the region of the active center at a distance of less than 1 nm, with two tyrosine and one tryptophan groups directly adjacent to the dinitrophenyl ring of the hapten.

A complex system of catalytic and regulatory subunits of aspartate transcarbamylase was studied by Matsumoto and Hammes (1975) using the method of fluorescent labels. Pyridoxamine phosphate (PP) covalently attached to the amino groups of the catalytic active center served as the fluorescent donor in the inductive resonance energy transfer to the mercury nitrophenyl group, i.e., to the acceptor attached to the SH residues of catalytic subunits. The active center appeared to be in the immediate vicinity of the SH group in the same subunits, but appeared to be separated from corresponding groups in other subunits of the catalytic trimer. The distance between PP and ANS attached to the active centers of different subunits was close to 2.6 nm. The fluorescent analogue of cystidine-5'-triphosphate in the regulatory center of the enzyme was quenched only slightly by the acceptors in the catalytic subunits and was reported to be separated by more than 4.2 nm from the active center. The validity of the distance calculations with $K^2 \approx \frac{2}{3}$ was confirmed by polarization experiments.

The complicated problem of determining the relative positions of allosteric and active centers of phosphorylase b was solved with the aid of a whole series of methods, including measurements of (1) the effect of Mn^{2+} in the active center on the NMR spectrum of AMP (the activating agent) and glucose-1-phosphate (the substrate), (2) the fluorescence quenching of a chromophore label on the SH group of the enzyme, and (3) the interaction between an iodoacetamide label on the SH group and the ligand protons as well as Mn^{2+}. The results of the complex investigations are presented in Figure 6.5 (Bennick *et al.* 1971).

Combined masurements of T_{1n} and T_{2n} for water protons and phosphoenolpyruvate analogues, as well as measurement of the broadening of the NMR spectral lines for the protons and phosphorus in phosphogluconate in the presence of pyruvate kinase and Mn^{2+}, enabled Nowak and Mildvan (1972) to develop a detailed model of the enzyme-substrate complex. The NMR measurements clearly showed the changes in the substrate upon inclusion of potassium ions in the active center.

The efficiency of combined application of the methods of NMR and spin labeling was demonstrated to its full extent in a study of the structure of proteins of small size using the toxin *Naja naja oxiana* (Tsetlin & Ovchinnikov 1986). The breadth of the available data allowed construction of a spatial model of the protein.

Two surface SH groups, Cys-316 and Cys-140, were modified by "short" and "long" spin labels to study the topography of a hydrophobic portion

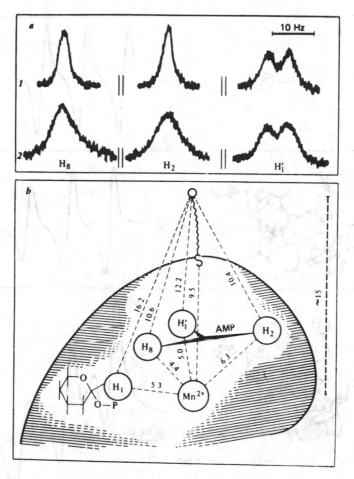

Figure 6.5. Results of an NMR investigation of spin-labeled phosphorylase *b*: (a) NMR spectra of $2.19 \cdot 10^{-2}$-M AMP in water; (1) in the presence of phosphorylase ($4 \cdot 10^{-4}$ M); (2) in the presence of spin-labeled phosphorylase ($3.7 \cdot 10^{-2}$ M). (b) Model of a complex of the enzyme with a substrate analogue and with the activator (AMP) (distances in angstroms) (Bennick *et al.* 1971).

of rhodopsin (Livshitz *et al.* 1988a). The rotation of the labels, as shown by the restricted proteolysis techniques, was hindered by hydrophobic portions of the protein polypeptide chain: Cys-322-Ala-346 and Cys-273-Ala-341. The photoinduced transition from the initial form of rhodopsin to metarhodopsin II led to some softening of the protein matrix, which was reflected in the change in the shape of the spin-labeling ESR spectrum and in the parameters of the saturation transfer ESR signal. The intensive

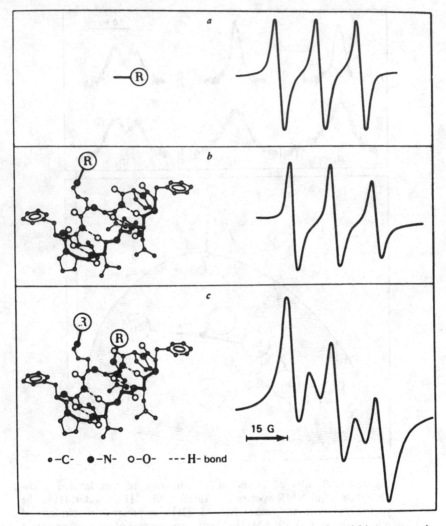

Figure 6.6. Determination of the conformation of ornithine groups in gramicidin S with the aid of spin labels at 42°C: (a) R is the nitroxide derivative of cyanuric chloride. (b, c) Derivatives of gramicidin S: (b) monoradical, (c) biradical (Ivanov *et al.* 1973).

light irradiation inducing the protein molecules to aggregate was monitored by spin-labeling techniques and by electron microscopy of rhodopsin prepared with the SH group modified by electron-scattering labels (Kalamkarov *et al.* 1977).

Recently, preparations of rhodopsin have been obtained with biosynthetically spin-labeled amino acids included in a hydrophobic portion of the protein polypeptide chain. The accessibility of the nitroxide fragments to dioxygen was traced by saturation transfer techniques (J. Hyde, *personal communication* 1989).

The first small polypeptide studied by spin labeling was cyclic decapeptide gramicidin (Ivanov *et al.* 1973). A label based on cyanuric chloride (SLXV) was attached to the side-chain ornithine NH_2, and doubly and singly labeled gramicidin derivatives were obtained. Besides the three principal lines, the solution spectra of the biradical derivatives had two lines (Figure 6.6), which indicates that there was an exchange interaction between unpaired electrons due to intramolecular encounters between the radical fragments. The distance between the radicals proved to be 1.25 nm. In fact, the characteristic features of the antibiotic structure are the rigid conformation of the polypeptide backbone, which is held in its place by four intramolecular hydrogen bonds, and the arrangement of the side-chain groups of the ornithine residues on the same side of the plane.

Data have been cited elsewhere on applications of spin and fluorescence labeling methods to study the structures and dynamics of other enzymes and proteins, including lysozyme, carboxypeptidase, serum albumins, ribo-nuclease, DNA polymerase, citrate synthetase, creatine kinase, aspartate aminotransferase, etc. (Berliner 1976; Likhtenshtein 1976a).

6.2. Conformational changes in proteins and enzymes

This chapter presents experimental data on the application of biological labels for detecting various types of conformational changes, including (1) comparatively large-scale denaturation changes; (2) predenaturational phenomena; (3) allosteric effects, which are transmitted from one enzyme subunit to another; (4) transglobular allosteric transitions; and (5) changes in the states of various segments of the active centers of enzymes when specific reagents act on the adjacent segments.

6.2.1. Large-scale and allosteric conformational changes

The thermal motions of spin labels in proteins are very sensitive to changes in the secondary and tertiary structures such of proteins. Such processes as the decomposition of hemoglobin into its subunits, the stretching of the polypeptide chains of hemoglobin, albumin, lysozyme, and aspartate aminotransferase upon ionization, the thermal denaturation of hemo-globin, the denaturation of proteins by urea, dioxane, and a number of other structural alterations generally result in release of the labels and changes in the rotational diffusion parameters of the nitroxide fragments by factors of 100 or more (Likhtenshtein 1976a). The experimentally observed changes in the mobility of a label upon alterations in the properties of the protein matrix largely depend on the structure (length the flexibility) of its "legs." In those cases in which labels with relatively large distances between the polypeptide skeleton and the nitroxide group ($\geqslant 1.2$ nm) are used, the molecular motion of the latter depends only slightly on the fine structure of the protein surface.

Labels that quench fluorescence were used to study conformational transitions of bovine and human serum albumins (BSA and HSA), which possess tryptophan and tyrosine fluorescent centers (Likhtenshtein *et al.* 1968; Likhtenshtein 1976b). The chromophore residues of the trichloro-triazine dyes 5BC, 2RP, and 4R1 served as the acceptor groups. The quenching parameter (I_f/I_f^0) of the single tryptophan residue of HSA depends strongly on the number of labels bound and on factors that disturb the protein conformation. As expected, the most substantial increases in the distances ($r = 0.5$–1.2 nm) between the tryptophan and the label groups, which reflect conformational changes causing loosening of the protein molecules, were observed under the action of 8-M urea. More modest changes in r (0.2–0.4 nm) occurred when the protein was deformed as a result of an acid–base transition in the protein groups. Similar phenomena have been observed in studies of energy migration between the tyrosine residues of albumin and the labels.

There is particular interest in quencher labels that contain paramagnetic ions (e.g., the copper-containing Procion dye 2RP). Application of such compounds permits simultaneous evaluation of the distance between the luminescent center and the label of the luminescence quenching and the distance between the labels of the dipolar broadening of the ESR spectrum.

The first really significant result from the use of spin labels was discovery of the allosteric conformational changes in hemoglobin in solution. Spin labels not only made it possible to determine that the change occurred in the vicinity of the β-93 SH group, at a distance of about 1.5 nm from the site of the reagent effect (the heme group), but also revealed the details of this phenomenon.

Allosteric effects were first reported by McConnell and associates (Ogawa & McConnell 1967; McConnell & McFarland 1970) in studies of spin-labeled horse hemoglobin. When oxygen is bound in the active center of hemoglobin, the ESR spectral components are markedly broadened, and splitting is observed. It is clear that portions of the protein are contracted as a result of the allosteric effect, which is transmitted from the heme group to the vicinity of the label. Experiments with the individual subunits have shown that addition of O_2 causes the transition of the hemoglobin β-subunits to a more compact conformation. Careful measurements of the dependence of the ESR spectrum of the label in hemoglobin on the O_2 pressure revealed a slight deviation from the isosbestic rule that was due to changes in binding contacts of O_2. Since the labels were better immobilized upon addition of O_2 in native hemoglobin $(\alpha_2\beta_2)$ than in the free chains, the authors concluded that the main allosteric effects are on the surfaces of the β-subunits in the areas of contact with the α-subunits. This conclusion was later confirmed by detailed investigations, including hybrid and genetically and chemically modified preparations of hemoglobin (Baldassare *et al.* 1970).

The use of spin labels that are derivatives of ATP, triphosphate, and chloramphenicol in regulatory centers of hemoglobin and mutant hemoglobin (leucine-12 instead of α-92 arginine) made it possible to monitor interactions between the regulatory and acceptor centers of the protein. A thorough quantitative analysis revealed a good fit between the experimental curves of carbon monoxide binding and the result predicted by the generalized concerted model (Ogawa & McConnell 1967; Gantchev *et al.* 1988).

6.2.2. *Transglobular conformational transition*

The studies by McConnell and associates, however, do not present a complete solution to the problem of allosteric interactions. The most dramatic events that occur during hemoglobin interaction with the substrate take place at the boundaries between the subunits and can be induced by disturbing the subunits without significant changes in their tertiary structure. Therefore it would be interesting to investigate conformational changes in the enzymes of structures other than quaternary.

That provided the reason for a series of studies aimed at experimental investigation of allosteric effects in lysozyme, myoglobin, aspartate aminotransferase, and myosin (Likhtenshtein 1976a). Differing in terms of molecular size, enzymatic function, and other properties, these enzymes make convenient systems for studying the general laws governing enzyme actions.

Hence, studies of the effects of specific inhibitors of lysozyme, particularly fragments of polysaccharide chains, are of special interest. According to X-ray diffraction analysis, the latter are embedded in clefts in the A, B, and C subsites and thereby prevent the formation of enzyme-substrate complexes. Addition of sucrose, glucose, and maltose, which are bound either very weakly or not at all to the active center, to lysozyme preparations labeled at histidine-15 affects the ESR spectrum of the label very little. However, addition of smaller quantities of the specific inhibitors NAG and NAG-NAG induces distinct changes in ESR spectra. The relative changes in the values of the h_0/h_{-1} parameter as functions of the concentration are in good quantitative agreement with the extent of binding to the substrates (Likhtenshtein *et al.* 1974; Likhtenshtein 1976a).

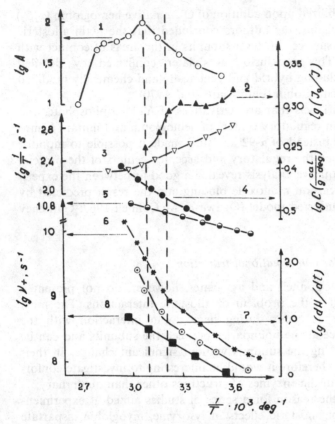

Figure 6.7. Arrhenius dependences of different characteristics of lysozyme enzymic activity (1), fluorescence quenching intensity I_0/I (2), relaxation rate of protons of water in the presence of the sample labeled by His-15 (3), partial heat capacity (4), frequency of rotational diffusion of spin labels (5), deuterium exchange (6), and lysozyme rotation (7) (Alfimova & Likhtenshtein 1979).

The increase in the rotational diffusion parameter of the spin label may indicate that there is some separation between the protein segments in the vicinity of the histidine group on the opposite side of the protein globule at a distance of 2.0 nm. According to X-ray diffraction data, NMR, and differential fluorescence spectra, this binding causes some contraction of the groups in the active center. Thus, there is a specific cooperative transition from the C subsite to the region of the histidine-15 group. The latter is located in the hydrophobic region between th Phe-3 and Asp-87 residues of lysozyme (Schmidt & Kuntz 1984).

Experimental results from studies of transglobular predenaturation transitions in lysozyme by a whole set of physicochemical methods are

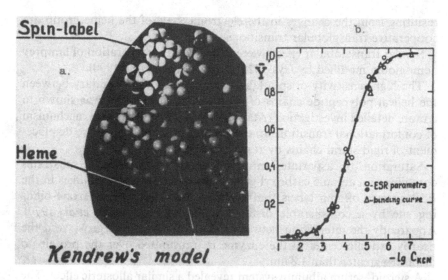

Figure 6.8. Transglobular conformational transition in sperm-whale myoglobin: (a) Scheme of disposition of nitroxide spin label and heme group in the Kendrew model. (b) Variations of the relative parameters of the ESR spectra of spin-labeled myoglobin ($Y = h_0/h_{+1}; h_0/h_{-1}$) during complexation with a specific inhibitor (cyanide) (Likhtcnshtein 1976a).

presented in Figure 6.7 (Alfimova & Likhtenshtein 1979). As can be seen, the methods of fluorescence quenching, spin labeling, and deuterium exchange reveal conformational changes at temperatures above 30°C, whereas the temperature dependences of the enzyme activity, the heat capacity, and the water proton relaxation change monotonically in this temperature region. The effects observed in the first series of experiments appear to have been caused by a shift of relatively large segments of the protein globule, with the shift not being affected by the parameters recorded in the second series of experiments.

Sperm-whale myoglobin is a low-molecular-weight, compact protein with a known structure. In one study, the three easily accessible myoglobin histidine groups, CD-6, EF-4, and CH-1, were modified with labels XI and XIV. According to the Kendrew model and the data of Atanasov *et al.* (1977), each of these groups is located at a distance of about 1.6–1.8 nm from the heme center and thus is not accessible to the direct action of the substrates at the heme active center. When myoglobin is titrated with the specific inhibitor cyanide, the curve describing the relative changes in the ESR spectral parameters of the label practically coincides with that describing the binding of cyanide to the active center (Figure 6.8). This result indicates that the conformational changes in the enzyme resulting from the changes in the electronic state of the heme group are cooperative transglobular transitions.

Similar transglobular effects were observed in a preparation of lamprey hemoglobin modified by Cys-142 groups (Artukh *et al.* 1990).

The high sensitivity of spin labels located near the boundary between the helical polypeptide chains of sperm-whale myoglobin was shown in a more detailed investigation (Atanasov *et al.* 1977). The main mechanism of conformational transitions in the protein probably is relative displacement of rigid spiral chains by restricted hinge-like movements.

Saturation of aspartate aminotransferase with specific substrates containing an anionic carboxyl group causes reproducible changes in the ESR spectra of spin labels that are separated from the substrate-binding site by a considerable distance (> 1.8 nm) (Timofeev *et al.* 1970). Apparently the interaction between the anionic substrate group and the nearby cationic center of the enzyme is transmitted over the protein to distances greater than 1.8 nm.

A steroid–serum albumin system revealed a similar allosteric effect. The presence of the steroids prednisolone, 6-methylprednisolone, hydrocortisone, Viadril, estradiol, etc., causes considerable changes in the ESR spectra of labels that are located at distances greater than 1.8 nm from the albumin binding site (Sergeev *et al.* 1974).

A complexation of antibodies with human γ-globulins was found to be accompanied by conformational changes in both proteins (Kaivarainen 1985; Nezlin & Sykulev 1988). In this case, spin labels were found to be more nearly immobilized on the γ-globulin molecules and less immobilized on the antibodies. This fact probably demonstrates the general ability of a protein globule to be contracted in one portion at the expense of extension in other portions. This effect can be transmitted from one molecule to another.

6.3. Molecular dynamic properties of proteins and enzymes

6.3.1. *General*

Our present conception of the molecular dynamics of proteins is based on three hypotheses advanced in the late 1950s and early 1960s.

First, Lumry and Eyring assumed that the interaction of substrates with enzymes was accompanied by a certain distortion of the structure (conformation) of the enzyme (Lumry & Eyring 1954), namely, by the change in the energy state of the system, which contributed to the favorable occurrence of the particular reaction. Later these concepts developed into the theory of "complementarity" of free energies of chemical and conformational processes (Lumry & Biltonen 1969). According to this concept, the energy redistribution to conformational energy upon binding of the substrate occurs in such a way that the energy profile of the process as a whole is eventually flattened.

Another basis for the modern concept of the molecular dynamics of proteins is the Linderström-Lang hypothesis of the mobile fluctuating state of the protein macromolecule. One of the manifestations of structural fluctuations is the ability of the inner peptide groups to exchange hydrogen atoms for water protons, with the process taking place in open conformations of the protein globule (Linderström-Lang & Schellmann, 1959).

The third concept is the Koshland induced-fit theory, according to which the structure of the enzyme can be induced to fit the structure of the substrate during their interaction. Hence, binding of the substrate to the enzyme must be accompanied by a change in the conformation of the enzyme (Koshland 1959).

Though these early hypotheses were based on speculative and indirect considerations, they nevertheless had great impact on the development of the science of enzymes, since they provided a clear formulation of the problem and, to a certain extent, outlined routes to its solution.

In the later stages of solving this problem, investigators have used quantitative data on the isotopic exchange of hydrogen between the aqueous phase and the inner portions of protein globules, the kinetics of enzyme catalysis, the reactivity of the functional groups inside the protein globule, and the proteolysis of proteins.

Analyses of the entropy and energy specificity of the enzyme catalysis and denaturation processes have led to the conclusion (Likhtenshtein 1966, 1976a,b) that the elementary acts of chemical conversion in the active sites of the enzyme are accompanied by cooperative changes in the structure of the surrounding water-protein matrix. Numerous data have been obtained on such changes in conformation during the functioning of proteins, providing indirect evidence in favor of their intramolecular mobility (Lumry & Gregory 1986). In particular, so-called transglobular conformational transitions have been discovered in enzymes consisting of a single protein subunit, which is the evidence for their intramolecular nature (Section 6.2.3).

At present, a new approach to the investigation of protein dynamics is being intensively developed. Its distinctive feature is application of the latest methods (in particular, NMR, X-ray diffraction, dynamic spectral methods, and physical labeling), making it possible to obtain detailed information on the mobility of individual functional groups in proteins and to compare the kinetic and dynamic characteristics of enzymatic processes with the results of theoretical calculations (Likhtenshtein 1976a,b, 1979a; 1988a; Permyakov & Burshtein 1977; Wütrich 1986).

6.3.2. Experimental data

The introduction of spin, fluorescent, triplet, and Mössbauer labels and probes, as well as radical pairs, into a protein molecule enables one to record the dynamics of different domains within the entire range of correlation times $\tau_c = 10^2 - 10^{-10}$ s (Chapters 1–5). Figure 6.9 is a conventional diagram of such an introduction. The interior labels, chromophores of tryptophan and tyrosine residues and ^{57}Fe atoms in the composition of the active centers of iron-containing enzymes, can also be employed. Several studies have been devoted to the effects of temperature and humidity on the dynamic states of proteins and enzymes, as well as their functional activities.

The first direct data on the intramolecular mobility of protein globules in the nanosecond region ($\tau_c = 10^{-7} - 10^{-9}$ s) and its amplitudes ($A = 0.05 - 0.1$ nm) were obtained by the methods of spin labeling and Mössbauer labeling (Frolov *et al.* 1973). Figure 6.10 shows the theoretical and experimental dependences of the nitroxide line width on the shift of the HFI constant for some spin-labeled proteins. As seen from this figure, a temperature increase leads first to narrowing of the line, followed by parallel increases in both the line width and the shift. This fact clearly indicates that in certain systems the model of anisotropic diffusion in a slow rotation region accurately describes the system. The diffusion parameters reflect the microviscosity of the environment (Likhtenshtein 1976a). This conclusion was independently confirmed by 2-mm-band ESR spectroscopy. As one can see from Figure 6.11, a monotonic parallel shift is observed for the HFI constant for the x, y, and z components, indicating slow-motion rotation of the nitroxide spin label attached to the His-15 group of lysozyme.

Studies of the intramolecular mobilities of bovine and human albumins over a wide temperature range (30–300 K) by combined use of methods that enable one to study the motion at frequencies of $v_c = 10^{-2} - 10^{11}$ s^{-1} have revealed the general picture of dynamic effects in these proteins (Likhtenshtein & Kotelnikov 1983; Krinichny *et al.* 1985; Likhtenshtein 1988a,b). It can be seen from Figure 6.9b that the general tendency is as follows: the lower the value of the characteristic frequency of the method, the lower the temperature at which the label mobility can be recorded. In a wide temperature range (130–300 K) the experimental data for surface labels lie about the Arrhenius straight line, with $E_{app} \approx 70$ kJ mol^{-1} and $\Delta S^{\neq}_{app} \simeq 160$ e.u. (Likhtenshtein 1985). Thus, the recorded mobility results from a gradual softening of the water-protein matrix, rather than an individual phase transition. The mobility of the spin label and diphenylamine radicals in the hydrophobic pocket of HSA has been described by the following parameters: $E_{app} = 20$ kJ mol^{-1}, $\Delta S^{\neq}_{app} = -23$ e.u.

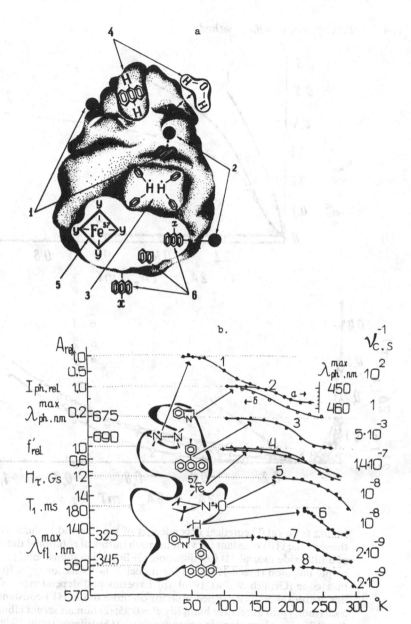

Figure 6.9. Study of protein intramolecular dynamics: (a) The principles of introduction of physical labels and probes into a protein globule: 1, spin labels; 2, spin probes; 3, radical pairs; 4, NMR probes; 5, Mössbauer labels; 6, luminescent labels and probes. (b) Temperature dependences of physical labels on dynamic parameters in serum albumin. Left: Conventional diagram of the label location. Right: Characteristic times of the appropriate methods; 1, radical-pair recombination; 2, tryptophan phosphorescence; 3, eosin-label phosphorescence; 4, Mössbauer-label T_{1e} parameter; 6, PMR; 7, tryptophan fluorescence; 8, eosin-label fluorescence (Likhtenshtein & Kotelnikov 1983; Likhtenshtein 1988a).

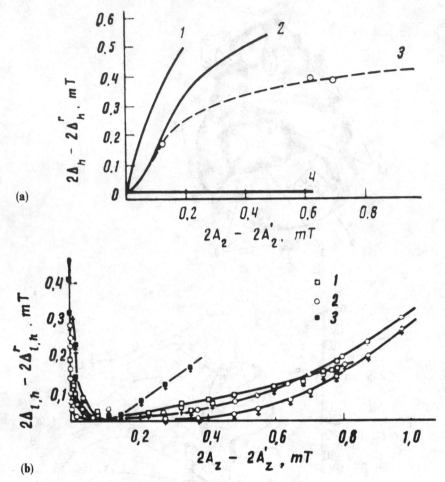

Figure 6.10. (a) Theoretical dependences of the line-width shifts (Δ) on those of the HFI constant for various mechanisms of rotational diffusion of a nitroxide radical: (1) free diffusion; (2) Brownian diffusion; (3) anisotropic diffusion (Freed 1976); (4) the model of fast rotation ($\tau_c \leqslant 10^{-9}$ s) in a cone (Griffith & Jost 1976). (b) Experimental dependences of the line-width shifts of IR-labeled proteins on those of the HFI constant. (1) α-chymotrypsin at relative humidity $P = 0.95$; (2) human serum albumin, $P = 0.95$; (3) α-chymotrypsin aqueous solution (Antsiferova *et al.* 1989a).

in the temperature range 80–280 K (Likhtenshtein 1985). Their mobility at $\tau_c < 10^{-7}$ s rises starting at about 200 K. This fact has been independently confirmed by the data on Rayleigh scattering of Mössbauer radiation. According to these data, the intensity of the elastic scattering of the radiation increases drastically at $T > 200$ K and at a degree of hydration $P > 0.1$ (Goldansky & Krupyansky 1988).

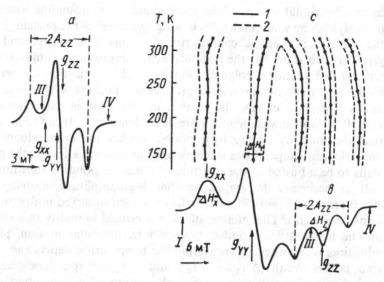

Figure 6.11. Temperature dependences of the parameters of the 2-mm-band ESR spectrum of spin-labeled lysozyme: (a) 3-cm-band spectrum, 150 K; (b) 2-mm-band spectrum, 150 K. (c) Temperature dependences of the x, y, and z components of the 2-mm-band ESR spectrum [positions (1) and line widths (2) of the components are indicated] (Krinichny *et al.* 1988).

However, it is evident that the general picture given here is approximate. More detailed measurements have revealed appreciable differences in dynamic properties between various portions of the protein globule that are hidden when proteins are studied at a wide range of temperatures and correlation frequencies. The side groups exposed to the solvent at physiological temperatures move with higher frequencies, and the groups that are closely packed into peptide blocks move with lower frequencies than would be expected from the averaged temperature dependence. It should be noted that the radical-pair and luminescence methods disclose a substantial heterogeneity in dynamic parameters that is characterized by an activation-energy distribution of $4-6 \, kJ \, mol^{-1}$ even in those cases in which the labels are localized in the same portion of the protein macromolecule. The heterogeneity of the macromolecule of serum albumin is evidenced by the recently discovered phenomenon of the long-wavelength shift in the fluorescence spectrum of the only tryptophan in HSA upon edge long-wavelength excitation (Demchenko 1986).

According to the localization of physical labels and probes, three zones can be distinguished: (1) the peripheral aqueous phase, which does not differ from the water bulk at a distance of 1.4–1.6 nm; (2) the zone formed by the functional groups of the protein and the surrounding water, and possibly also by water molecules in wide "grooves" of the protein globule; the molecular dynamic characteristics of this zone correspond to a glycerine-like state, i.e., the state of viscous amorphous liquids and soft natural and synthetic polymers above the glass-transition temperature; (3) the zone that contains narrow grooves and pockets on the surface and also polypeptide chains; the motion in this zone is characterized by $v_c \approx 10^7 \text{ s}^{-1}$ and a weak temperature dependence on v_c. It can be presumed that the mobility of the hydrophobic pocket in serum albumins at physiological temperatures is functionally important, allowing the pocket walls to be adjusted to the structure of the corresponding substrate.

It is necessary to emphasize that high-amplitude mobility with $A > 0.1$ nm and $\tau_c = 10^{-2} - 10^{-11}$ s has not been observed in dry preparations of proteins. This motion starts at a critical humidity that depends on the temperature. Another type of intramolecular motion, phonon vibration, is recorded by measuring the temperature dependence of the spin pocket width ($\Delta H_{1/2} \sim 1/T_{2e}$) and $1/T_{1e}$ of spin labels and the probability of recoilless resonance absorption of a γ-quantum (f') of Mössbauer labels. This motion does not depend significantly on the humidity of the protein preparations (Likhtenshtein 1986).

Rotational motions of individual fragments of immunoglobulins, with $\tau_c \approx 15-35$ ns, were found in experiments with preparations modified by spin and fluorescent labels. Such motion was also observed in complexes of immunoglobulins with receptors (Kaivarainen 1985; Nezlin & Sykulev 1988).

The first direct data on intramolecular mobility of hemoprotein globules were obtained by the Mössbauer labeling method (Belonogova *et al.* 1978; Bauminger *et al.* 1983). The experiments were carried out on dry and moistened powders and in myoglobin single crystals, which excluded any motion of the macromolecule as a whole. The [^{57}Fe] atoms were incorporated into the heme group in hemoglobin and myoglobin. Thus, it became possible to trace the displacement of the [^{57}Fe] atom tightly bound to the porphyrin ring relative to the protein globule.

Since the heme ring is bound to the protein matrix by numerous hydrophobic contacts, it is evident that any motion of the heme must be accompanied by displacement of the helical polypeptide chain. As shown by these experiments, in dry and moistened samples at temperatures below 200 K the vibrations of the [^{57}Fe] correspond to "ordinary" solid-state vibrations, with $\langle \Delta x \rangle_c$ being equal to 0.001 nm. However, as the relative moisture content exceeds a certain critical value ($P/P_S > 0.65$), a new type of motion arises, with $\langle \Delta x \rangle_c = 0.02$ or 0.05 nm, and a frequency of $v_c \geqslant 10^7 \text{ s}^{-1}$. These values characterize the dynamics of intramolecular vibrations. The increase in mobility recorded by γ-resonance spectroscopy correlates well with the data on the isotopic exchange in myoglobin and on the spin-labeling method.

The vibrations of the heme group begin at the same moisture content as the exchange in the inner region of the globule. Under conditions in which the exchange involves the portions directly adjacent to the heme group, the latter becomes involved in a diffusive motion, with $v_c \geqslant 10^7 s^{-1}$. The latest investigations have confirmed the basic inferences drawn in the works cited earlier (Parak *et al.* 1983; Goldansky *et al.* 1983).

The flexibility of the cavity of the myoglobin active site is evidenced by the mobility of a rather bulky spin label, a derivative of isocyanide attached to the heme group in the single crystal. At room temperature the mobility parameters are $v_c = 10^7 - 10^8 s^{-1}$, $E_{app} = 12 \, kJ \, mol^{-1}$, and $\Delta S^{\neq}_{app} = -60$ e.u. (Likhtenshtein 1976a; Likhtenshtein & Kotelnikov 1983; Doster *et al.* 1989; Smith *et al.* 1989).

The values of mean displacements ($\langle \Delta x \rangle_c = A$) have been calculated (Frauenfelder 1989) for separate myoglobin atoms on the basis of X-ray diffraction data. These results agree well with the findings obtained by physical labeling techniques and with those obtained in subsequent works devoted to Mössbauer spectroscopy of myoglobin.

For example, according to the authors of the works cited, $A = 0.022$ nm for the atoms of the heme group, and $A = 0.02-0.03$ nm for the motions of most of the atoms in the polypeptide chain. For the more mobile end of the globule and the cavity of the active site, $A = 0.035-0.040$ nm, and for the sheath of the globule, $A = 0.045$ nm. The temperature dependences of the quantities A and $\langle \Delta x \rangle_c$ and of the rate constants for electron transfer between the hemes of hemoglobin have also been found to be similar (Mayo *et al.* 1983; Peterson-Kennedy *et al.* 1984).

The dynamic behavior of the protein macromolecule can be approximately described in terms of the following model (Likhtenshtein 1985): The protein globule is made up of rigid blocks (α-helical portions, β-structures, closely packed side groups) and a softer, amorphous region (disordered, loose portions and water molecules). The blocks are bound by relatively flexible hinges and fairly weak hydrogen, hydrophobic, and polar interactions at the contacts between the blocks and the amorphous regions. Thus, the block system is embedded in a viscous amorphous medium. The protein globule is surrounded by a glycerin-like "sheath" composed of water molecules and exposed surface protein groups.

The globule may be in a state in which the blocks are closely packed (the closed state) or in an open state. The number of open states is relatively small. In an open state, the blocks may have a looser structure than in the closed state. This may manifest itself, for example, in the ability to undergo a hydrogen exchange between the peptide portions and water.

Like many other amorphous bodies, the external sheath is characterized by a high value of the viscosity thermal coefficient. As the temperature increases, the sheath softens substantially, and the motion of the blocks is then determined by the interactions in a system composed of blocks and an internal amorphous region. The dynamics of the amorphous region are characterized by a low value of the viscosity temperature coefficient.

The concepts of proteins as systems possessing the properties of amorphous bodies have recently been developed on the basis of data on the temperature dependence of the heat capacity in the region of helium temperatures (Goldansky *et al.* 1983). Analogous concepts have been worked out for protein globules and rubber-like systems (Lumry & Gregory 1986). The viscous properties of protein bodies have been taken into account in a number of theoretical works in which the chemical reaction has been treated as the result of the combination of (1) the thermal-disorder-activated motion of the nuclei of the reaction complex along the reaction coordinate at a thermal velocity and (2) the diffusive motion of the nuclei system of the surrounding medium. From this standpoint it is possible to account for the effects of the solvent viscosity on the rate of enzymatic reaction and on the intramolecular dynamics of proteins recorded by Mössbauer spectroscopy (Goldansky *et al.* 1983).

6.3.3. *Dynamics and functional activities of proteins*

It is of interest to compare the data on the dynamic state of myoglobin with the results of kinetic studies of ligand binding in CO-deoxymyoglobin. As seen from Figure 6.12, the following correlations are observed in water-glycerin solutions: The region of the motions of fluorescent and spin labels with $\tau'_c = v_c \approx 10^7 \, s^{-1}$ near 230 K coincides with the region of variation of the apparent activation energy of the CO association with the heme group from 80 up to $12 \, kJ \, mol^{-1}$. Approximately in this temperature range the slope of the temperature dependence $N(t)$ changes in the Arrhenius coordinates, where $(N_{(t)})$ is the function of myoglobin molecules that have no time to be complexed in the reaction between the CO molecules in solution and the heme during the time of $10^{-2} \, s$. The rate-determining step under these conditions is apparently the movement of CO from the solvent into the active-site cavity.

The temperature dependence of the quantum yield of photodissociation of carboxymyoglobin, I_{ph}, is correlated with the data from fluorescence and Mössbauer spectroscopy (Likhtenshtein & Kotelnikov 1983). Below 200 K the quantum yield is close to zero. In the same temperature region, the quantum yield begins to increase; I_{ph} and f' drop sharply, the value of $\Delta\lambda_{max}$ of fluorescence spectra increases (and hence the conformational mobility of the heme increases), and the dynamics of the water-protein

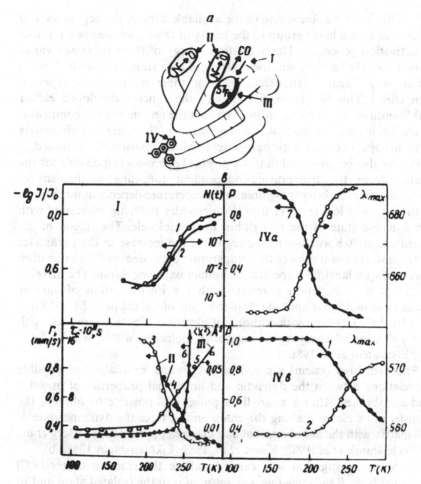

Figure 6.12. Temperature dependences of (a) the dynamic parameters of physical labels II–IV and (b) the kinetic parameters of the processes of carboxylation and decarboxylation (I) of sperm-whale myoglobin. Experimental data: I/I_0 is the relative quantum yield of photodissociation; $N(t)$ is the fraction of nondissociated MbBO molecules 10^{-2} s after a short light flash (Likhtenshtein 1988b).

matrix arc intensified. If may be presumed that the mobility of the matrix under these conditions makes it possible for the CO to escape from the active-site cavity after the elementary photodissociation act is completed. In the 140–190 K temperature range, the method of phosphorescence labeling records several dynamic processes in the matrix, e.g., slow reorientations of the electrical dipole assemblies around the excited chromophore $(v_c \approx 10^3\,\mathrm{s}^{-1})$, and motion of the O_2 molecule, with

$v_{tran}^{-1} \approx 10^{-7}$–$10^{-8}$ s. According to the available data, in this region the CO associates with a heme group in the cavity of the active site via a number of activation processes. The activation energy of these processes varies from 10 to 50 kJ mol^{-1}, and the entropy of activation from 45 to 120 e.u. There is a linear relationship between these quantities (compensation effect). This fact is consistent with the concept developed earlier (Likhtenshtein 1976a,b) according to which the presence of the compensation effect is the evidence that rearrangements of the water-protein matrix have an impact on the kinetic parameters of the chemical process studied.

It may also be presumed that the physical processes responsible for the change from the temperature-independent tunneling mechanism of reassociation of carboxymyoglobin to a temperature-dependent mechanism in the region of low temperatures (Frauenfelder 1989) are associated with the dynamic state of the myoglobin macromolecule. The region of the transition at 70 K is close to the region of the increase in the parameter $\Delta H_{1/2}$ observed in studies of the temperature dependence of the saturation curves of spin-label ESR spectra for human oxyhemoglobin. The increase in $\Delta H_{1/2}$ was shown to correlate with the intensification of various motions: slow, high-amplitude displacements of radical pairs ($A = 0.5$ nm, $v_c = 10^{-2}$ s^{-1}), faster reorientations of electrical dipole assemblies, with $v_c \approx 1$ s^{-1}, and relatively rapid jumps of O_2 molecules, with $v_c = 2 \cdot 10^{-4}$ s^{-1} (Likhtenshtein *et al.* 1986a).

Photosensitive systems are convenient objects for analysis of possible correlations between the dynamic and functional properties of proteins and membranes. After a short light pulse, it is possible to observe the transfer of an electron along the chain and to trace the dynamic state of the matrix with the aid of internal and external physical labels (Berg *et al.*, 1979a,b; Parak *et al.* 1983; Knox *et al.* 1989; Likhtenshtein 1988a,b).

Such an investigation was carried out on the reaction center (RC) extracted from *Rhodopseudomonas spheroidos* in the isolated state and in the composition of the membranes of chromophores. Spin, Mössbauer, fluorescent, and phosphorescent labels were introduced into the various portions of the protein matrix; they were covalently bound to the surface groups of the RC adsorbed by the hydrophobic segments of the protein and membrane, and the [57Fe] atoms were introduced by way of biosynthesis into iron-containing proteins.

Then, in the same samples, the dependence on temperature and moisture content was measured for the label mobility and the rate constants of the electron transfer between the components of the photosynthetic chain. The results from one of the experiments are illustrated in Figures 6.13 and 6.14. It can be seen from these figures that the emergence of an electron from the primary photosynthetic cell, which is favorable for photosynthesis (transport from the primary acceptor A_1 to the secondary acceptor A_2),

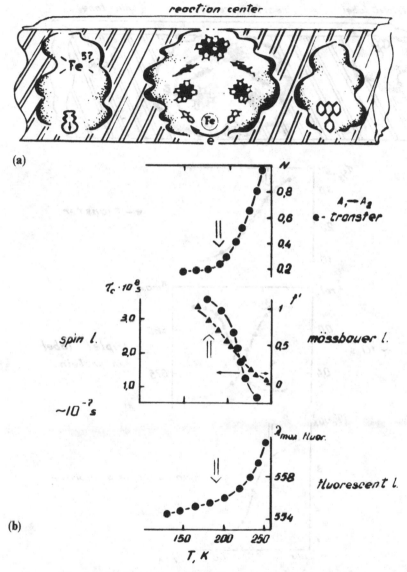

Figure 6.13. Temperature dependences of the parameters of physical labels on chromatophores from *R. rubrum*: (a) Schematic diagram of the locations of spin, fluorescent, and Mössbauer labels. (b) Experimental data: N is the efficiency of direct electron transfer from the primary (A_1) to secondary (A_2) acceptor. Arrows indicate the beginning of the dynamic process, with $\tau_c \leqslant 10^{-7}$ s (Berg *et al.* 1979a,b; Knox *et al.* 1989; Likhtenshtein 1988b).

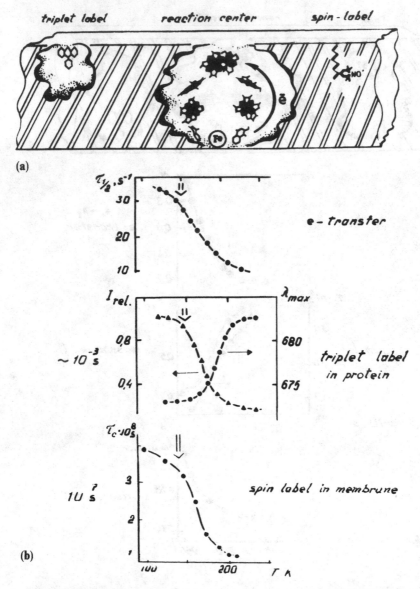

Figure 6.14. Temperature dependences of the parameters of physical labels on chromatophores from *R. rubrum*: (a) Schematic diagram of the locations of spin and triplet labels. (b) Experimental data: $\tau_{\frac{1}{2}}$ is the characteristic time of electron transfer from the reduced primary acceptor to the oxidized primary donor (Berg *et al.* 1979a,b); Kotelnikov *et al.* 1983; Likhtenshtein & Kotelnikov 1983).

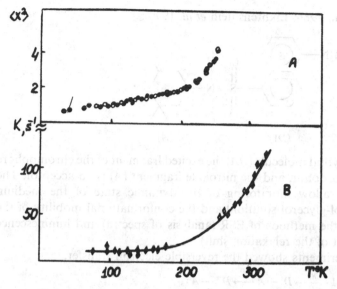

Figure 6.15. Electron transfer (k) and intermolecular dynamics in hemoglobin: $\langle x^2 \rangle$ (in nm^2) of human deoxyhemoglobin ^{57}Fe-enriched in the α and β chains (a), and $\langle x^2 \rangle$ of the electron transfer rate in the $\{\alpha[\text{Fe(III)}$ (H$_2$O)]; β(Zn)$\}$ hybrid hemoglobin (b) (Mayo *et al.* 1983; Peterson-Kennedy *et al.* 1984; Likhtenshtein *et al.* 1990).

takes place only under conditions in which the labels record the mobility of the protein moiety of the membrane with $v_c \geqslant 10^{-7}\,\text{s}^{-1}$. The rate of another undesirable kinetic process, the recombination of the primary products of the photoseparation of the charges of the reduced primary acceptor (A_1^-) and oxidized primary donor (D_1^+), falls from 10^3 to $10^2\,\text{s}^{-1}$ when dynamic processes with $v_c = 1-10^{+3}\,\text{s}^{-1}$ occur.

Thus, the process of transforming sunlight energy into chemical energy of spatially separated charges occurs only in a certain dynamic state of the water-protein matrix.

The correlation between the functional activity and dynamics of the water-protein matrix in the nanosecond range of correlation times was also observed for the reactions of electron transfer in hemoglobin (Figure 6.15) and of hydrolysis of cinnamoyl α-chymotrypsin (Likhtenshtein & Kotelnikov 1983; Likhtenshtein 1988a,b). In the latter case, the dependence of the parameters of nanosecond dynamics on relative humidity at room temperature was correlated with the corresponding dependence of the enzyme stability against thermal denaturation (Roslyakov & Churgin 1972).

The donor–acceptor pairs of the following type also appeared to be convenient objects for analyzing the activity–dynamics relationship

(Bystryak *et al.* 1986; Likhtenshtein *et al.* 1990):

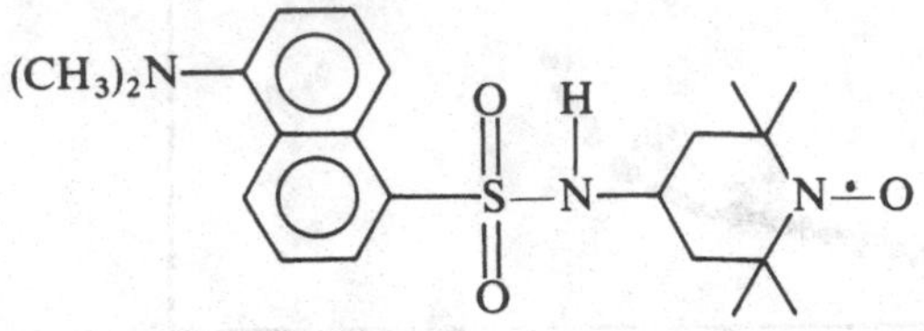

(D) (A)

In such a hybrid molecule (DA) the excited fragment of the chromophore (D) serves as a donor, and the nitroxide fragment (A) is an acceptor. The same groups allow monitoring of the dynamic state of the medium (water-alcohol-glycerol solution) and the conformational mobility of the molecule by the methods of ESR (analysis of spectra) and luminescence (measurement of the relaxation shift).

These experiments showed the reversible electron transfer,

$$D—A \xrightarrow{\;hv\;} D^s—A \longrightarrow D^+—A^-,$$

to be temperature-independent in the range 77–330 K. Irreversible photo-reduction of the nitroxide fragment takes place only when the electric dipoles in the bulk of the solution relax, with $\tau_c < 10^{-8}$ s. Such mobility seems to provide energy stabilization of the photoseparated charges D^+-A^- at the expense of solvation. A correlation was observed between the rate of irreversible reduction of the nitroxide fragment of the D–A molecule incorporated into the hydrophobic pocket of HSA and the rotational diffusion of the fragment in the nanosecond region (Rubtsova *et al.* 1992). Similar results were reported for another model protein system, spin-labeled α-chymotrypsin (Belonogova & Likhtenshtein 1991). In the latter system, the acceptor, a nitroxide fragment, was located on a methionine residue in the vicinity of the active center, and the excited tryptophan group was a donor.

What is the role of intramolecular dynamics in the realization of the functional and catalytic properties of proteins? A very precise mutual orientation of the reacting molecules is known to be necessary for efficient chemical conversion in polyfunctional active centers of enzymes in the course of every step of the conversion. Therefore, the necessary condition for a gain in the rate due to orientation is that some substrate nuclei be mobile, while the others are not. This can be achieved by alteration of the shape of the catalyst during the reaction without significant loss of essential contacts between the substrate and the binding groups (dynamic adaptation).

Processes in which proteins participate that are accompanied by the necessary conformational changes, such as complex formation between antigens and antibodies, allosteric transitions, mechanical transformations, etc., seem to take place only because of the ability of the protein globules to quickly and reversibly change their conformation (Likhtenshtein 1976a,b; Lumry & Gregory 1986).

The experimental values of frequencies of large-scale conformational changes of functional significance range from 10^2 to $10^5 s^{-1}$, while the aforementioned processes (photodissociation, reassociation, electron transfer, hydrolysis of the substrate-enzyme complex) require high-frequency $(v_c = 10^7 – 10^9 s^{-1})$ and low-amplitude $(A = 0.01–0.1 nm)$ dynamics. Therefore, the overall large-scale conformational transitions appear to result from a great number of fast and reversible low-amplitude movements.

6.4. Physical labeling as a tool for studying the electron transfer mechanism

6.4.1. General

Electron transfer processes play decisive roles in photosynthesis, dinitrogen fixation, oxidation, hydroxylation, and other biological reactions of great importance. Electron transfer is thus a convenient and simple model reaction for developing a quantitative theory of elementary chemical conversions.

According to contemporary theory, the rate of electron transfer (k_{tr}) in fixed donor–acceptor pairs $(D–A)$ is primarily determined by the following factors:

1. The degree of overlap of the wave function of the system before (Ψ_i) and after (Ψ_f) the transfer:

$$k_{tr} \sim |V|^2 \sim |S|^2 \sim |\Psi_i \Psi_f|^2, \qquad (6.1)$$

   where V and S are the resonance and overlap integrals, respectively. Both values depend on the distance (r) and mutual orientation of the donor and acceptor groups.

2. The thermodynamic preference of the transfer characterized quantitatively by the free energy of the process (ΔF_0); the value of ΔF_0 depends on the polarity of the medium.

3. The efficiency of adjustment of the nuclear environment to the new electronic configuration of the donor and acceptor, which is a function of the free energy of the environmental reorganization (ΔF_r) and of the rate of nuclear relaxation [the extinction of normal modes of vibration, collective vibration (phonons), classic reorientation, etc.].

According to the theory of electron transfer (Levich *et al.* 1970; Marcus & Sutin 1985),

$$k_{tr}^0 = v \cdot æ \exp(-\Delta F_r^{\neq}/RT),$$
$$\Delta F_r^{\neq} = (\Delta F_r + \Delta F_0)^2/4\Delta F_r RT, \tag{6.2}$$

where $\Delta F_r^{\neq}$ is the activation free energy of reorganization, v is the frequency factor, $æ$ is the coefficient of nonadiabaticity ($æ \sim |V|^2$), and ΔF_0 and ΔF_r depend on the dielectric constant (ε_0) and the structure of the *D–A* complex.

In solutions, one can express the overall apparent rate constant of electron transfer upon collision between the donor and the acceptor molecules (k_{tr}^{app}) under kinetically controlled conditions as

$$k_{tr}^{app} = z\alpha_e f_g \tau_{rs} \cdot k_{tr}^0, \tag{6.3}$$

where z is the collision frequency, α_e and f_g are the electrostatic and geometric steric factors, respectively, τ_{rs} is the residence time of the *DA* complex, and k_{tr}^0 is the rate constant for intracomplex electron transfer.

As follows from Chapters 1–5, the method of biophysical labeling can contribute to solutions for the following problems, solutions that are essential for the development of a general theory of electron transfer in such complicated systems as proteins: (1) estimation of the parameters of delocalization of spin and electronic density (conductivity) over a protein globule; (2) evaluation of the parameters of local polarity of proteins; (3) determination of the frequency of collisions between protein macromolecules and the values of the electrostatic and steric factors; (4) measurement of the parameters of intramolecular dynamics of proteins and analysis of the correlation between these parameters and the electron transfer rate.

6.4.2. Delocalization of spin density and local polarity in proteins

For developing the theory of electron transfer in proteins, an important prerequisite is an evaluation of the potential of protein globules to conduct spin and electron density. Are the protein globules "usual" dielectrics, or do they include specific conducting zones such as aromatic bridges or polypeptide "paths"? To elucidate the problem, one must evaluate the efficiency of spin exchange and triplet–triplet energy transfer between two centers separated by a protein matrix. The exchange processes may be considered as idealized electron transfers and treated with an electron transfer model.

Experimental results on the efficiency of energy transfer from chromophores in proteins in the excited state (tryptophan residues, fluorescent labels) to acceptor groups (heme groups of proteins, spin labels, chromophores) indicate that the experimentally measured values of the rate constants of triplet–triplet energy transfer (k_{ex}) and the exchange integrals (J) are in all cases 5–13 orders of magnitude lower than those expected for conducting media and correspond most closely to non-conducting media within one order of magnitude (Kotelnikov *et al.* 1981, 1986).

Thus, the available experimental data show that in the first approximation, none of these proteins exhibits strong anomalous electronic conductivity. In analyzing the data on electron transfer in proteins, evidently one can use the concepts of outer-sphere electron transfer and, in particular, equation (6.2). It should be noted that fine differences in the conducting properties of proteins can be beyond the sensitivity of this approach.

As shown in Chapters 1 and 3, the polarity of the environment is reflected in the parameters of the ESR spectra of spin labels and probes (superfine splitting A_{zz}, a_{iso}, g-factor) and luminescence spectra (relaxation shift in the fluorescence and phosphorescence spectra, $\Delta\lambda_{max}$).

An analysis of the available data indicates that the interior parts of protein globules and dry protein preparations feature low values of the apparent dielectric constant ($\varepsilon_{app} = 3$–4); the values of ε_{app} for the active-center pockets and hydrophobic clefts lie in the range $\varepsilon_{app} = 10$–20. The periphery of a protein globule is characterized by $\varepsilon_{app} = 30$–50. Strictly speaking, the value of ε_{app} can be different from the real dielectric constant ε_0 because of the specificity of donor–acceptor interactions.

Estimations of the parameters of polarity of protein globules will be useful for quantitative analysis of the factors affecting the rate of electron transfer in proteins.

6.4.3. Collisions between molecules: steric factor

The collision between molecules, including protein globules, is an important stage in many biochemical reactions and in the processes of biological signal transmission. As shown in Chapter 2, measuring the broadening (ΔH) and the spin–lattice relaxation time ($1/T_{1e}$) of the ESR signals of nitroxide spin labels and probes in the presence of other paramagnetics in solution permits one to determine the rate constant of the exchange interaction in collisions in the range of $k_{ex} = 10^6$–10^{10} $M^{-1}s^{-1}$. This range can be widened to $k_{ex} = 10$–10^{10} $M^{-1}s^{-1}$ by using dynamic quenching of the excited triplet states and photochrome labeling, particularly triplet labeling (Section 4.6).

The foregoing approaches are sensitive enough to study collisions with paramagnetic macromolecules, such as spin-labeled proteins and metal proteins.

The value of k_{ex} has been measured for the exchange interaction between two molecules of spin-labeled lysozyme in aqueous solution. It was determined by analyzing the dependence of $1/T_{1e}$ for nitroxide fragments on the protein concentration and was equal to $2 \cdot 10^8 \, M^{-1} s^{-1}$ at 20°C (Kulikov *et al.* 1983).

The accessibility of heme to an excited triplet eosin probe from the solution decreases in the following series: cytochrome c, leghemoglobin, myoglobin, which agrees with the data on the structure of these proteins (Vogel *et al.* 1986). Under conditions of strong exchange, $k_{ex} = f_{app} \cdot k_d$, where k_d is the constant of diffusional collisions, and f_{app} is the effective steric factor. The f_{app} value depends on the value of the geometric steric factor f_g, which is the ratio of the area occupied by heme to that of the whole protein. According to theory (Temkin & Jacobson 1984), $f_g = f_{app}^2$ for strong exchange. For cytochrome c the experimental value of f_{app}, 0.012, is close to that calculated from the X-ray structural model, 0.02. These data have been used to separate the factors that influence the rate of outer-sphere electron transfer between the oxidized and reduced forms of cytochrome c in solution (Vogel *et al.* 1986). At high ionic strength (1 M), the experimental value of the electron transfer rate constant k_{tr} is $5.5 \cdot 10^3 \, M^{-1} s^{-1}$ (Gupta, Koenig, & Redfield 1972).

The value of f_{app} ($1.5 \cdot 10^{-1}$) obtained earlier was then used to determine the $\ae \cdot \exp(-\Delta F_r^{\neq}/RT) = 2 \cdot 10^{-1}$ product for $k_d = 5 \cdot 10^9 \, M^{-1} s^{-1}$, $\tau_{rs} = 3 \cdot 10^{-11} \, s$, $\alpha_{ee} = 1$, and $\nu_0 = 6 \cdot 10^{12} \, s^{-1}$ [see equation (6.3)]. Estimation of the transition coefficient $\ae$ for collision of the heme edges carried out on the basis of the spin density data has led to a value of about 10^{-6}. Hence, $\Delta F_r^{\neq} \sim 4 \, kJ \, mol^{-1}$, and $\Delta F_r \sim 16 \, kJ \, mol^{-1}$. The value of $\Delta F_r^{\neq} < 6 \, kJ \, mol^{-1}$ has recently been obtained on the basis of X-ray diffraction data (Churg & Warshel 1986).

Thus, the electron transfer between molecules of cytochrome c is more likely to be characterized by substantial nonadiabaticity and relatively slight reorganization of the medium. Such a combination appears to be optimal for fairly fast electron transfer in relatively long-lived specific complexes with residence times of $\tau_{rs} = 10^{-2} - 10^{-4} \, s$ and at the same time for preventing electron transfer in nonspecific collision complexes with $\tau_{rs} = 10^{-10} - 10^{-11} \, s$.

6.4.4. Mechanisms of dynamic adaptation at electron transfer

As shown in Chapters 1, 3, 4, and 5, the method of physical labeling can be used to determine certain local dynamic parameters of the environment, namely, microviscosity, η_m, estimated by measuring the parameters of rotational diffusion of nitroxide labels and probes and those of Mössbauer spectroscopy ($f', \Delta r$), and the value of the relative relaxation shift in fluorescence and phosphorescence spectra (P). This approach permits one to obtain quantitative correlations between the electron transfer rate (k_{tr}) and the parameters of the specific molecular dynamics of the processes in the medium.

The dynamics of protein globules and solvent can influence the rate of electron transfer by two alternative mechanisms (Likhtenshtein *et al.* 1990). The first mechanism, referred as a "spontaneous" one, is the process that starts with thermal fluctuation of the surrounding particles to arrive at the configuration that corresponds to the final state of the system, with the electron being transferred. In this case, the correlation between k_{tr} and the parameters reflects the dynamic state of the environment in the bulk space (Section 5.3.5).

In the other, "induced" mechanism, the initial step of the reaction is electron transfer with the formation of an intermediate state in which the solvating molecules are not equilibrated. In such a case, the relaxation of the medium would proceed not only by thermal random effects but also by means of the effect of the electric field on the transferred charges, and the k_e value would be expected to correlate with the relaxation shift parameter (P).

As mentioned in Section 3.5.3, the Arrhenius dependence of the apparent values of relaxation times, τ_{rel}, calculated on the basis of the experimental values of the relaxation shift [equations (3.3) and (3.18)], indicate two sites. One of them, a high-temperature site, appears to reflect spontaneous reorientation of the environmental particles; the other, a low-temperature one, results from both spontaneous and induced relaxation. That is why extrapolating the first site to the low-temperature region and the use of the experimentally obtained value of τ_{rel} allow calculation of the relaxation parameter P_v, which is a quantitative characteristic of spontaneous dynamics in the bulk space over all temperature regions of the relaxation shift.

Experiments have shown the rate constants of electron transfer (k_{tr}) to depend on certain dynamic parameters of the medium in the hybrid donor–acceptor molecule in a model solution (Bystryak *et al.* 1986), in the donor–acceptor pair on α-chymotrypsin and serum albumin (Belonogova & Likhtenshtein 1991), and in Zn-substituted hemoglobin (Bauminger *et al.* 1983; Peterson-Kennedy *et al.* 1984). A satisfactory linear correlation is observed between k_{tr} and the parameters of spontaneous dynamics (η_m, τ_c, and P_v). Therefore, in the systems under consideration, the stage of electron transfer appears to be preceded by the spontaneous formation of a configuration of the protein group and the solvent that is close to the final state of the electron transfer.

Thus, the method of biophysical labeling can aid in the solution of a number of structural and molecular dynamic problems of enzyme catalysis, namely, (1) deciphering the structure of the active centers, (2) detecting large-scale, allosteric, transglobular, and local conformational transitions in proteins, (3) determining the parameters of translational, rotational, and intramolecular dynamics, (4) evaluating electron conductivity and steric factors in protein globules, and (5) studying the mechanisms of dynamic adaptation.

7

Structure and dynamics of membranes

The role of membrane participation in the fundamental biochemical and biophysical processes that produce the vital activities of all organisms cannot be overestimated. Biological membranes act as barriers to separate cell and subcellular structures; they also form ensembles of enzymes that catalyze chemically connected reactions, and they contain specific transportation channels for certain physiologically active compounds. Membranes perform transmembrane transport, regulate enzyme activity, and transmit electric and other biological signals. Practically all biologically essential phenomena, such as intercell interaction, cell division, mobility, differentiation, pathological processes, etc., are in some way associated with the peculiarities of structure and dynamics of biological membranes.

According to contemporary concepts, biomembranes are formed from lipid molecules arranged, as a rule, in bilayers. Biomembranes, with their great variety of ingredients, including protein and glycoprotein macromolecules and small compounds (electron and proton carriers, antioxidants, regulators of rigidity, etc.), "resemble a crowded swimming pool" (Jost & Griffith 1980).

The relative rigidity of the superficial portions of membranes, consisting of polar phosphatidyl and glyceryl fragments, and the flexibility of the hydrocarbon "tails" provide a unique combination of lateral and longitudinal stability for the frame and mobility for the inner membrane.

Specific modifications of individual parts of biological and model membranes by a whole set of labels and probes (Figure 7.1) allows the biological labeling method to be extremely effective; it is not infrequently the only tool for in-depth research in molecular membranology. The achievements and limitations in this field have been outlined in a number of monographs and reviews (McConnell & McFarland 1970; Griffith & Jost 1976; Likhtenshtein 1976a; Cherry 1979; Seelig & Seelig 1980; Devaux 1988; Hoffman & Restall 1984; Watts 1985; Beddard 1986; Waggoner 1986; Dobretsov 1989; Edidin 1987; Oldfield 1988; Robinson 1988; Hemminga & de Jager 1989; Keana, 1984; Marsh 1989, 1990; Bammel et al. 1990).

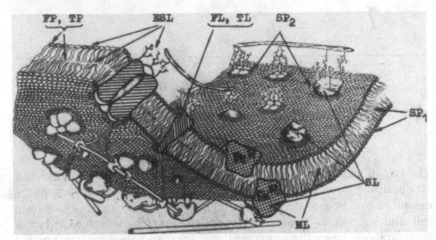

Figure 7.1. Conventional diagram of locations of biophysical labels in various portions of the membrane. SL, FL, TL, and ML are spin, fluorescent, triplet, and Mössbauer labels, respectively; ESL is the electron-scattering label; SP, FP, and TP are spin, fluorescent, triplet probes, respectively.

The most commonly used spin and fluorescent labels and probes are those of the SLXIX, SLXX, LPXVI, and LPXXVI–XXVIII types. Recently a number of sophisticated spin labels and probes have been synthesized that contain ^{15}N and ^{2}H in their nitroxide fragments (Park & Trommer 1989):

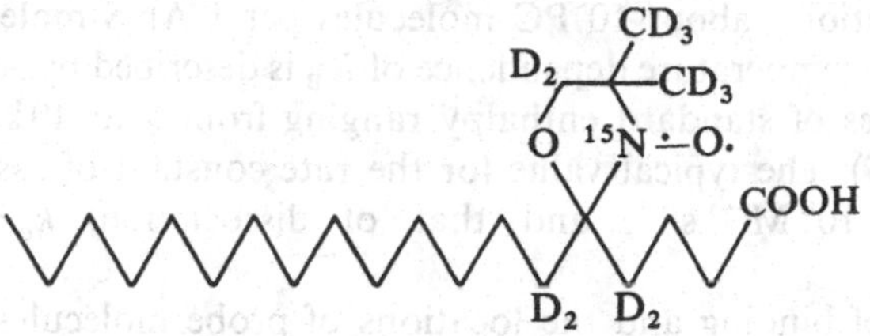

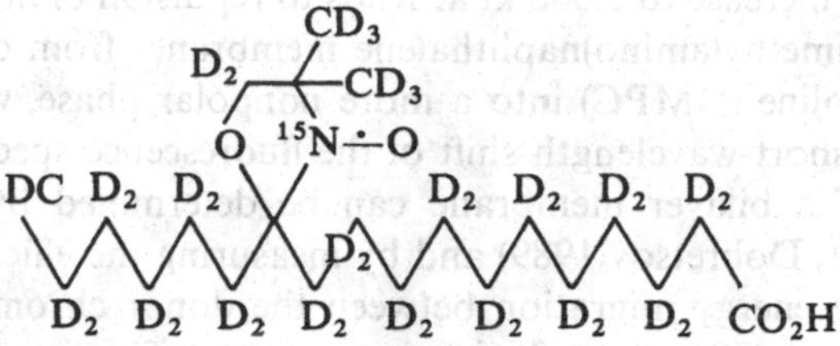

$$
\begin{array}{c}
\text{CD}_3 \\
\text{D}_2 \diagdown \diagup \text{CD}_3 \\
\text{O} \quad {}^{15}\text{N} \overset{\cdot}{\text{-}} \text{O.} \\
\text{X} \quad \text{COOH}
\end{array}
$$

$$\text{D}_2 \quad \text{D}_2$$

$$
\begin{array}{c}
\text{CD}_3 \\
\text{D}_2 \diagdown \diagup \text{CD}_3 \\
\text{O} \quad {}^{15}\text{N} \overset{\cdot}{\text{-}} \text{O} \\
\text{DC} \ \text{D}_2 \ \text{D}_2 \qquad\qquad \text{D}_2 \ \text{D}_2 \ \text{D}_2 \ \text{D}_2
\end{array}
$$

$$\text{D}_2 \quad \text{D}_2 \quad \text{D}_2 \quad \text{D}_2 \quad \text{D}_2 \quad \text{D}_2 \quad \text{D}_2 \quad \text{D}_2 \quad \text{CO}_2\text{H}$$

This chapter reviews the fundamentals and main results in the field of application of physical labeling to membranology.

7.1. Model membranes

Model membrane systems serve as a "proving ground" in which new physical and chemical approaches, including biophysical labeling, have been worked out. Models with individual enzymes, receptors, regulatory and structural proteins, and other compounds incorporated into the membranes allow studies of the structural and functional properties of these systems under relatively simple conditions.

7.1.1. Structure of model membranes: localization of labels and probes

Among a great variety of possible lipid systems, the following structures are most widespread: micelles, monolayer, bilayer, and multilayer liposomes, and bilayer vesicles. A vesicle, which looks like a ball covered with a lipid bilayer, with a liquid, water-containing "core," represents the best model for biological membranes. Researchers have developed a whole collection of fluorescent and spin probes that bind efficiently to model membranes.

The value of the association constant between fluorescent dyes and membranes, K_B, ranges from 10^5 to $6 \cdot 10^4 \, \text{M}^{-1}$. In the case of ANS, for example, the highest affinity for the dye ($K_B = 0.15 \cdot 10^4 \, \text{M}^{-1}$) was observed in liposomes of phosphatidylserine (PS), with phosphatidylcholine (PC) being characterized by the lowest affinity ($K_B = 5.6 \cdot 10^4 \, \text{M}^{-1}$). Under saturation conditions, about 10 PC molecules per 1 ANS molecule can be observed. The temperature dependence of K_B is described by Arrhenius plots, with values of standard enthalpy ranging from 5 to 40kJ mol^{-1} (Dobretsov 1989). The typical value for the rate constant of association is $k_a = 2 \cdot 10^8 - 7 \cdot 10^9 \text{M}^{-1} \text{s}^{-1}$, and that of dissociation, $k_d = 3 \cdot 10^2 - 7 \cdot 10^5 \text{s}^{-1}$.

The degrees of binding and the locations of probe molecules depend not only on the membrane structure but also on external conditions. For example, a pressure increase to 2,000 kbar leads to repulsion of molecules of 6-propionyl-2-(dimethylamino)naphthalene membranes from dimyristoylphosphatidylcholine (DMPC) into a more nonpolar phase, which is accompanied by a short-wavelength shift of the fluorescence spectrum.

The thickness of a bilayer membrane can be determined by X-ray analysis (Kuhn 1971; Dobretsov 1989) and by measuring the efficiency of inductive resonance energy migration between the donor chromophore and the acceptor (e.g., ANS and acriflavine, respectively). The sizes, shapes, and thicknesses of bilayers of micelles and liposomes have been studied by electron microscopy, including the method of mercarbide electron-scattering labels (MESL) (Section 5.4). n-Dodecylmercaptan (DDM) has been used to label hydrophobic parts of micelles and liposomes (Figure 7.2) (Levchenko *et al.* 1979; Likhtenshtein *et al.* 1986b; Marakushev *et al.* 1986); the presence of a nonpolar fragment allowed this reagent to be embedded into hydrophobic areas. Sulfhydride groups that were arranged into polar regions then reacted with MESL. As a result, electron-scattering ball-like formations containing four to six or more MESL molecules could be observed in micrographs of samples of modified dodecylsulfonate micelles (DDSM) and samples of lipid vesicles. The diameters of these formations of DDSM (ca. 3.0 nm) are in satisfactory agreement with the X-ray data.

The thickness of the bilayer in membranes was also determined by an approach based on introducing phosphatidylethanolamine (PE) modified by MESL into the polar parts of liposomes and introducing labeled 11-mercaptoundecanoic acid into the middle of the lipid bilayer (Figure 7.3).

The effects produced by paramagnetic ions (Mn^{2+}, Ni^{2+}, Cd^{3+}) on the spin relaxation parameters and the line positions of the NMR spectra of the nuclear ^1H, ^{13}C, and ^{31}P of the lipid polar fragments were employed to determine the ratio (P) of the exterior to the interior lipid number in vesicles (Shapiro 1988). The values of P in liposomes of PC, sphingomyelin, and phosphatidyl inositol (PI) are 2.02, 2.03, and 2.5, respectively, in good agreement with the X-ray data.

In elegant classical experiments, McConnell and associates (McConnell & McFarland 1970; Griffith & Jost 1976) introduced doxyl spin labels

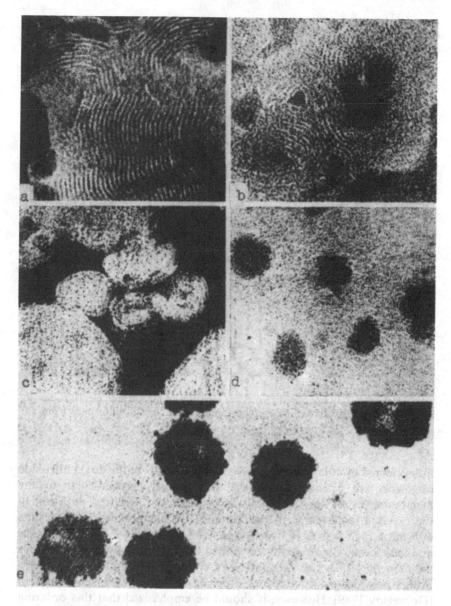

Figure 7.2. Electronic micrographs of liposomes, from phosphatidylethanolamine (PE). Magnification × 300,000: (a) liposomes, negative contrasting by 2% phosphotungstic acid (PTA); (b) liposomes MESL-modified via scheme (5.17), without contrasting (bold); (c) liposomes from PC and PE (1:4), negative contrasting by PTA (1%); (d) liposomes from PC and PE (4:1) modified via scheme (5.17), negative contrasting by PTA (0.2%); liposomes from PC and *n*-dodecylmercaptan (5:1) modified directly by MESL, negative contrasting by PTA (0.2%) (Levchenko *et al.* 1979).

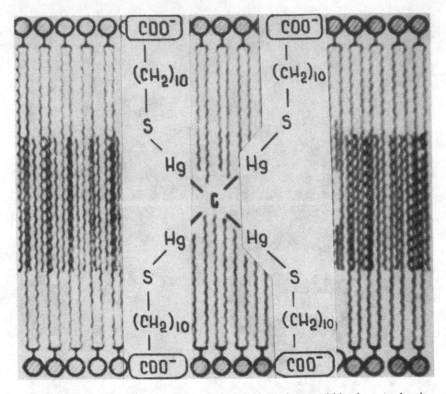

Figure 7.3. Proposed model for location of mercarbide electron density label derivative of 11-mercaptoundecanoic acid in liposomes from PC (Marakushev *et al.* 1986).

into oriented multilayer membranes. The *p*-orbitals of the doxyl nitroxide fragments are oriented along the axis that is perpendicular to the membrane plane. For the steroid spin probe, the preferred direction of the *z*-axis orientation is in the plane of the membrane. These results indicate a high degree of ordering in the membranes of lipid molecules and have been independently confirmed by deuterium nuclear resonance data (Seelig 1977). The orientation of chromophore molecules in membranes provides evidence for an ordering of the lipid molecules as well (Dobretsov 1989). However, it should be emphasized that this ordering is not ideal because of dynamic processes and structural defects. The angular orientations in membranes appear to have a Gaussian distribution (Griffith & Jost 1976).

The position of a nitroxide fragment in the chemical formula of a spin probe does not always reflect a priori its position in the membrane. Thus, an electron spin-echo study has shown that nitroxide fragments of doxyl stearic probes in micelles from hexakis(ethyleneglycol)monododecyl ether are arranged in such a way that a fragment in position 10 is immersed a distance of about 1.1 nm from the micelle surface, with the depth of immersion of the N—O fragment s in positions 5 and 16 being equal to 0.6–0.7 nm (Kevan & Baglioni 1990). By the method of ion-relaxators (Section 2.2.4) it was found that nitroxide fragments of stearic spin probes in liposomes from egg phosphatidylcholine lay at depths of 0.3 nm (4-doxyl probe) and 0.5 nm (16-doxyl probe). The tendency of the nitroxide fragments to localize in the superficial portions of the membrane may be explained in some cases by the significant dipole moment of the N—O group (Cherepanova *et al.* 1990).

The method of physical labeling has contributed significantly to estimations of the polarity of various portions of membranes. The parameters of the ESR spectra of spin probes, namely, the values of hyperfine splitting A_{zz} and a_{iso}, depend on the polarity of the environment, e.g., on the dielectric constant ε_0 and in particular on the hydride bond between the N—O fragment and the water molecule (Section 1.6). By means of a whole set of spin probes, with the doxyl fragments in various positions on the probes, it was shown that the superficial portions of lipid membranes were characterized by a polarity that corresponded approximately to that of ethanol, while the middle of the membrane was nonpolar (Griffith & Jost 1976).

Similar results have been obtained by other independent methods. The micropolarity of membranes has also been estimated by techniques based on determining the association constant between methyl viologen and an aromatic compound, the shifts of the absorption spectra of chromophores, the maximum of the fluorescence spectrum of *trans-trans*-1-(4)-*N,N*-dimethylaminophenyl-4-(4-nitrophenyl)-1,3-butadiene, and the lifetime of the 4-methoxy-4-nitrostilbene excited state (Shin *et al.* 1988). The micropolarity of the environment for all three probes is also approximately equal to the polarity of ethanol. Yet another indicator of membrane micropolarity may be provided by a set of parameters on the vibronic structure of pyrene (Matsuzaki *et al.* 1989).

Data on the dynamic quenching of polar-probe fluorescence of compounds such as ANS provide evidence that the superficial layers of membranes are significantly polar (Haynes 1974; Dobretsov 1989). The value of the quenching rate constant, $k_q = 3.3 \cdot 10^8 \, M^{-1} s^{-1}$, and that of the association rate constant, $K_B = 0.15 \, M^{-1}$, were determined from data on the quenching of probe fluorescence by water molecules. The concentration of water in the carbonyl layer found by this method appeared to be 7.7 M. Considering that the concentration of bulk water equals 55 M, the aforementioned value for the membrane seems to be an overestimate.

The density f of electrostatic charges in membranes (the number of charges per 100 lipid molecules) measured with charged fluorescent probes depends on the lipid structure, the pH, and the ionic strength (Dobretsov 1989). Binding of probes of the ANS type increases as f increases, but to a lesser extent than would be expected in the framework of the model, which suggests that charges are distributed uniformly over the membrane surface. Evidently the discrete distribution of charges must be taken into account in quantitative calculations.

This peculiarity is clearly observed in experiments on absorption of paramagnetic ions of ferricyanide and Mn^{2+} (Kulikov *et al.* 1979). Addition of these ions to a lecithin membrane spin-labeled with a nitroxide fragment in the polar head of the lipid is accompanied by a monotonic decrease in the integral intensity of the spin-label ESR signal, without a change in the ESR line shape. Therefore the positive and negative ions obviously are bound in the immediate vicinity of the nitroxide fragment oriented along the membrane plane. Such an orientation leads to "quenching" of the ESR signal due to the strong spin–spin interactions. The planar disposition of the phosphatidylcholine head of the lipid in the membrane has been confirmed by NMR techniques with the use of paramagnetic shift reagents (Shapiro 1988) and by theoretical calculations as well (Pullman & Berthold 1974).

The conclusion regarding the structural heterogeneity of the model lipid membranes has been sustained by two further results: the anomalous broadening of the fluorescent-probe spectrum, and the dependence of the spectral depolarization on the wavelength of the incident light in a membrane of dimyristoylphosphatidylcholine (DMPC) (Demchenko 1986; Dobretsov 1989).

The values of the longitudinal modulus of elasticity for the lecithin membrane are antithetic to the degree of binding of the hydrophobic spin probes (Hianik *et al.* 1986).

Fine differences in the structural dynamics of the exterior and interior portions of a liposome dispersion of dipalmitoylphosphatidylcholine (DPPC) were detected by measuring the temperature dependences of the ESR spectra of 2-, 5-, and 16-doxyl stearic probes. As shown, a more ordered structure of the lipid bilayers corresponds to a zone with a large radius of curvature, with less ordered structures related to zones of small radii of curvature.

The ability of phospholipid PC vesicles to conduct proton flux has been demonstrated using pH-sensitive spin probes (Section 1.6) (Khramtsov *et al.* 1989). The values of the permeability coefficient were 2, $1.4 \cdot 10^{-4}$, and $\leqslant 10^{-7}\ cm\ s^{-1}$ for HCl, HNO_3, and H_2SO_4, respectively.

7.1.2. Molecular dynamic properties and conformational transitions in model membranes

The method of physical labeling is an adequate and effective tool for studying vibrational, rotational, and translational mobility in condensed phases. The remarkable possibilities of this method have been to a great extent put into practice in investigations of membranes.

The values for the characteristic time of rotational diffusion (τ_R) in liposomes of dyes measured by fluorescence depolarization techniques range from 20 to 60 ns, which correspond to an apparent viscosity of several poise (Chapman & Benga 1984; Waggoner 1986; Edidin 1987; Van Ginkel 1989). The method of multifrequency fluorescence gives approximately the same values (Lakowicz *et al.* 1989).

When the ^2H derivatives of hexatriene were used as double fluorescent and NMR probes, the data on the rotational diffusion of the probe obtained by the technique of depolarization agreed well with those obtained by NMR (Keintanar *et al.* 1986).

Rochev *et al.* (1985) introduced ferrocene derivatives into monolayer and bilayer membranes to serve as Mössbauer probes:

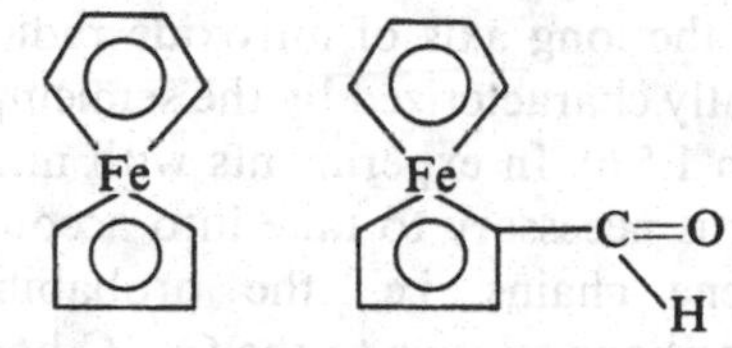

Measurements were made in the lecithin membrane of the temperature dependence of the Mössbauer spectral parameters, the intensity f', and the line width Γ. In monolayer microsomes these probes recorded the reorientational mobility of the deep portion of the membrane, which had a correlation time $\tau_c \leqslant 10^7$ s and an amplitude $A \geqslant 0.005$ nm starting from $T = 160$ K. At 240 K the mobility spread over all parts of the membrane. In the bilayer membranes the mobility of the interior portion was characterized by an activation energy $E_a \approx 10\,\mathrm{kJ\,mol^{-1}}$ and by a room-temperature microviscosity μ_m of about 15 poise (P). For the hydrophobic portion of the membrane, $E_a \approx 20\,\mathrm{kJ\,mol^{-1}}$, and $\mu_m \simeq 10$ P. Introduction of a cholesterol molecule to the bilayer membrane increased the microviscosity to 70 P.

More detailed conclusions on the dynamics of various fragments of lipids in liposomes have been reached by using spin-labeling and deuterium NMR methods (Figure 7.4) (McConnell & McFarland 1970; Hubbell & McConnell 1971; Griffith & Jost 1976; Likhtenshtein 1976a; Watts 1985; Marsh 1989, 1990). These conclusions are as follows:

    1. The rotational motion of doxl fragments is characterized by axial

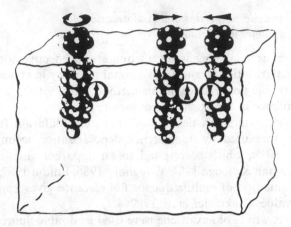

Figure 7.4. Schematic representation of location of nitroxide spin probes in a membrane.

symmetry relative to the axis $\bar{R}$, which is perpendicular to the membrane plane, with the p-orbital of the nitroxide being parallel to $\bar{R}$.

2. The degree of ordering in the long axis of nitroxide radicals in the membrane is conveniently characterized by the semiempirical order parameter S (Section 1.5.6). In experiments with nitroxide derivatives of fatty acids it is necessary to take into account the flexibility of polymethylene chains, i.e., the probability of trans–gauche isomeric transitions relative to the C—C bonds. If the probability of the energetically unfavorable gauche conformation P_g is small and constant for all the C—C bonds of the chain, then

$$\log S_n = n \log(1 - P_g) + \log S_0 + C,$$

where S_n is the order parameter, n is the number of methylene carbons between the nitroxide fragment and the terminal group of the chain, S_0 is the order parameter of the radicals considered as rigid rods, and C is the iteration constant.

Hydrophobic residues like acetylcholine rotate very rapidly ($\tau_c = 10^{-11}$ s, NMR data).

3. In low-viscosity membranes without cholesterol added and at high temperature, rotation takes place in the fast region of correlation times ($\tau_c \leqslant 10^{-9}$ s), and ESR spectra with three components are observed. Another indication of anisotropic fast motion of spin probes in the membranes is the occurrence of internal extremes in the experimental ESR spectrum of nitroxide $A_\perp$ (Figure 1.17).

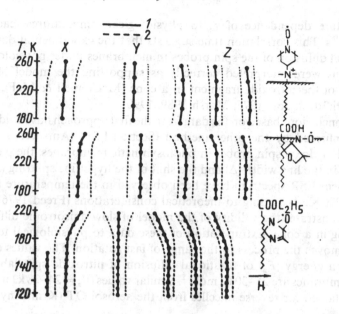

Figure 7.5. Temperature dependences of the positions (1) and line widths (2) of the 2-mm-band spectra of nitroxide spin probes (right) in lecithin liposomes (Krinichny *et al.* 1988).

The interpretation of another dynamic parameter of the nitroxide ESR spectrum, the value of the shift of the hyperfine splitting ΔA_{zz}, appears not to be unequivocal, and therefore it is animatedly discussed (Freed 1976; Griffith & Jost 1976; Dudich *et al.* 1977; Antsiferova *et al.* 1989a). The important point is that the shift may be caused either by fast rotation ($\tau_c \leqslant 10^{-9}$ s) of the nitroxide fragment within a restricted cone with an angle $\theta \leqslant 90°$ or by slow anisotropic rotation with $\tau_c = 10^{-7}$–10^{-8} s. The experimental shape of the ESR spectrum can be independently described in the framework of either model.

To discriminate between alternative models of the spin-probe rotation in membranes, two approaches have been developed (Krinichny *et al.* 1985, 1987a; Antsiferova *et al.* 1989a; Krinichny 1990). In the first, the temperature dependence of the high-resolution ERS spectrum is analyzed for the nitroxide spin probe in the membrane. As seen in Figure 7.5, the z- and x-components of the 2-mm ESR spectrum of the 5-doxyl stearic spin probe in the lecithin membrane shift monotonically within the temperature range of 220–260 K, while the y-component undergoes a substantially smaller shift. Thus, within this temperature range, the nitroxide rotation is essentially anisotropic, with a correlation time $\tau_c \leqslant 10^{-7}$–10^{-8} s, the Freed (1976) model. Extrapolation of the Arrhenius temperature dependence of τ_c to physiological temperatures leads to $\tau_c \leqslant 10^{-9}$ s. This correlation time suggests that the experimental data on rotational diffusion of the spin probes in membranes under physiological conditions were interpreted correctly as supporting the model of fast rotation of the nitroxide fragment in a cone (McConnell & McFarland 1970; Griffith & Jost 1976; Marsh 1989, 1990).

The conclusions based on the data from the first approach have evidently been confirmed by the other method (Section 1.5.6) (Antsiferova *et al.* 1989a). For doxyl spin probes in photosynthetic membranes, the parallel increase in the line width (Δ) and the shift of the hyperfine splitting (ΔA_{zz}) in the 3-cm ESR spectrum have been observed in the temperature range of 150–200 K. According to theoretical considerations (Freed 1976), this fact demonstrates the validity of the model of slow anisotropic diffusion occurring in a cone. Extrapolation of these data to physiological temperatures moves the process to the range of fast rotation. The values of the activation energy (E_a) of rotational diffusion of nitroxide spin labels in lipid membranes are 25–30 kJ mol^{-1}. Similar values ($E_a = 24$–28 kJ mol^{-1}) were obtained for reverse micelles from the aerosol OT (Krinichny *et al.* 1989).

One of the remarkable achievements of the spin-labeling method has been the determination of the flexibility profiles of model and biological membranes (Griffith & Jost 1976; McConnell 1978; Seelig 1977). The values of the order parameters obtained by two independent methods (spin probes and deuterium NMR) decrease with increasing distance between the nitroxide or deuterium fragment and the membrane. The two methods give close, though not identical, results. Some discrepancies seem to be caused by the differences in the chemical structures of the probes and in the characteristic times of the ESR and NMR methods.

Physical labeling has led to such achievements as the discovery and quantitative study of the phenomenon of lateral diffusion of lipids and other molecules in membranes. This process is monitored by a number of independent techniques: by quenching of fluorescence and phosphorescence by other molecules, by measuring luminescence of exciplexes and excimers and delayed annihilation fluorescence, and by the methods based on fluorescence correlation spectroscopy and on fluorescence recovery after photobleaching (FRAP) (Chapters 2–4). The lateral diffusion of lipid molecules in membranes has been modeled by collisions of the nitroxide fragments of lipid spin labels (Figure 7.4). The frequency of collisions is determined by analyzing the intensity of spin–spin exchange interactions.

The values of the translational diffusion coefficient D_{tr} in liposomes from egg lecithin, DMPC, and DPPC measured by the fluorescence techniques (excimer formation of pyrene and other aromatic probes) are equal to $(3$–$14)\cdot 10^{-8}$ cm^2 s^{-1}, with the activation energy of the diffusion equal to 7–8.8 kJ mol^{-1}. The microviscosity of the membranes was found to be 0.5–0.6 P (Dobretsov 1989; Lemmetyinen et al. 1989). Similar values of D_{tr} were obtained by other methods. The value of D_{tr} for diffusion of gramicidin in PC liposomes determined by the FRAP method at room temperature was $3.5 \cdot 10^{-8}$ cm^2 s^{-1} (Wu 1977a,b). The method based on monitoring the kinetics of the recovery of fluorescence after microphotolysis of the phosphatidylcholine derivative of anthracene gave a D_{tr} value of $3.5 \cdot 10^{-8}$ cm^2 s^{-1} in multilayers from PC and about half that value in monolayer liposomes.

Several methods have been developed for quantitative investigation of lateral translational diffusion in membranes with the use of spin labels. In the first approach, developed by McConnell (McConnell & McFarland 1970), small drops (about 1 mm in diameter) of a spin-labeled probe (e.g., spin-labeled PC) are inserted into films of oriented multilayers of dihydrostearylphosphatidylcholine (DHSPC) or lecithin. Because radical diffusion proceeds parallel to the plane of the film, the spectrum changes from a singlet characteristic of large local radical concentrations to a triplet. The coefficient of translational diffusion is approximately equal to 10^{-8} cm^2 s^{-1} at 25°C, which is of the same order as D_{tr} for the spin-labeled steroid (Träuble & Sackman 1972).

The most detailed information on diffusion of lipid spin probes in membranes was obtained by the ELDOR method (Hyde & Feix 1989). These authors developed an elegant approach based on simultaneous use of two types of probes with ^{15}N and ^{14}N in nitroxide fragments of doxyl stearic spin probes. The efficiency of the spin–spin exchange interaction at collisions of the nitroxide fragments of different probes in DMPC liposomes via the mechanism of saturation transfer (Section 1.5) was monitored by the ELDOR technique. The main result of that work was to determine the sequence of weakening of the exchange interaction between the ^{14}N and ^{15}N fragments located at the following positions of the doxyl stearic probes: 12:12 > 16:12 ~ 12:16 ≫ 12:5 ≫ 16:5 ≫ polar head:any position. As seen from the sequence, the nitroxide fragments remote from the membrane plane (12:12 and 16:12) collided most frequently. Nevertheless, translational motions in the direction perpendicular to the membrane plane were intense enough. These motions were characterized by relatively strong dependences on temperature and pH. The protonation of the 5-doxyl stearic spin probe was accompanied by immersion of the nitroxide fragment into the lipid bilayer. The influence of the variations in pH and temperature on the frequency of collisions between the fragments in position 16 was found to be noticeably weaker.

A more thorough investigation indicated kinetic heterogeneity of relaxation processes in phospholipid model membranes. Thus, three dynamic processes with characteristic times of 21, 60, and 350 ns were revealed in liposomes of egg DPPC in pyrene excimer formation. The decay of the probe fluorescence is, as a rule, described by a set of exponents. that are an order of magnitude different from each other (Binder & Dittes 1987).

Another peculiarity of membrane dynamics manifested itself, in particular, in experiments that monitored the kinetics of the fluorescence polarization anisotropy of diphenylhexatriene in DMPC. The peculiarity is the noticeable difference in the apparent microviscosities of rotational and lateral diffusions. In the framework of the Stokes-Einstein law, lateral (as compared with rotational) diffusion is impeded by a factor of about 10. The foregoing observation may be explained by the structural heterogeneity of liposomes.

As noted in Chapter 4, owing to the long lifetime of the excited triplet state, the method of triplet labeling is very sensitive to the rare collisions between triplet probes (delayed annihilation of fluorescence) or between a triplet probe and a quencher of phosphorescence. That gives one an advantage in being able to use very small, noninvasive concentrations of the probe and to study dynamic processes in rigid membranes, e.g., to determine the rare collision frequency in a membrane at temperatures below the transition temperature (T_c) and the frequency with which triplet probes and "quenchers" come to the membrane surface as a result of diffusion across the membrane (Mekler *et al.* 1982, 1983; Likhtenshtein *et al.* 1986b; Mekler & Likhtenshtein 1986; Mekler & Umarova 1988; Likhtenshtein 1990b).

The methods of spin and fluorescent labeling are widely used for the recording of structural transitions in membranes induced by temperature. The general tendencies in this field are as follows:

1. The most drastic changes in such parameters as correlation times (τ_c) of rotational diffusion of fluorescent and spin labels, the degree of the probe binding with the membrane, the frequency of collisions, the coefficient of translational diffusion (D_{tr}), etc., have been observed in the temperature region of the phase transition (T_c) between the solid-crystal and liquid-crystal states of the membrane ($\tau_c < 10^{-8}$ s at $T > T_c$, and $\tau_c > 10^{-8}$ s at $T < T_c$). The values of the apparent activation energy of rotational diffusion are $E_{app} = 12$–$30\,kJ\,mol^{-1}$ at $T < T_c$, and $E_{app} = 28$–$80\,kJ\,mol^{-1}$ at $T > T_c$.

2. A number of additional "prephase" and "postphase" structural transitions in membranes have been reflected in the nonmonotonic character of the temperature dependences of the dynamic parameters of the probes at temperatures below and above the transition temperature. Such transitions are, as a rule, found to occur at temperatures that are 10–15°C lower than T_c.

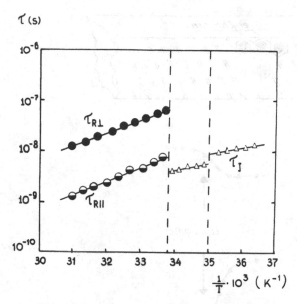

Figure 7.6. Temperature dependences of the parameters of anisotropic rotation of 6-doxyl-DMPC spin probes in DMPC bilayers (Marsh 1989).

3. The temperature dependences of the dynamic parameters of the probes are not completely identical for different portions of a given membrane.

The physical meaning of the foregoing kinetic peculiarities seems to be as follows: the realization of an elementary act of rotational and translational diffusion of a probe requires some free space. At temperatures above T_c the required space is formed by a cooperative disordering, "melting," of an ensemble of lipid molecules. The cooperation of the processes is manifested in relatively large values of the apparent energy (E_{app}) and entropy ($\Delta S_{app}^{\neq}$) of activation of diffusional processes (Likhtenshtein 1976a). At temperatures below T_c the free volume of rotation appears to be attained at the cost of some ordering, "crystallization," of surrounding molecules that leads to the low values of E_{app} and $\Delta S_{app}^{\neq}$. The other source of nonmonotonicity of the Arrhenius dependence and other "anomalies" may be structural inhomogeneity of liposomes caused by differences in liposome sizes and shapes and microstructural differences in a liposome's various portions.

Figure 7.6 shows the typical dependence of the parameters of rotational diffusion of a spin probe in a membrane on temperature. The dependence indicates abrupt discontinuities at temperatures of 13°C and 25°C.

The temperature dependences of the parameters of 2-mm ESR spectra of spin probes with nitroxide fragments located in various parts of the

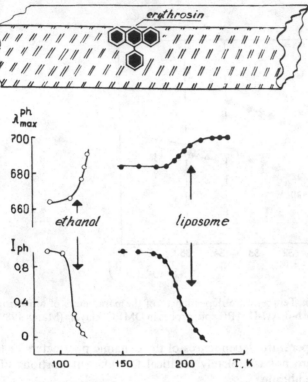

Figure 7.7. Temperature dependences of intensities (I_{ph}) and relaxation shifts ($\Delta\lambda_{max}^{ph}$) in phosphorescence spectra of Erythrosin in lecithin liposomes (Kotelnikov *et al.* 1983).

lecithin membrane indicate a noticeable increase in the fragment mobility ($\tau_c \leqslant 10^{-7}$ s) in the lipid phase at $T \geqslant 240$ K and on the hydrophobic surface of the membrane at $T \geqslant 220$ K (Figure 7.5) (Krinichny *et al.* 1987b).

As follows from the temperature dependences of the relaxation shift ($\Delta\lambda_{max}^{ph}$) and intensity (I_{ph}) of phosphorescence spectra of Erythrosin in the lecithin liposome, the dynamic processes in the environment with characteristic times $\tau_c \approx \tau_{ph}^* \approx 10^{-3}$ s activate at $T \geqslant 180$ K (Figure 7.7) (Kotelnikov *et al.* 1983). The excited triplet state of the Erythrosin probe is quenched by molecules located in both aqueous and lipid phases. Therefore the probe is located in a superficial portion of the membrane. Electric relaxation with $\tau_c = 10^{-8}$–10^{-9} s was observed in the vicinity of the neutral fluorescent probe ANS monitored with relaxation shift techniques in the range of -20°C to $+20$°C (Shcherbitskaya & Demchenko 1989).

Another remarkable achievement of the spin-labeling method has been the discovery of a new dynamic process in bilayer vesicles that is referred to as a "flip-flop" transition (McConnell & McFarland 1970). The process involves reorientation of lipid molecules by 180° along an axis perpendicular to the membrane plane of the transition. The kinetics of the process are followed by monitoring the reduction rate of the nitroxide group of the probe by ascorbic acid located on the opposite side of the membrane. The characteristic time of the flip-flop transition of a PC spin label with a nitroxide fragment in the polar head located in egg lecithin vesicles was found to be $6.5\,h^{-1}$ at 30°C.

Lipid transport between various vesicles has been monitored using spin probes (ESR) and shift paramagnetic reagents (NMR) (De Kruif & Wirtz 1977; Shapiro 1988). In the case of vesicles from PC, the value of the characteristic transport time was $(2-4)\cdot10^4\,s$. An amalgamation of the vesicle membrane was observed by an elegant approach involving loading one part of the vesicle with a spin probe and the other with Mn^{2+} ions, which are reagents that broaden the probe ESR spectrum (McConnell 1978).

7.1.3. Mixed and protein-lipid model membranes

Biological membranes in living cells are very complicated mixtures of various lipids, proteins, and other ingredients. Variation in the lipid composition in membranes of artificial mixed systems is a powerful factor in the regulation of their molecular mobility, permeability, absorption of physiologically active compounds, regularity of enzyme activity, and ability to transfer biological signals. The method of physical labeling has proved to be an effective approach in studying the microstructure and dynamics of mixed membranes.

One can use spin probes in mixed membranes composed of lipids with different "melting" temperatures to determine a transition temperature T_c in order to obtain a suitable phase diagram (McConnell 1978). Three states (liquid, solid, and mixed) have been established. The mixed state includes a liquid phase enriched with dieladoylphosphatidylcholine, for example, and a solid phase with concentrated DPPC. Clusters of spin-labeled molecules of PC about 50 nm in size can be observed by ESR spectroscopy.

The process of lipid microcrystallization can be visualized by electron-scattering labeling combined with electron microscopy (Levchenko *et al.* 1979; Likhtenshtein *et al.* 1986a,b). Micrographs of mixed liposomes containing DMPC (75%) and PE (25%), with an amine group modified by MESL at $T_c > 15°C$ according to scheme (5.17), show stripped electron density formations typical of the "usual" liquid-crystal state of membranes.

At $T = 5°C$, which is lower than T_c, ensembles of 15–20 closely arrangd MESLs appear in the micrographs.

The mutual influences of lipid molecules in mixed membranes have been studied by spin labeling and by ^2H and ^{31}P NMR (Seelig & Seelig 1980; Sixl & Watts 1982; Marsh *et al.* 1983). In mixtures of PC, PE, and PS the conformational and dynamic states of the hydrocarbon tails and phosphate groups of lipids did not change significantly as compared with their states in individual lipids. Introduction of an unsaturated lipid into the mixture decreased the value of the order parameter S by 35%. Noticeable decreases in the quadrupole splitting of the NMR signals of the $-N^+(CH_3)_3$ and $-CH$ groups of the acetylcholine head of DMPC upon increasing the lipid concentration indicated the influence of the adjacent lipid on the conformational state of the acetylcholine head.

Methods of spin and fluorescent probes have been used to determine the electrostatic characteristics of the surface of the mixed membrane of PC and phosphatidic acid. The electrostatic state of the membrane surface, monitored by the distribution of the positively charged fluorescent probe and by the quenching of the negatively charged fluorescent probe 2-4-hexadecylaminonaphthyl-6-sulfonate under an attack of positively charged nitroxide radicals, has been found to be homogeneous.

The biological significance of cholesterol has generated a great number of investigations of the effects produced by this compound on the structure, dynamics, and functional activity of model membranes (Likhtenshtein 1976a; McConnell 1978; Owicki & McConnell 1980; Dobretsov 1989; Yeagle *et al.* 1990).

An increase in the cholesterol concentration in lecithin and other membranes monotonically enhances the order parameters of spin probes embedded into various portions of bilayer membranes.

Marrot *et al.* (1987) have studied the influence of the orientation of cholesterol and androsterone on the parameters of the ESR spectrum of the ^{15}N-derivative of the 16-doxyl stearic probe and of the ^2H NMR spectra of α, β, and CH_3 groups of the glycerol residue of the probe. For the cholesterol molecule, only one orientation with the acyl group oriented toward the center of the membrane was found to exist. For the androsterone molecule, two orientations were observed, and vertical diffusion across the membrane was characterized by $\tau \approx 10^{-7}$ s.

A combined study of the effects of steroids and lipid composition on bilayer dynamics was carried out in a series of ESR and fluorescence experiments (Korstanje *et al.* 1990). The main conclusion of those authors was that "changes in the membrane molecular order and reorientation dynamics have to be considered separately and are not necessarily correlated by the common concept of membrane fluidity" (Chapman & Benga 1984). For example, the incorporation of 20% of steroids into the dioleophosphatidylcholine bilayers enhanced the orientation order of the cholestane spin label, but induced only slight (if any) changes in the rotation parameters of the spin labels.

A new possibility for using triplet labeling to determine the localization and dynamics of physiologically active compounds in membranes was demonstrated using a biological carrier of electrons and protons, ubiquinone Q-10, in liposomes from DPPC (Mekler *et al.* 1984a). The efficiency of quenching of the excited triplet states of Erythrosin and other triplet probes by a Q-10 molecule in membranes and in a model solution was measured experimentally. The rate constants for quenching (k_q) of excited Erythrosin and anthracene molecules by Q-10 in a 30% ethanol–water solution via the electron transfer mechanism were $2 \cdot 10^9 \, M^{-1} s^{-1}$ and $1.7 \cdot 10^9 \, M^{-1} s^{-1}$, respectively. In membranes the ubiquinone molecules appeared to be efficient quenchers not only of the intramembrane probes but also of Erythrosin, which was located in the superficial portion of the membrane. In the latter case, $k_q = 1.5 \cdot 10^7 \, \text{mmol} \, g^{-1} s^{-1}$ of lipids. This value is typical for diffusion-controlled processes in lipid membranes. Fairly high efficiency of quenching was obtained at temperatures below $T_c = 30°C$. Thus, molecules of Q-10 can efficiently diffuse to the polar portion of the membrane to take part in electron and proton transfer. This conclusion was indirectly confirmed when the well-known reduction of the nitroxide was accompanied by redox processes in which ubiquinone participated (Kerimov *et al.* 1979).

The processes of rotational and translational diffusion of protein macromolecules incorporated into model membranes can be studied under simple and controlled conditions. The coefficient of rotational diffusion (D_R) has been measured by two independent methods: (1) by monitoring depolarization of phosphorescence of triplet labels and (2) by measuring the efficiency of saturation transfer in the ESR signals of spin labels, the ST ESR method (Sections 1.5.3 and 3.3).

The necessary prerequisites for application of both methods are a tight binding of the label to the macromolecule under study and rotational correlation times within the range $\tau_R^m = 10^{-2}$–10^{-6} s.

An important problem to be solved in each particular case is to find the orientation of the label relative to the rotational axis of the macromolecule in the membrane. The rotation is, as a rule, an essentially anisotropic process, because the contacts of the hydrophobic portion of the membrane prevent the macromolecule from being turned upside down.

Monitoring the kinetics of depolarization of labeled macromolecular phosphorescence in the membrane allows direct determination of the rotational correlation time. Stationary versions of this technique enable one to estimate the value of the order parameter only in a region close to the lifetime of the triplet excited state.

When the value of τ_R^m is determined by comparing the ESR data on the system under investigation to the standard system, namely, an isotropically rotating ball-like macromolecule with τ_R^{app}, it appears to be equal to

$$\tau_R^m = \tau_R^{app} \sin \theta/2,$$

where θ is the angle between the z axis of the nitroxide radical and the axis of the macromolecular rotation (Marsh *et al.* 1983). The maximum value of $\tau_R^m = \tau_R^{app}/2$ is realized in the case of $\theta = 90°$. If θ is small, the value of τ_R^m calculated in this manner becomes lower than the correct value.

The τ_R value for rotational diffusion of protein "band 3" from Erythrosin in DMPC membranes determined by the ST ESR technique was found to be $25\,\mu s$ (Sakaki *et al.* 1982). This value increased upon addition of cytoskeletal proteins. Similar values of τ_R were obtained for spin-labeled bacteriorhodopsin in liposomes formed of the PC derivatives with various lengths of hydrocarbon chains (Kusumi & Hyde 1982; Ahl & Cone 1982). At a lipid:protein ratio of 150:1 the values of τ_R for liposomes from lipids with n—CH_2 groups ($n = 12$–18) ranged from 27 to $8\,\mu s$, passing their maximum at $n = 16$. The rhodopsin molecules in the membrane were found to be arranged into dimers as the smallest ensemble. Judging by the fast rotation of cytochrome b_5 in the DMPC liposome, this protein is a monomer in the membrane (Vaz *et al.* 1982). The rotational diffusion coefficient of cytochrome P-450, with the SH group modified by ^{15}N and ^2H derivatives of the nitroxide spin label in unilammelar liposomes from octylglycoside, was determined by the ST ESR method (Schwarz & Pirrwitz 1988). The value of the coefficient suggests that the protein macromolecules aggregate in the membrane.

The values of the translational diffusion coefficients (D_{tr}) for bacteriorhodopsin, rhodopsin, acetylcholine receptor, and Ca^{2+}-ATPase in synthetic lipid membranes were determined by the FRAP method; they ranged from 1.3 to $3.4 \cdot 10^{-8}\,cm^2\,s^{-1}$ (Criado *et al.* 1982; Vaz *et al.* 1982; Waggoner 1986; Edidin 1987). These values increase gradually as the protein:lipid ratio increases. It should be noted that the D_{tr} data for proteins are surprisingly close to those for low-molecular-weight compounds. For pyrene, for example, D_{tr} measured by monitoring the excimer fluorescence is equal to $4 \cdot 10^{-8}\,cm^2\,s^{-1}$.

In protein-lipid membranes, most lipids (20–35 per macromolecule) interact directly with the protein surface (Jost & Griffith 1980). As an example, we can consider the data on the system consisting of DMPC liposomes and embedding the protein from the bacteriophage M13 shell (Datema *et al.* 1988). The dynamic state of the lipid phase was traced by means of deuterium NMR and spin probes that were derivatives of palmitic and 14-doxyl stearic acids. Insertion of the protein into the membrane enhanced the relaxation time T_{2n} for deuterium, especially in the 16th position; however, that did not affect the order parameter of ^{2}H in the 2nd, 9th, or 16th position of the probe. A new component from the immobilized lipid spin probes appeared in the presence of the protein in the membrane. At the same time, it was not observed in the NMR spectrum of the ^{2}H probes. To explain these results, the authors proposed a model according to which the lipids exchange with a frequency of about 10^7 s in the lipid phase, and some lipids are in contact with the protein surface. Similar results were obtained for myelin and cytochrome c oxidase in proteolipid complexes and for Na$^+$,K$^+$-ATPase in the lipid bilayers from α-palmitoyl-β-oleoyl-α-L-PC (Marsh 1989, 1990).

Combined use of the spin-labeling method, ^{1}H and ^{31}P NMR, differential scanning calorimetry, and X-ray scattering allowed Aripov *et al.* (1984) to obtain fairly complete information on the effects of cytotoxins V$_c$5 and V$_c$1 from the Asian cobra *Naja naja oxiana* in liposomes from PS, PC, DPPC, cardiolipin, and their mixtures on the dynamic state of the liposome lipid bilayer. Specific modifications of individual lysine groups, both hydrophilic (Lys-35, 23 and 50) and hydrophobic (Lys-12, 18, 41 and 58) showed the depth of immersion of the macromolecule into the membrane to be about 2.4 nm. Experiments with applications of 1-oxyl-2,2-dimethyloxazoline derivatives of keto stearic acid containing nitroxide fragments in the 5th, 12th, and 16th positions showed that immersion of the toxins into the membrane causes phase separation of the lipid, formation of cytotoxin-lipid complexes, and disordering of the rotational axes of the radical probes.

7.2. Biological membranes

This section deals with the data on biological membranes isolated by the spin-labeling method. The results of studies on membranes in intact tissues and organisms will be considered in the next chapter.

7.2.1. Erythrocyte membranes

Owing to their relative availability and stability and the important biological role of erythrocytes, these systems have been subjected to numerous physicochemical investigations, including a number using the methods of physical labeling.

Spin and fluorescence labeling approaches have indicated significant similarities between the strutural and dynamic properties of erythrocytes and model bilayer liposomes (Cherry 1979, 1986; Fowler & Branton 1977; Gavish *et al.* 1988; Blatt & Sawyer 1985). The values of erythrocyte viscosity estimated by the aforementioned approaches were on the order of several poise, with the values of activation energy (E_a) ranging from 16 to $25 \, \text{kJ mol}^{-1}$, and those of the rotational correlation time (τ_R) being on the order of 10 ns.

Lateral diffusion of lipid fluorescent labels in the intact erythrocyte cell is characterized by translational diffusion coefficients $D_{tr} = (31–210) \cdot 10^{-10} \, \text{cm}^2 \, \text{s}^{-1}$ at 12–37°C (Bloom & Webb 1983). Values of $D_{tr} = (22–190) \cdot 10^{-10} \text{cm}^2 \, \text{s}^{-1}$ were estimated under these conditions for erythrocyte ghosts.

Determination of τ_R for the erythrocyte protein "band 3" labeled by eosin by the method of phosphorescence depolarization indicated a whole set of exponential decays at $\tau_R = 160 \, \mu s$ and $\tau_R = 3{,}400–64{,}160 \, \mu s$ (Edidin 1987). These values are large compared with those for protein band 3 in a model lipid membrane of similar viscosity. Relatively low values for the D_{tr} of lateral diffusion of this protein in ghosts of human erythrocytes ranged from $0.06 \cdot 10^{-10}$ to $4 \cdot 10^{-10} \text{cm}^2 \, \text{s}^{-1}$ at 12–37°C, which also suggests their aggregation. A similar value of D_{tr} ($0.45 \cdot 10^{-10} \, \text{cm}^2 \, \text{s}^{-1}$) was estimated for the ghosts of mouse erythrocytes, with this value increasing up to $25 \cdot 10^{-10} \, \text{cm}^2 \, \text{s}^{-1}$ upon removal of spectrin.

Erythrocyte membranes efficiently bind both positive and negative fluorescent dyes. About $2 \cdot 10^6$ molecules of the dye

per one erythrocyte are bound, with $K_e = 2 \cdot 10^6 \, \text{M}^{-1}$ (Dobretsov 1989). Hydrophobic properties of spin probes were found to be essential for binding. Thus, the relationship between the parameter Δ (the ratio of the amplitude of the bound spin probe to that of a nonbound probe) and the parameter P (the coefficient of partition of the probe between heptane and water) is experimentally described by a simple equation: $\Delta = 7.2 \cdot 10^{-4} P$ (Gendel et al. 1981; Gendel & Krugliakova 1986). A probe's hydrophobicity significantly affects the rate of its reduction by ascorbate localized in the aqueous phase. The relative rate of the reduction is $V = 1 - 0.45\Delta$.

Studies of the absorption of ions with electrostatic charges of different signs and of the polycation protamine allow one to detect the formation of clusters of electrostatic charges on the surface of the erythrocyte membrane. Aggregation of glyceraldehyde-3-phosphate dehydrogenase molecules modified at the SH group by MESL has been observed on electron micrographs (Levchenko et al. 1979). Total labeling of the erythrocyte surface by MESL according to scheme (5.17) led to images that corresponded to aggregation of proteins in ensembles distributed randomly over the membrane.

Erythrocyte membranes are suitable models for investigation of various biological processes. Thus, in this system the transmembrane potential, E_{tm}, has been found to be induced in the presence of K^+ ions, with E_{tm} ranging from 9 to 20 mV, depending on the experimental conditions and the source of erythrocytes (Hoffman & Laris 1974). As has been shown (Cherry 1986), induction of the transmembrane potential causes aggregation of protein band 3 in erythrocyte ghosts. A similar aggregation effect was observed upon the incorporation of melitin, a protein that induces an influx of hemoglobin from erythrocytes.

The molecular mechanisms of the effects of erythrocyte freezing and thawing and of cryoprotector action were studied by introducing spin labels and probes into the protein and lipid phases of the membrane (Tsymbal & Moiseev 1985). The most abrupt changes in the microstructure of the membrane were produced by freezing the erythrocytes near the temperatures of intracellular water crystallization (-8 to $-12°C$); these were followed by conformational transitions in the lipid bilayer ($T = -16$ to $-30°C$). Addition of cryoprotectors (polyethylene glycol, dimethylsulfoxide, ethyleneglycol, glycerin) significantly increased the width of the temperature transition, prevented abrupt crystallization of ice, and stabilized the microstructures of the protein and lipid portions of the erythrocyte membrane.

The method of physical labeling made it possible to study the kinetics and mechanisms by which dyes, labeled dextrins, and polyethylene glycols penetrate into the erythrocyte cell (Gavish *et al.* 1988; Dobretsov 1989). A paramagnetic anticancer preparation of chloromethylnitrosourea was found to be covalently bound to the SH groups of erythrocytes (Lassmann *et al.* 1987). Preliminary blocking of these groups by Ca^{2+} enhanced the membrane permeability to hydrophobic probes and diminished that to hydrophilic compounds.

7.2.2. Sarcoplasmic reticulum

The Ca^{2+}-ATPase of the sarcoplasmic reticulum (SR) is one of the main membrane enzymes that regulate the Ca^{2+} ion concentration during muscle contraction. The enzyme consists of a hydrophilic portion and a hydrophobic tail, the latter being embedded into the SR membrane. The objectives of investigations of the SR using physical labeling were as follows: (1) locating the ATPase center; (2) monitoring the conformational changes at various steps of the ATP-dependent transport of Ca^{2+} across the membrane; and (3) obtaining information on lipid–lipid and lipid–protein interactions in SR (Champeil *et al.* 1980; Livshitz *et al.* 1988a; Marsh 1989; Zhdanov 1981).

Localization of the triplet probe eosin in the ATPase center of the enzyme was established by incorporating the probe into the center, followed by tracing the phosphorescence by hydrophilic (NR) and hydrophobic (anthracene derivatives) probes. The eosin phosphorescence in the enzyme active center was quenched by radical SLIX and ferricyanide at rates 140- and 2-fold less efficiently than these quenchers performed in aqueous solution.

Experiments with triplet probes (derivatives of perylene, benzanthracene, and pentacene), which produced delayed annihilation fluorescence (DAF) upon rare collisions, with a characteristic time of about 1 ms, indicated that there were no lipid zones in the SR that were isolated by the surrounding protein: the probe molecules diffused without collisions for a distance of not less than 30 nm (Mekler *et al.* 1983, 1984a).

The method of ESR saturation transfer using spin labels (ST ESR, based on analysis of saturation curves, Section 1.5.3) was employed for a detailed study of Ca^{2+}-ATPase in model vesicle membranes from DMPC and egg lecithin, as well as native SR membranes (Livshitz *et al.* 1988a). The values for the rotational correlation times τ_R of spin labels attached to the three SH groups of the enzyme were found to be 0.32, 1.2, and 10 μs, respectively. The latter appears to correspond to the rotational diffusion of the protein globule in the membrane as a whole. Experiments in the oriented layer of the SR failed to reveal any preferred orientation of the nitroxide fragment of the label with respect to the membrane plane. Quantitative correlations were observed among the parameters of label mobility monitored by the ST ESR method and the parameters of the Ca^{2+} concentration, the lipid composition, and the enzyme activity. Analysis of the dipole–dipole interaction of spin labels and Mn^{2+} ions in the active center showed the distance between the label and the center to be 1.5–1.7 nm. This conclusion was confirmed independently by other authors (Champeil *et al.* 1980; Zhdanov 1981). Analysis of the dipole–dipole interaction between labels located on different macromolecules provided evidence that 60% of the macromolecules are aggregated in the SR membranes.

Dynamic contacts between the Ca^{2+} molecules in the membrane were monitored by means of the triplet label *N*-(3-pyrene maleimide) attached covalently to the SH groups of the enzyme. The intensity of DAF upon contact increased with the ratio of triplet label:protein up to a ratio of 3:1.

Protein–label interactions in membrane preparations of Ca^{2+}-ATPase have been studied by spin labeling (Marsh 1989). The values of the order parameter *S* for the 5-doxyl stearic acid probe and the rate of reduction of the probe by ascorbate were noticeably higher than those in model membranes from egg lecithin. The experimental temperature dependence of the order parameter was found to coincide with that for the kinetic parameters of ^1H–^2H exchange in the Ca^{2+}-ATPase protein globule. These data demonstrated the essential role of lipid–protein interactions in the catalytic and regulatory functions of the sarcoplasmic reticulum.

7.2.3. Rhodopsin membranes

Light is transformed to a biochemical signal in the photoreceptor membrane, where the protein rhodopsin is a main protein component. The primary step in the transformation is absorption of light quanta by the pigment retinal, followed by isomerization, with the formation of a whole cascade of intermediate forms, including meta-rhodopsin. Spin, fluorescence, and elecron-scattering labeling methods have been used to monitor conformations, the degree of aggregation of the rhodopsin globules, and the dynamic state of the lipid layer of the rhodopsin membrane under various conditions, including those present in the course of the photoreceptor's functioning.

As early as 1970, the value of the rotational diffusion time (τ_R) for rhodopsin macromolecules in mammalian photoreceptor membranes and the microviscosity of the lipid portion of the membrane were estimated at $20\,\mu s$ and 0.7–$0.6\,P$, respectively, by measuring the rhodopsin luminescence depolarization after flash photolysis (Cone 1972). These results were further confirmed by ST ESR techniques (Kusumi & Hyde 1982). When the membrane preparation was diluted by lipids up to a lipid:protein ratio of 50:1, the value of τ_R for rhodopsin was measured as $10\,\mu s$, which suggested a dimer structure for the rhodopsin molecules in the native membrane. Micrographs obtained after two SH groups of rhodopsin were modified indicated that the spots of electron density were distributed uniformly (Kalamkarov *et al.* 1977; Levchenko *et al.* 1979). The sizes of the spots exceeded that which could be expected for a monomer molecule. The value of the coefficient of translational diffusion for rhodopsin in mammalian photoreceptor membranes ($D_{tr} = 3 \cdot 10^{-9}\,cm^2\,s^{-1}$) was lower by a factor of 10 than in synthetic membranes (Edidin 1987).

Intense irradiation caused some softening of the bovine rhodopsin membrane and noticeable aggregation of the rhodopsin macromolecules (Kalamkarov *et al.* 1977; Livshitz *et al.* 1988b). According to spin and electron-scattering labeling methods and measurement of the molecular masses of the aggregates, the degree of aggregation after illumination was three to four times larger than in the intact membranes.

A detailed investigation of conformational and aggregational changes in mammalian rhodopsin membranes was carried out by Livshitz *et al.* (1988b). Cys-140 and -316 groups of rhodopsin protruding to the aqueous phase were modified by two maleimide derivatives of the spin label, each with a short and a long "leg." The spin label's mobility on the Cys-140 group was found to be higher than on Cys-316. The light-induced transition rhodopsin → metarhodopsin II decreased the immobilization of both types of the spin label. Two independent approaches based on determination of τ_R for the tightly bound short-leg label and on analysis of spin–spin interactions between the long-leg labels indicated noticeable aggregation of the rhodopsin molecules in the dark within the pH ranges from 7.2 to 5 and from 7.7 to 7.5. Attempts to detect such an effect in rhodopsin preparations isolated with the help of a detergent were not successful.

Splitting of the rhodopsin C-end fragment Cys-322-Ala-345 and the loop Glu-273-Ala-244 enhanced the spin label's mobility in the vicinity of the Cys-140 and Cys-316 groups; this result suggested a cooperative relationship between these portions of the protein.

Considering the location of the photoactive retinal fragment in the interior hydrophobic portion of retinal and the Cys-140 and Cys-316 groups on the hydrophilic surface of the protein, the aforementioned data evidently showed that phototransformation of the retinal fragment was accompanied by a transglobular conformational transition.

Investigation of the protein–lipid interaction in the rhodopsin membrane by the spin-labeling method revealed their several types (Baroin *et al.* 1979; Davoust *et al.* 1983). Lipid spin probes, such as 14-doxl stearic acid, that were being "squeezed" between the proteins were characterized by a rotational correlation time $\tau_R \geqslant 10^{-7}$ s, whereas labels stuck to the protein surface rotated with $\tau_R \leqslant 10^{-7}$ s. The former type of probe was exchanged with those located in the lipid bilayer with a frequency of $\nu'_{ex} \sim 10^5 – 10^6$ s^{-1}. Elegant experiments were carried out using rhodopsin labeled covalently by the ^{15}N derivative of nitroxide and PC labeled by nitroxide with ^{14}N, with the frequency of dynamic spin exchange interaction between the spin label and the spin probe being found to be $\sim 10^7$ s^{-1}.

7.2.4. Microsomes

A hydroxylating system that includes cytochrome P-450 and the electron donors NADPH–cytochrome reductase and cytochrome b$_5$ is the most important physiological component of microsomes of the endoplasmic reticulum. The cytochrome P-450 molecules in the microsomal membrane have been studied by a whole arsenal of physicochemical methods involving spin, fluorescence, triplet, and electron-scattering labeling. Most of the early results in this field have been reviewed elsewhere (Cherry 1979; Edidin 1987).

The main conclusion derived from an analysis of the available data is that the greater part of the cytochrome P-450 molecules in the intact microsomal membrane are aggregated. Nevertheless, some discrepancies exist in the estimates of number of macromolecules in a single aggregate. According to the FRAP method, the majority of the cytochrome P-450 molecules are associated as dimers; the ST ESR spectroscopy data

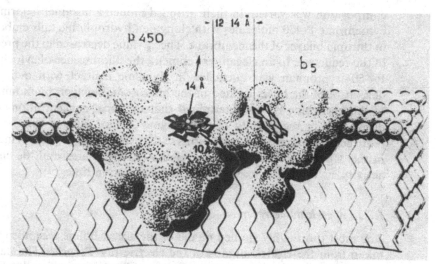

Figure 7.8. Schematic representation of location of cytochromes P-450 and b_5 in microsomal membranes.

probably can be explained within the framework of aggregates of 6–12 molecules (Schwarz *et al.* 1982; Schwarz & Pirrwitz 1988). However, it was impossible to avoid a priori some uncertainty about the orientation of the nitroxide fragment of the spin label with respect to the membrane surface.

When membranes were prepared with cytochrome P-450 molecules each labeled with two to three molecules of mercarbide electron-scattering labels, the electron micrographs showed that the electron-density spots were of a size corresponding to two to three macromolecules in a single ensemble (Alterman *et al.* 1981). The average size of the spot increased by as much as a factor of 2 in the presence of molecules of NADPH–cytochrome P-450 reductase. In a preparation of cytochrome P-450 isolated with the aid of detergents, spots were observed that clearly were two to three times larger than those in the intact membrane.

Figure 7.8 shows a model of the interaction between cytochrome P-450 and cytochrome b_5 that was suggested in order to account for the data on the location of the heme group of cytochrome P-450 in a microsomal membrane (Section 6.1.4) and data on the location of the cytochrome b_5 heme (Coon *et al.* 1981).

Kiselev *et al.* (1990) have studied the influence on rotational diffusion (τ_R) of the lipid composition, the complexation of cytochrome P-450 (from LM$_2$) with its reductase, and the redox state of cytochrome P-450. The values of τ_R were found to range within 110–180 μs when the lipid composition was varied. In their proposd model, a hexamer is formed of cytochrome P-450 molecules with elongated hydrophobic tails embedded in the lipid bilayer of the membrane. The τ_R value decreased in the presence of the reductase. In an elegant experiment, the fluorescence of cytochrome P-450 tryptophan was quenched at dynamic contact with a bromine derivative of the lipids located in the microsomal bilayer; by monitoring of the quenching it was determined that the local concentrations of Br derivatives in the vicinity of the enzyme surface differed for oxidized, reduced, and enzyme-substrate forms of the cytochrome P-450. An increase in the Mg^{2+} ion concentration caused a parallel increase in the enzyme activity, as well as increased rotational correlation times.

7.2.5. *Acetylcholine receptor*

The most thoroughly studied model of the acetylcholine receptor (AR) is taken from the electric organs of the electric ray *Torpedo*. The data on the dynamic properties of AR were generalized in a review by Edidin (1987). Attempts to observe translational diffusion in the intact membrane of AR by the FRAP technique failed. The ST ESR method showed no translational diffusion either; the value of τ_R measured by this method exceeded $100\ \mu s$. Nevertheless, in model synthetic liposomes the value of the translational diffusion coefficient $D_{tr} = (2-3) \cdot 10^{-8}\ cm^2\ s^{-1}$ suggested considerably higher mobility of AR (Clarke *et al.* 1983; Criado *et al.* 1982).

Detailed information has been obtained on the mechanism of binding of neurotoxin I from *Naja naja oxiana* with AR (Tsetlin & Ovchinnikov 1986). Those authors applied a combination of physicochemical methods, including specific modification of the lysine and histidine groups by nitroxide spin labels, to determine the distances between the labels by the ESR method, the distances between the labels and certain residues of the neurotoxin I protein globule by the NMR method, and the distances between the labels and the tryptophan groups by fluorescence quenching techniques. The results allowed those authors to suggest a three-dimensional model of neurotoxin I.

That model served as a basis for studies of the mechanism of neurotoxin binding with AR. The ESR spectrum of a specially labeled group on the neurotoxin and the effects of the ferricyanide paramagnetic ions and Ni^{2+} on the spectrum indicated that the Lys-26, Lys-46, and His-31 groups of neurotoxin are located in the depths of the neurotoxin-AR complex, while the Lys-1 and Lys-15 groups are on the boundary of the complex. Additional information was obtained by specific modification of the toxin's lysine residues by fluorescent dansyl probes. The relaxation shift of the fluorescence maximum ($\Delta\lambda_{max}$) indicated that the dansyl fragments on the Lys-1, Lys-15, and Lys-25 groups protruded into the aqueous phase.

7.2.6. *Membranes of chromatophores of photosynthetic bacteria*

The main functional component of membranes of chromatophores in photosynthetic bacteria (MCPB) is the reaction center (RC), which is a specific photoenzyme that catalyzes transformation of the energy of solar quanta to that of photoseparated charges: a positive bacteriochlorophyll (P^+) and a negative ubiquinone anion (A^-). Cytochrome c_{555} is an electron donor for the RC. The method of a paramagnetic probe-relaxator (Section 2.4.4) was employed to study the P^+ location in membranes of chromatophores from *Rhodospirillum rubrum*. In the preparations without cytochrome c_{555} the effects of the hydrophilic probe-relaxator ferricyanide on the saturation curves of the ESR signal from P^+ were adequately described by the model, suggesting that P^+ was immersed in the RC of the protein globule to a depth of about 8 nm (Kulikov *et al.* 1981). In the presence of cytochrome c_{555} the pigment P^+ was not influenced magnetically by ferricyanide but was accessible to the influence of a Co^{2+} acetylacetone probe-relaxator located in the superficial portion of the membrane lipid bilayer (Alakhverdiev *et al.* 1991). In the latter case, the nearest approach of the probe-relaxator to P^+ was about 0.10–0.12 nm. The distance r between the radical and the paramagnetic ion was estimated based on the effects of the spin–spin dipole–dipole interaction on the parameters of the radical saturation curve (Section 2.4); the value was ~ 1.8 nm for the distance between the heme group of cytochrome c and the P^+ pigment of the RC. The results obtained for the complex in native membranes agreed fairly well with the data from X-ray analysis of the isolated complex (Michel & Deisenhofer 1986).

Photosynthetic membranes are suitable models for studing correlations between molecular dynamics and functions of biomembranes (Berg *et al.* 1979a,b; Kotelnikov *et al.* 1983; Chamorovsky *et al.* 1986; Likhtenshtein 1988b; Knox *et al.* 1989).

When spin, fluorescent, triplet, and Mössbauer labels were introduced into chromatophore membranes from *R. rubrum* and the temperature, humidity, and composition of the solvent were varied, a number of dynamic effects were revealed (Figure 7.9). At temperatures above 30 K and 80 K, the spin–lattice relaxation rate, $1/T_{1e}$, and the spin pocket width, $\Delta H_{\frac{1}{2}}$, respectively, increased for 14-doxyl stearic acid. This result suggests a phonon process (Section 1.5.7). At $T > 80$ K the ESR signal line width narrowed because of low-amplitude wobbling of the nitroxide fragment, which averaged the dipole–dipole interaction of the electron and nuclear spins of surrounding protons. A further temperature increase led to defreezing of high-amplitude rotation of the nitroxide fragments, which could be monitored by the decrease in the hyperfine splitting of the ESR spectrum (A_{zz}) and by the alterations in the parameters of the ST ESR spectrum (Section 1.4.1).

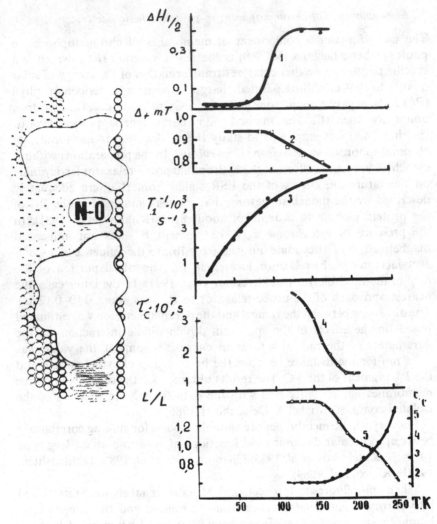

Figure 7.9. Temperature dependences of the dynamic parameters of 5-doxyl stearic spin labels in the photosynthetic membrane of purple bacteria: 1, $\Delta H_{\frac{1}{2}}$ spin pocket width; 2, Δ_r line width; 3, T_i^{-1}; 4, τ_c; 5, parameters of the ESR signal of the second harmonic absorption out of phase (Kochetkov *et al.* 1984; Likhtenshtein 1986).

Similar consequences for the dynamic events following the temperature increase were observed qualitatively for spin-labeled membrane proteins. However, in the latter case, wobbling and rotational mobility of the nitroxide fragments were detected at temperature shifts of 60–70°C to a higher-temperature region. At physiological temperature, experiments using spin, fluorescence, and Mössbauer labeling independently indicated high-amplitude dynamic processes in the nanosecond region (see Figures 6.11 and 6.12) that correlated with the kinetics of electron transfer in the RC.

7.2.7. *Other membranes*

Na^+, K^+-ATPase is a membrane protein consisting of α- and β-subunits. The enzyme catalyzes transportation of Na^+ and K^+ ions across the membrane, coupled with ATP hydrolysis. When ST ESR techniques were applied to a protein preparation modified by a tightly bound nitroxide spin label with an active vinylketone group, they showed the rotational diffusion correlation time to be 25 μs (Esmann *et al.* 1989). The latter value was evidence for a dimeric form $(\alpha, \beta)_2$ of the enzyme in the membrane.

A correlation was established between the catalytic activity of the Na^+, K^+-ATPase from brain and the order parameter S of hydrophobic lipid spin probes (Boldyrev 1988). The Arrhenius dependences of both catalytic and dynamic processes contained discontinuities at temperatures above 20°C. Treatment of the membrane preparation by phospholipase to hydrolyze the lipid polar "head" linearized the Arrhenius dependence.

The dynamic affinity of spin-labeled lipids for the Na^+, K^+-ATPase surface in the membrane decreased in the following sequence: cardiolupin, phosphatidylserine > stearic acid > phosphatidylglycerol $\simeq$ phosphatidylcholine $\simeq$ phosphatidylethanolamine (Esmann *et al.* 1985).

As early as the 1960s it was observed that the ESR spectra of lipid nitroxide spin probes were sensitive to biochemical reactions in the mitochondrial membrane (Koltover *et al.* 1968, 1971; Gendel *et al.* 1968). Addition of the ATP substrate led to increases in the solubilization parameter L of hydrophobic spin probes. Energization of the membranes was accompanied by a decrease in the accessibility of the nitroxide groups for spin exchange interactions with ferricyanide distributed in the aqueous lipid phase.

A carbodiimide spin label,

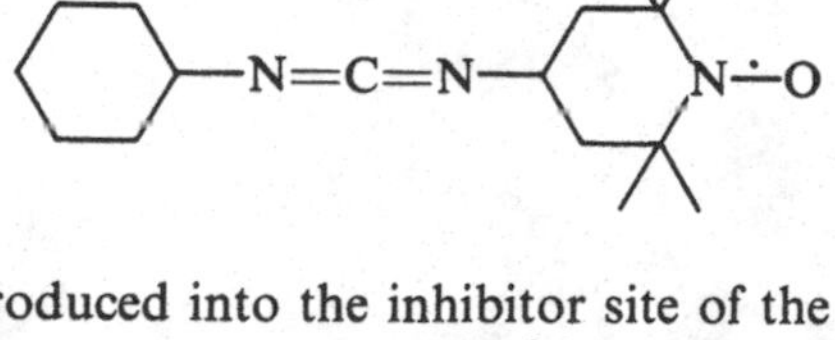

was introduced into the inhibitor site of the mitochondrial ATPase, and the paramagnetic complex Mn^{2+}-ATP was accepted by the catalytic site of the enzyme (Azzi *et al.* 1973). The distance between these two sites was estimated to be about 20 nm (Taylor *et al.* 1969). The Arrhenius plots for both types of parameters (the rotational diffusion correlation time of the iodoacetamide spin label attached to the SH groups of the enzyme, and the ATP activity) were characterized by discontinuities at temperatures of about 20°C and 30°C.

The data presented in this chapter have demonstratd that biophysical labeling is an efficient tool for the study of microstructure and molecular dynamics in both model and, particularly, native membranes. In these sophisticated heterogeneous structures, in which "direct" physicochemical methods often fail, specific modifications of various parts of membranes by physical labels and probes can facilitate solutions to a number of structural and dynamic problems. Among the unique possibilities for uses of the method, the following have been most effectively employed: (1) experimental estimation of microviscosity and micropolarity of certain portions of membranes; (2) determination of quantitative parameters of rotational and translational diffusion of lipid and protein components; (3) study of component location and distribution; (4) monitoring of phase and conformational transitions and protein–lipid interactions; (5) evaluation of membrane potential and local pH. An analysis of the correlations between these parameters and the functional activities of membranes (catalytic and regulatory activity, permeability, transportation, transmission of biological signals, etc.) will allow us to learn still more about the fine mechanisms of membrane functioning.

8

Nucleic acids and other biological systems: biological assays

8.1. Nucleic acids

Applications of physical labeling to study the structures and conformational changes of nucleic acids are based on the principles presented in Chapters 1–5 (Sukhorukov *et al.* 1967; Likhtenshtein 1976a; Mill *et al.* 1986; Ulanov *et al.* 1976; Bobst 1979; Kamsolova & Postnikova 1981; Spartenstein *et al.* 1989). However, at the present time the use of this approach is still greatly hampered by a number of specific features of the nucleic acids, primarily by difficulties in selectively and noninvasively modifying them chemically. These problems are most strongly experienced in work with DNA, and to a lesser degree in work with RNA and tRNA. Actually, tRNA is a very convenient object for chemical modification with labels because we know its base sequence and because it contains a clearly defined aminoacyl fragment and sections with minor nucleotides.

8.1.1. Modification of nucleic acids with physical labels

The following methods have been developed for modification of nucleic acids (Bobst 1979; Piette & Hsia 1979; Kamsolova & Postnikova 1981; Gnewich & Sosnovsky 1986; Spartenstein *et al.* 1989):

    1. Convalent attachment of the nucleotide bases to the functional group with the use of spin labels:

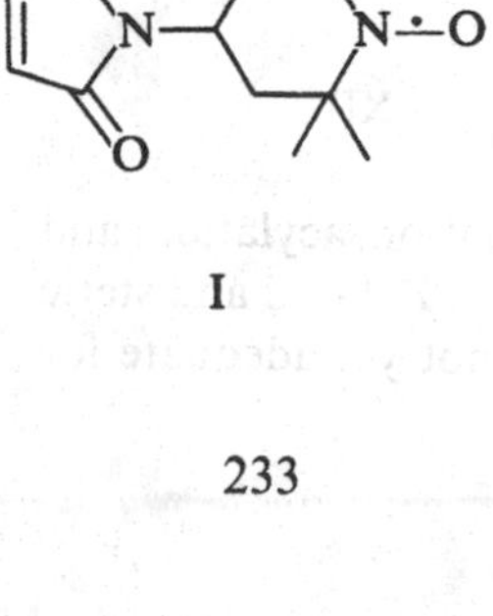

         I 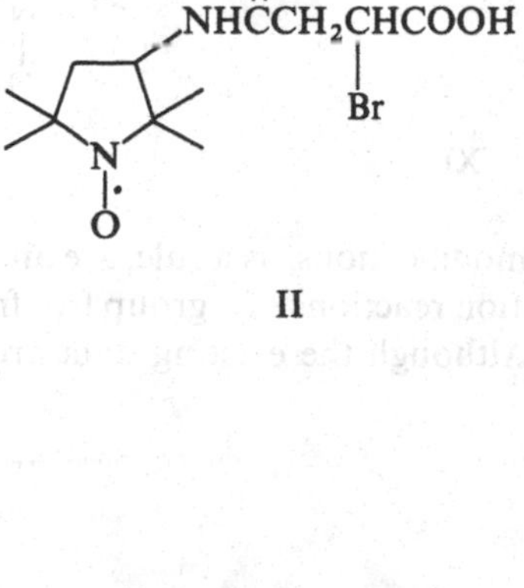   II 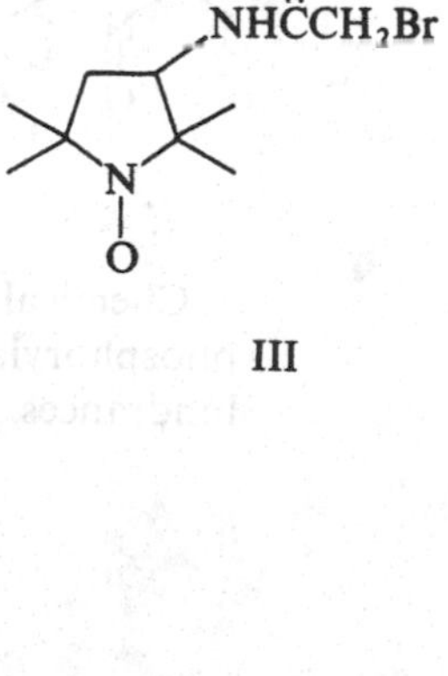   III

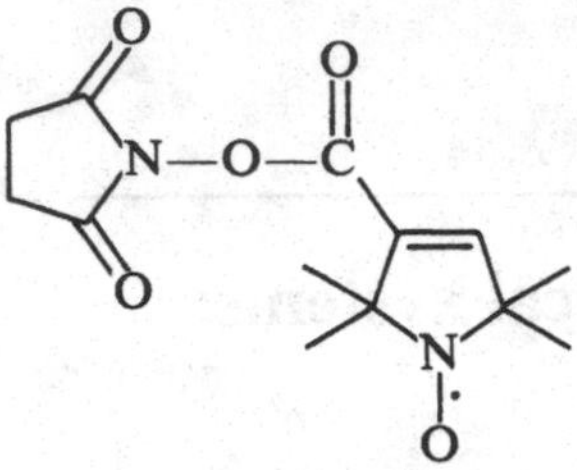

IV

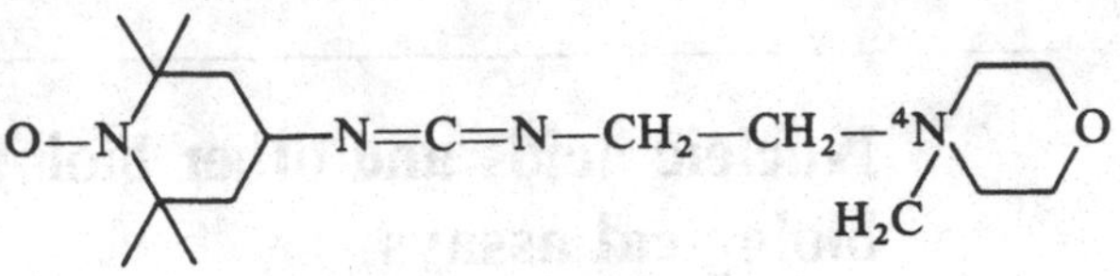

V

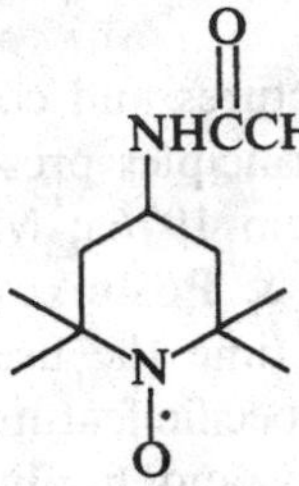

VI

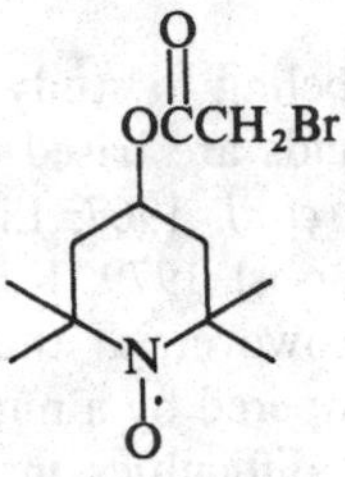

VII

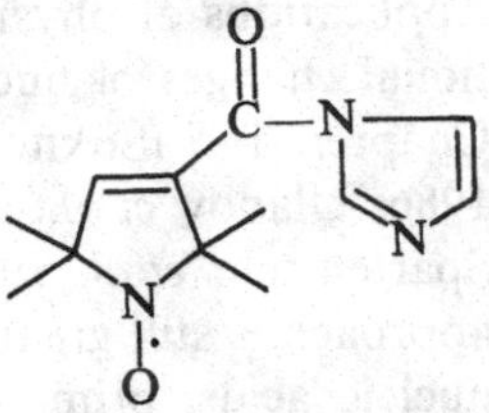

VIII

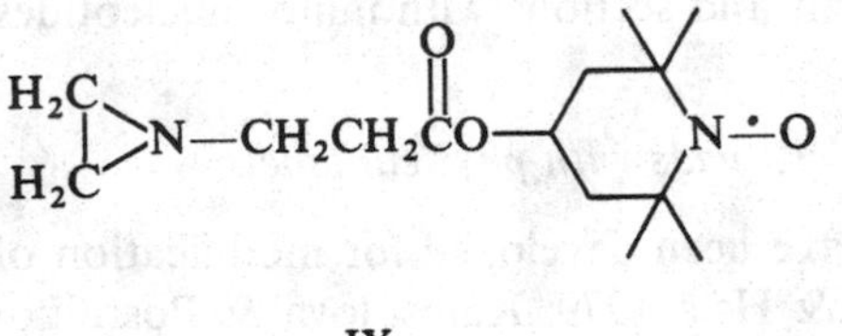

IX

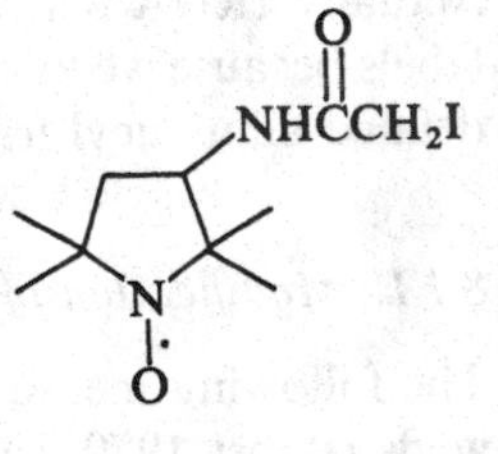

X

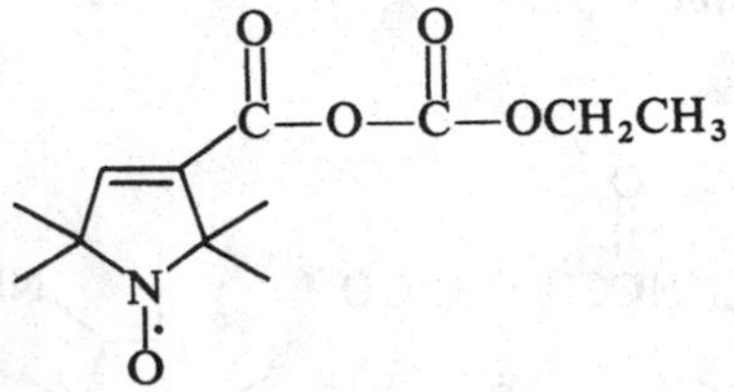

XI

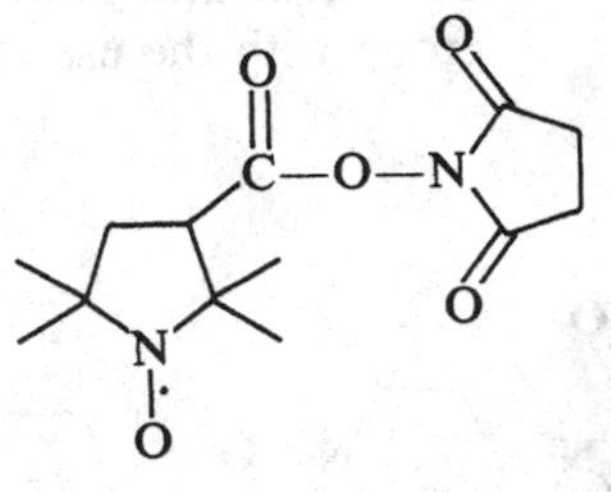

XII

Chemical modifications, as a rule, are alkylation, acylation, and phosphorylation reactions of a group free from H-bond and steric hindrances. Although the existing data are not yet adequate for serious generalization, it is still clear that guanine and adenine are the most reactive bases and that the reactivity decreases along the following series: individual nucleotides > polynucleotides > RNA > DNA.

2. Covalent attachment to hydroxyl groups on the ribose glucosyl rings. A number of forms of this type of modification have been developed: (*a*) periodate oxidation of the rings, followed by reactions of aldehyde groups formed with an amine derivative of the labels and reduction with $NaBBH_4$; (*b*) acylation of free OH groups (e.g., label VIII).

3. Alkylation or acylation of the SH group of the 4-thiouracil or NH_2 group on the aminoacyl residue of tRNA.

4. Enzymatic synthesis of nucleic acids with the use of a labeled base (e.g., uridine-5'-diphosphate spin label polymerized with the polynucleotide phosphorylase from *Micrococcus luteus*).

5. Intercalation of spin and fluorescent probes between bases in the double-helix portions of nucleic acids:

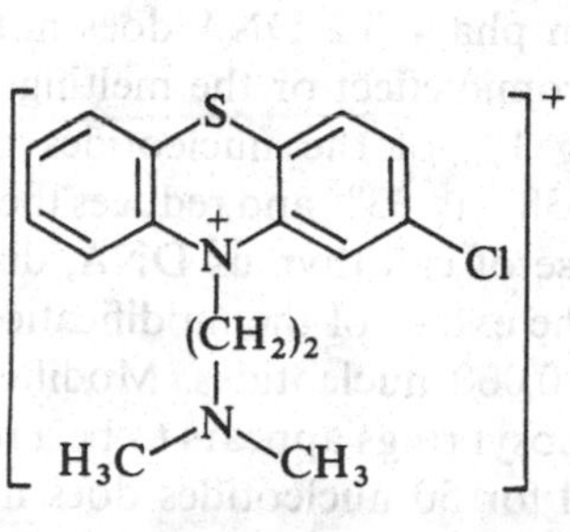

XIV

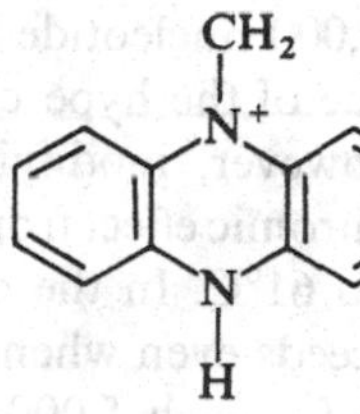

XV

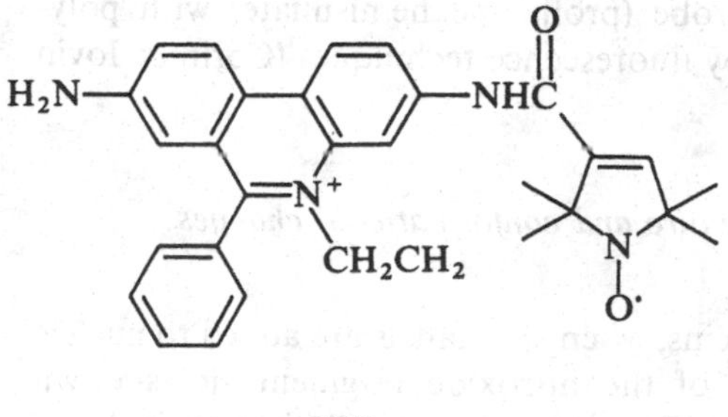

XVI

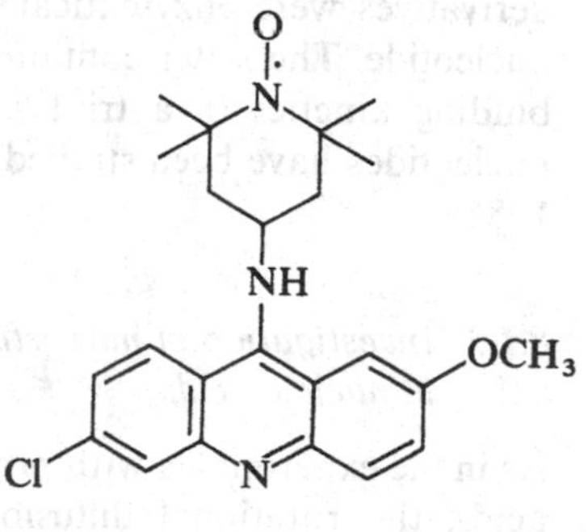

XVII

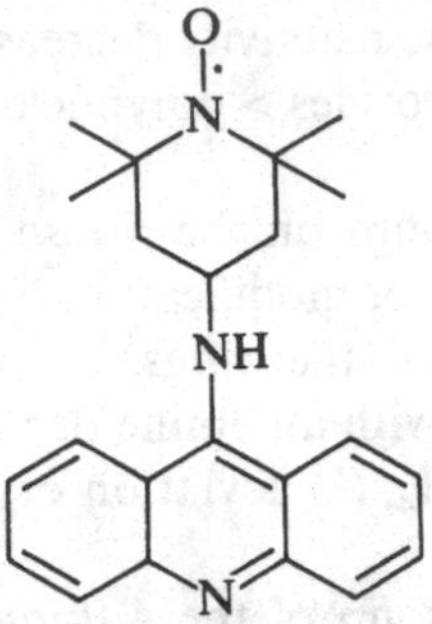

XVIII 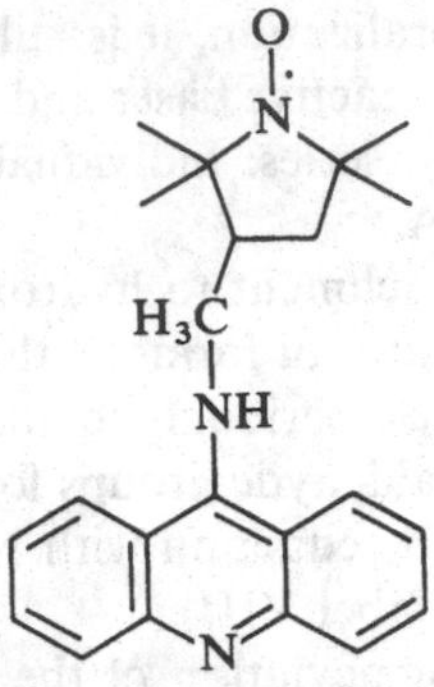   XIX

6. Complexation with paramagnetic (Mn^{2+}) or Möessbauer ($^{57}Fe^{n+}$) atoms.

The effects of physical labels and probes on the structural and functional properties of modified nucleic acids depend on the degree of modification and the nature of the object under study. Thus, binding of approximately one label to every 1,000 nucleotides in phage T2 DNA does not significantly affect the value of the hyperchromic effect or the melting point of the preparation. However, modifying 1% of the nucleotides of DNA decreases the hyperchromic effect from 38% to 33% and reduces the melting point from 82.5°C to 61°C. In the case of calf-thymus DNA, disruption of the structure proceeds even when the extent of the modification corresponds to one label for each 5,000–10,000 nucleotides. Modification of hydroxyl groups on the ribose and glucosyl rings appears to be a relatively noninvasive procedure: one spin label for 50 nucleotides does not affect the physicochemical properties of the preparation. Attachment of spin labels to the SH group of 4-thiouracil of tRNA from *Escherichia coli* does not prevent binding of tyrosine and methionine.

It should be noted that in recent work (Bobst *et al.* 1988), deoxyuridine derivatives were enzymatically introduced into 26-mericoligodeoxyribo-nucleotide. The latter contains an endonuclease EcoRI binding site. The binding kinetics of a triplet probe (proflavine hemisulfate) with polynucleotides have been studied by fluorescence techniques (Corin & Jovin 1986).

8.1.2. Investigation of microstructure and conformational changes in nucleic acids

As in the experiments with proteins, when spin labels are added to nucleic acids, the rotational diffusion of the nitroxide fragment slows down considerably ($\tau_c = 10^{-9}–10^{-10}$ s). The extent of immobilization is deter-

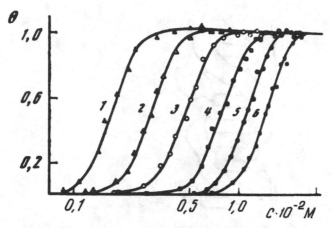

Figure 8.1. Isotherms of adenosine binding by polycarbohydrate at various temperatures. (1) 3°C; (2) 8°C; (3) 12°C; (4) 17°C; (5) 22°C; 1-M NaCl, 0.1-M tris-HCl, pH 7.5 (Sukhorukov & Petrov 1988).

mined mainly by the structure of the nucleic acid, the length and flexibility of the label, and the ambient conditions (pH, temperature, ionic strength, and degrees of binding of various compounds).

Thermal pH-induced transitions of nucleic acids at their melting temperatures (τ_m) are accompanied by a drastic decrease in the value of the τ_c parameter. Strong spin–spin exchange effects were observed in the case of a spin-labeled polyuridyl acid with a biradical fragment SL-U-U-SL when the temperature was increased to 85°C (SL is spin label V) (Levental *et al.* 1972).

The ESR spectra of spin-labeled preparations of nucleic acids are sensitive to fine conformational changes as well (Mill *et al.* 1986). For calf-thymus DNA, for example, the transition from the *E* form to the *C* form was accompanied by a decrease in the τ_c parameter for label IX attached to DNA. The decrease appeared to result from some contraction of the DNA helix.

The degree of immobilization of a spin label on the OH groups of the sugar moieties decreases along the following series: poly(A) < poly(C) < poly(U) (Sukhorukov & Petrov 1988). It was found that the formation of double and triple helixes led to a significant increase in the value of the τ_c parameter. This fact allowed those authors to quantitatively investigate the complexation of polynucleotides, both complementary [double helix of poly(U): 2 poly(U):adenosine; 2 poly(U):AMP; 2 poly(U):poly (A); poly(A):poly (U)] and noncomplementary [poly(U):poly(C); poly(U)poly(A); poly(A):uridine]. They clearly demonstrated the cooperative character of the mononucleotide and mononucleoside binding (Figure 8.1). After

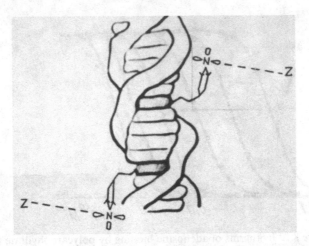

Figure 8.2. Disposition of the intercalated spin label on DNA (Mill *et al.* 1986).

analyzing these results, they suggested that nonspecific stacking inter-actions contribute significantly to the stability of the DNA double helix.

Spin labeling has been used to study the mechanism of intercalation of biologically active dyes in DNA (Bobst 1979; Chignell 1979; Hsia & Piette 1969; Sinha & Chignell 1975). In carrying out their experiments, Mill *et al.* (1986) assumed that intercalation of a dye molecule between two bases in the DNA helix could force the DNA fragment to elongate by about 0.35 nm. If the dye bound to the macromolecular surface (in a narrow cleft, for example), significant elongation could not be expected. When molecules of spin-labeled ethidium bromide and acriflavine were in-troduced into the labeled DNA macromolecule (one label per 10 pairs of nucleotides), the average distance between the labels increased by 0.8–0.9 nm. This result favors the intercalation mechanism of binding.

When spin labeling and absorption spectroscopy were combined to investigate oriented preparations of DNA labeled by the nitroxide deriva-tives acridine and rubomycin, two types of dye binding were indicated: strong binding by intercalation (Figure 8.2) and weak binding presumably resulting from electrostatic interactions with phosphate groups. The stacking interaction from the association of 2-aminopurinoriboside and its 5-mono-phosphate and diphosphate derivatives was studied by means of spin label-ing and luminescence quenching (Gajewska *et al.* 1982).

The high sensitivity of the ESR parameters of spin-labeled nucleic acids has been exemplified in a number of works. When tRNA became elongated upon binding, the degree of immobilization increased for the spin label covalently attached to the 4-thiouridine group of tRNA (Janick *et al.* 1990).

Similar effects were observed for the interaction between the following pairs: spin-labeled poly(U):polysine; protein from phage T4:singly stained DNA; spin-labeled polynucleotides:monolayers of cells.

Local sections on the boundary between the dihydrouridine and aminoacyl fragments were compared for different types of tRNA from *E. coli* by adding spin label VII to the SH group of 4-thiouridine. The τ_c values for methionine, phenylalanine, tyrosine, and valine tRNAs were 0.77, 0.5, 0.77, and 0.21 ns, respectively (Hara *et al.* 1970). The plot of $\log \tau_c$ as a function of $1/T$ for valine tRNA located on the α-MH groups showed a distinct discontinuity at 51°C. The value of the transition temperature was close to the corresponding value obtained spectrophotometrically.

The use of paramagnetic probes, e.g., nitroxide, has revealed a number of interesting features of the temperature dependence of τ_c in samples of calf thymus DNA of different water contents (Sukhorukov *et al.* 1967). In dry preparations the probes were found to be immobilized ($\tau_c \geqslant 10^{-7}$ s). Addition of two water molecules per one nucleotide led to a significant increase in the probes' mobility.

The intramolecular dynamics of DNA have been studied by FRAP techniques (Scaletfar *et al.* 1990). The rotational wobbling of ethidium bromide intercalated in the DNA double helix reportedly had characteristic times of about 100 ms in aqueous solution and times of milliseconds in agarose gel.

In conclusion, developments in the chemistry of nuclic acids lead us to believe that in the near future the assortment of physical labels and probes will be expanded and that there will be increased use of these methods in further studies of the structrual and conformational changes in nucleic acids.

8.2. Polysaccharides

Polysaccharides included into immunoglobulins and other glycoproteins, cotton fibers, celluloses, dextrins, etc., play an important role in biology. Despite the fact that there have been few investigations in this field using physical labels, the data available demonstrate the potential of this labeling method for studies of microstructure, intramolecular dynamics, and conformational transitions of polysaccharides.

The methods of polysaccharide modification by labels are, to some degree, similar to those described in Section 8.1 for the ribose and glucosyl rings of nucleic acids. In several studies, nitroxide labels were covalently attached to the OH groups in the reactions of acylation and acetylation (Nezlin *et al.* 1970; Aplin *et al.* 1976; Lee & Grant 1980; Marupov *et al.* 1981; Nezlin & Sykulev 1988). Fluorescent and phosphorescent chromophores were subjected to nucleophilic substitution, which is fairly well developed in the chemistry of fibers. Another modification procedure included activation of bicarbonate by bromcyan, followed by reaction with an amino radical.

A number of specific methods of modification are based on obtaining aldehyde groups by periodate oxidation of the C—C bonds or of hydroxyl groups of sialic acid. Another procedure involves oxidation of —CH$_2$—OH groups by glucosidase, followed by reaction with an amino derivative of the nitroxide spin label and reduction by NaBH$_4$. Chlorodiimide derivatives of physical labels can be used for modification as well.

Spin-labeled dextrins with intercalated radicals and biradicals of the type

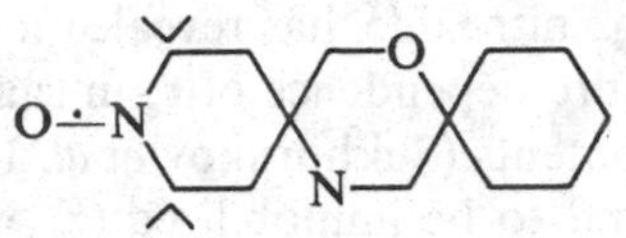

have been obtained by direct interaction between reagents.

8.2.1. Glycoproteins

Carbohydrate fragments of glycoproteins have a particular structural and functional significance. These fragments provide for specific interactions between subunits, sensitivity of proteolysis, permeability across the membrane, intercell recognition, etc.

Spin-labeled immunoglobulins (Ig) have been studied in thorough detail. Oligosaccharides, which constitute up to 2.5–12% of the total molecular mass of Ig, are attached to heavy chains of the proteins (Nezlin & Sykulev 1988) in a ratio of up to five oligosaccharide chains per protein chain. When spin labels were introduced into the carbohydrate portions of immunoglobulins by the periodate procedure cited earlier, the number of labels per macromolecule ranged from 1 to 30, and the ESR spectra of spin-labeled preparations indicated two fractions of the radical: an immobilized fraction with the order parameter $S = 0.81$ and a mobilized fraction with $\tau_c = 10^{-9}$ s. Spin labels attached to the protein portion of Ig by acylating fragments (VI, SLXV) were found to be immobilized ($S = 0.71$).

The values of the rotational correlation time, τ_c, for a spin-labeled macromolecule of human Ig and for individual fragments (Fab) were about 26 ns at room temperature. This value was close to that determined by an independent method of depolarization of fluorescence. In all cases, the experimental values of τ_c were much lower that those expected for rotation of the Ig macromolecule as a whole and even for rotation of individual Fab. These results supported a model in which both oligosaccharide and protein portions took part in local mobility relative to the Ig globule (Kaivarainen 1985; Nezlin & Sykulev 1988; Arutyunyan *et al.* 1989).

Similar results were obtained for spin-labeled Ig from various sources and other glycoproteins (ovalbumin, fetuin, glycoforin, glycoconjugates of erythrocyte membranes).

The ESR spectra of spin labels on Ig were found to be sensitive to interactions with antigens. Thus, complexation with dansyl led to decreases in the correlation times of spin labels on carbohydrates. On the other hand, aggregation of Ig or complexation with C1q complement led to immobilization of the label. An analogous effect was observed upon complexation of glycoforin and lectin.

An important, though not yet resolved, problem is precise determination of the locations of oligosaccharides relative to other fragments of glycoproteins and detailed elucidation of the mechanism of allosteric transglobular transitions. These problems can be solved, in principle, using combinations of spin, fluorescent, triplet, and electron-scattering labels.

8.2.2. Cotton fibers and cellulose

Preparations of spin- and luminescence-labeled cotton fiber and cellulose from various sources have provided the basis for studies of the structures and molecular dynamics of these objects (Bobodzhanov & Likhtenshtein 1974; Marupov *et al.* 1981; Yusupov *et al.* 1988; Islomov *et al.* 1989).

Treating cotton fibers and α-cellulose with the spin label SLXV resulted in covalent binding of the label to hydroxyl groups in amorphous parts of the fibers. An approach based on estimation of the effect of a paramagnetic molecule (ferricyanide) on the parameter of spin relaxation of the label indicated that the nitroxide fragment was immersed in the cotton microfiber to a depth of about 0.1 nm. In an aqueous medium, two fractions of spin label were observed, with correlation times $\tau_c \geqslant 10^{-7}$ s and $\tau_c \approx 10^{-9}$ s. The values of the rate constants for spin exchange (k_{ex}) between the nitroxide fragment of the label and a ferricyanide ion in solution for mobilized and immobilized fractions were $7 \cdot 10^8$ and $3.7 \cdot 10^7 \, M^{-1} s^{-1}$, respectively.

Measurements of the temperature dependences of the parameters of the spin label ESR spectrum in the 3-cm and 2-mm ranges and V'_2 in the 3-cm range (Section 1.4.1), as well as those of fluorescence and phosphorescence spectra, revealed dynamic effects in cotton fiber preparations, which will be described next (Marupov *et al.* 1981; Yusupov *et al.* 1988; Krinichny *et al.* 1986).

In dry samples of spin-labeled cotton fibers within the temperature range 70–320 K, the parameter of hyperfine splitting (A_{zz}) in the spectrum of the 2 mm band of ESR was found to remain practically unchanged. A gradual decrease in A_{zz} began when the temperature was increased above 320 K. Experiments in the 2-mm range confirmed the z- and y-component shifts at $T > 320$ K and showed an additional low-field shift of the x-component at temperatures above 200 K. The latter effect cannot be explained by any dynamic process and is likely to have resulted from local mechanical tension of the H-bonds between the nitroxide and the adjacent hydroxyl group. The data obtained are in agreement with the model suggesting rotation of the fragment

along the O–C axis, which is parallel to the x axis of the nitroxide. At temperatures $T > 90$ K, narrowing of the line width (Δ) of the spectrum probably was caused by low-amplitude wobbling of the nitroxide fragment.

In wet preparations of cotton fibers the temperature dependences of the A_{zz} and Δ parameters at $T = 70$–200 K were found to be similar to those in dry preparations. Within this range were detected a decrease in the intensity and long-wavelength shifts in the fluorescence and phosphorescence spectra of eosin. These effects testified to the emergence of relaxation processes with correlation times less than 10^{-3} s in the vicinity of the chromophores. At temperatures above 200 K, a decrease in the A_{zz} parameter accompanied by an increase in Δ suggested intensification of the high-amplitude rotational mobility of the label, with $\tau_c = 10^{-7}$–10^{-8} s (Section 1.5.6). In both dry and wet preparations the lack of merging of individual components of the 2-mm-band ESR spectrum appears to be a strong argument in favor of the model of slow anisotropic mobility of the spin-labeled nitroxide fragments.

At $T > 303$ K the 3-cm band of the ESR spectrum of wet spin-labeled cotton fibers theoretically can be divided into two components corresponding to mobile and immobile fractions of the label (Antsiferova *et al.* 1989a), the former fraction being equal to 6–28% for various preparations. The ESR spectra of partially hydrated samples were fairly well described in theory, suggesting the presence of two fractions: totally hydrated and nonhydrated. The ratio between the fractions can be calculated from A_{zz} as a function of the water content by comparison with the corresponding theoretical calibration curve (Antsiferova *et al.* 1989b).

The influences of various plasticizing liquids on the mobility of the nitroxide fragment were studied in spin-labeled cotton fibers (Islomov *et al.* 1989). High-amplitude rotational mobility of the radicals was observed in water, ethanol, and chloroform at temperatures of 0, -2, and $-30°C$, respectively. The plasticizing ability of the solvents decreased along the following series: ethanol > water > chloroform. The values of τ_c at 60°C were 7.7, 8.9, and 16 ns, respectively.

The values of the parameters of the ESR spectra and their temperature dependences were different for cotton fibers from different sources (Marupov *et al.* 1981). The general tendency was found to be as follows: the more mobile the fractions of the label in the preparation, the lower the value of the tearing stress of the cotton fiber.

Analysis of the changes in the A_{zz} and τ_c parameters and their temperature dependences in spin-labeled cotton fibers under UV irradiation allowed Kostina *et al.* (1987) to formulate a mechanism for radiative destructive action. According to this mechanism, irradiation led to an increase in the rigidity of the amorphous portion of the fiber as a result of covalent linkage of polycarbohydrate chains. At the same time, the number of microdefects and the molecular mobility in this portion of the fiber were increased.

Dynamic effects in oxypropylcellulose were monitored by determining the fluorescence parameter of the pyrene excimer in the picosecond range of correlation times (Yamazaki *et al.* 1987).

The experimental approaches and results cited earlier should prove useful in investigations of a wide range of biological fibers.

8.3. Spin-labeled, physiologically active compounds

Even a short review of all spin-labeled molecules of biological importance would exceed the limits of this book. Thus, the description here will be confined to consideration of the main types of such compounds. More detailed information can be found in a number of monographs and reviews (Rozantsev 1970, 1990; Chignell 1979; Rozantsev & Zhdanov 1986; Zhdanov 1981; Hideg 1990; Keana *et al.* 1990).

Amino acids and peptides. Syntheses have been reported for several types of spin-labeled analogues of amino acids, e.g.,

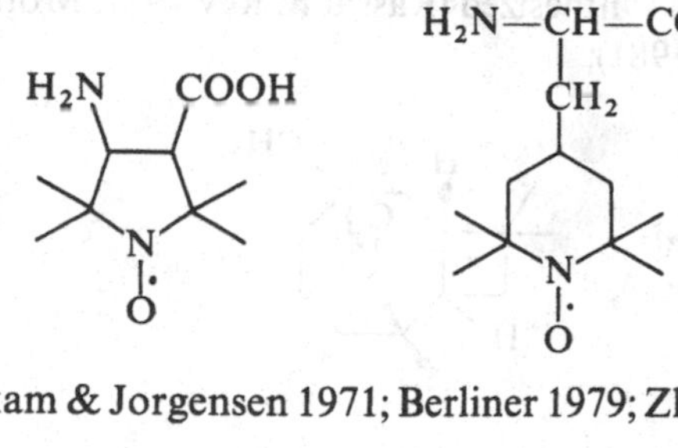

(Weinkam & Jorgensen 1971; Berliner 1979; Zhdanov 1981; Hideg 1990).

Steroids. Formulas for a number of radicals and biradicals, analogues of steroids, are as follows (Dodd & Lewis 1975; Keana 1978; Zhdanov 1981; Volodarsky 1988):

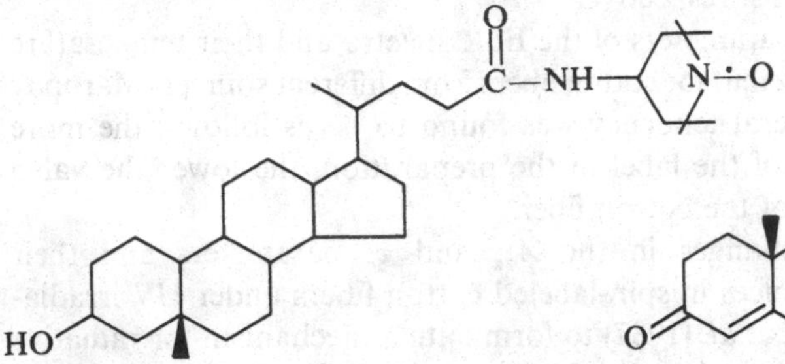

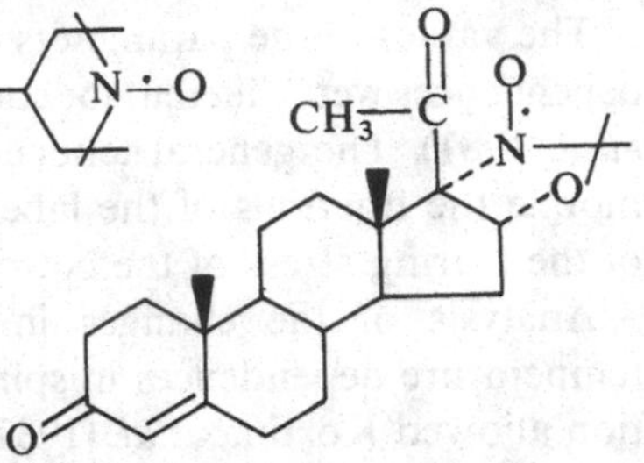

Phosphoorganic compounds. The assortment of spin-labeled phosphoorganic compounds is fairly wide. The most typical compounds are as follows (Rozantsev 1970; Sosnovsky 1990):

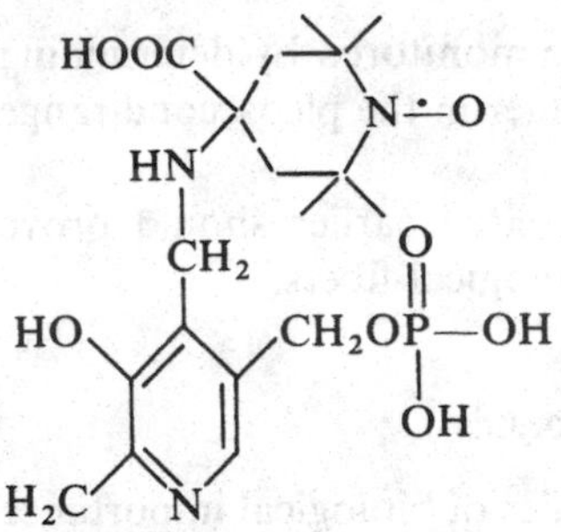

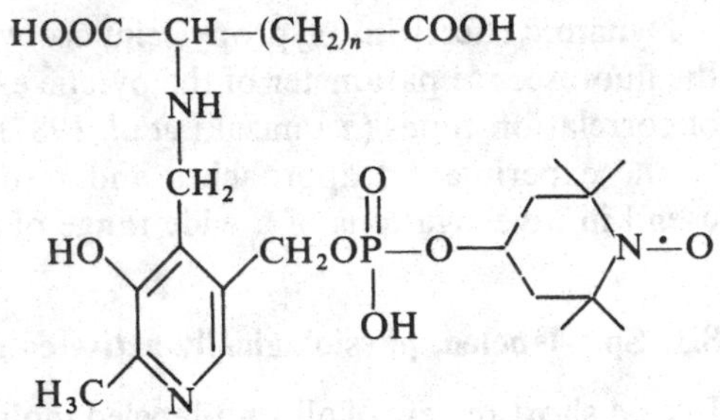

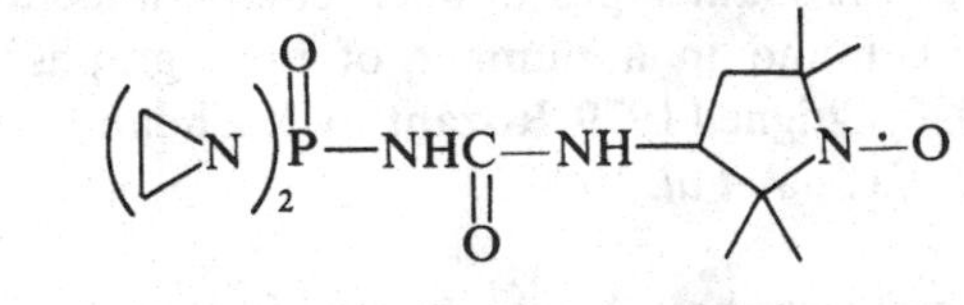

Alkaloids and terpenes. Spin-labeled analogues of camphor, carrofillen, and isomincyledin have been synthesized (Rassat & Rey 1971; Motherwell & Roberts 1972; Zhdanov 1981):

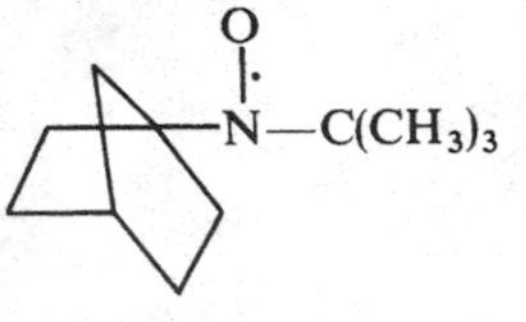

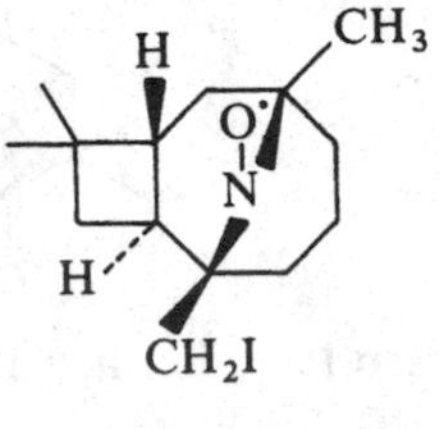

Nucleotides and their fragments. Owing to the great biological significance of nucleotides, quite a number of corresponding spin-labeled analogues have been synthesized (Wenzel & Trommer 1977; Robey *et al.* 1979; Park & Trommer 1989):

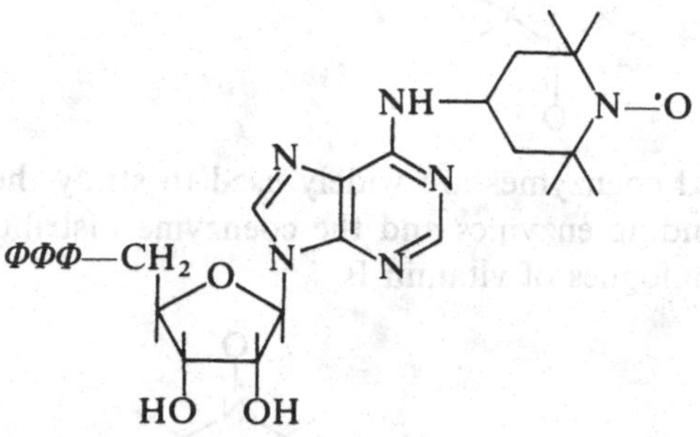

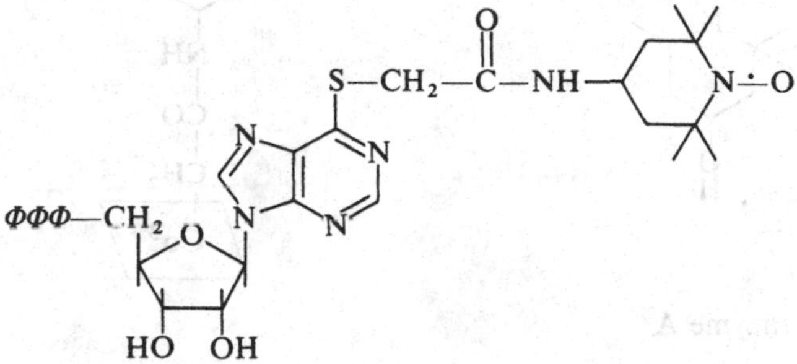

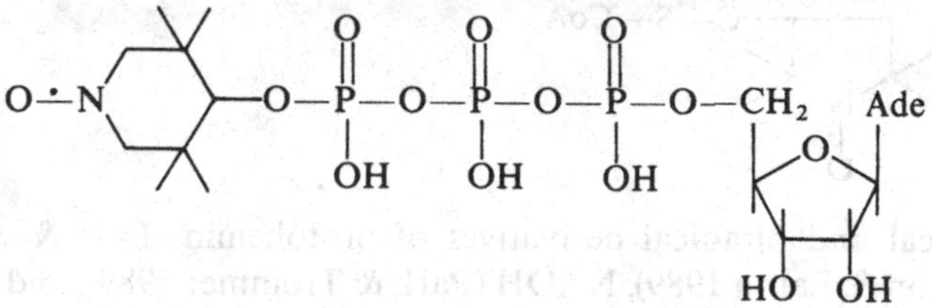

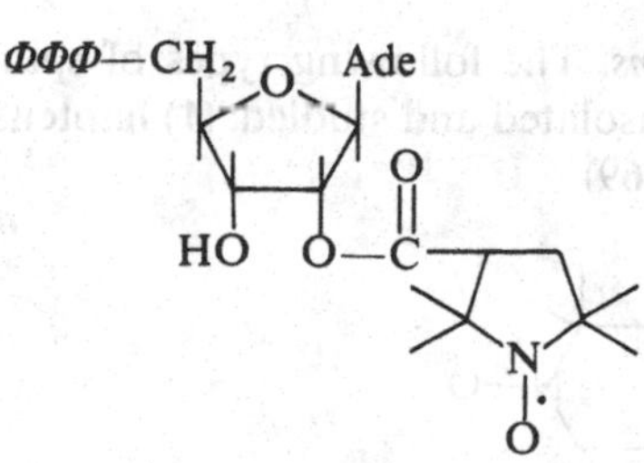

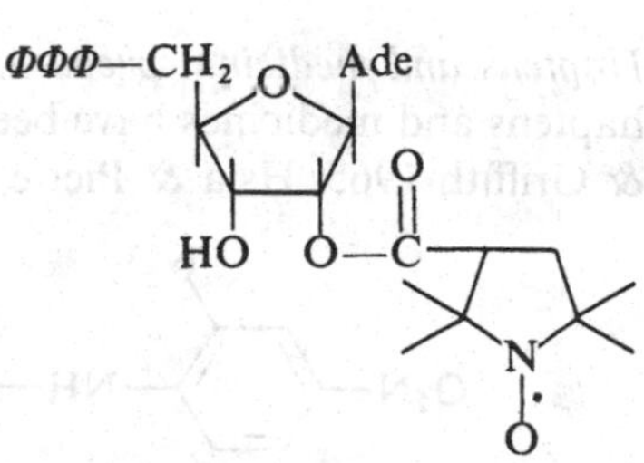

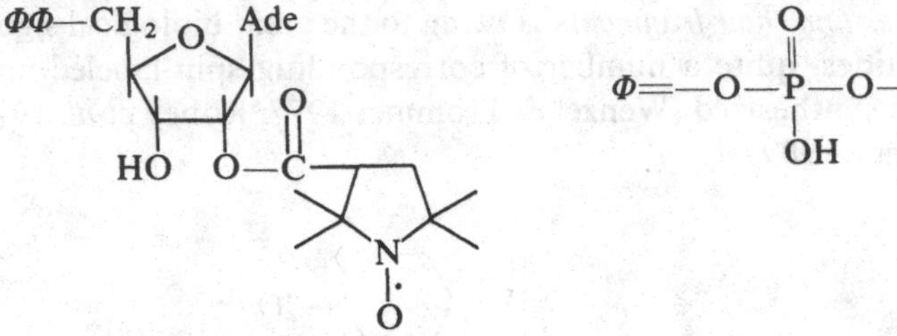

$$\Phi\Phi—CH_2 \quad O \quad \text{Ade}$$

$$\Phi{=}{-}O{-}\overset{\overset{\displaystyle O}{\|}}{P}{-}O{-}$$

$$\underset{OH}{}$$

Coenzymes. Spin-labeled coenzymes are widely used to study the active centers of the corresponding enzymes and the coenzyme distribution in bioobjects, including analogues of vitamin B_{12},

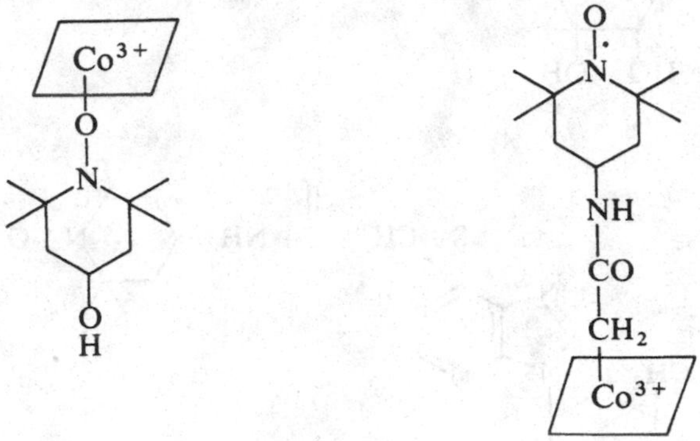

acetyl coenzyme A,

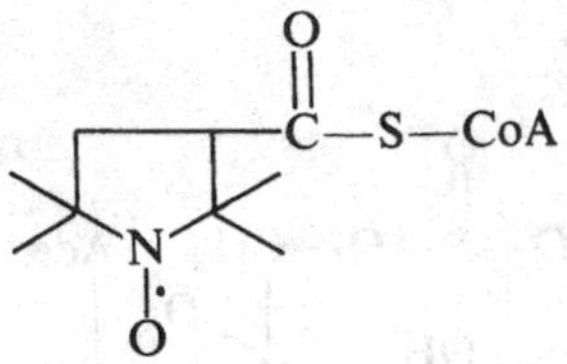

and radical and biradical derivatives of protohemin (Law & Asakura 1976; Eaton & Eaton 1989), NADH (Park & Trommer 1989), and vitamin B_6 (Zhdanov 1981).

Haptens and medicinal preparations. The following types of spin-labeled haptens and medicines have been isolated and studied: (1) haptens (Stryer & Griffith 1965; Hsia & Piette 1969)

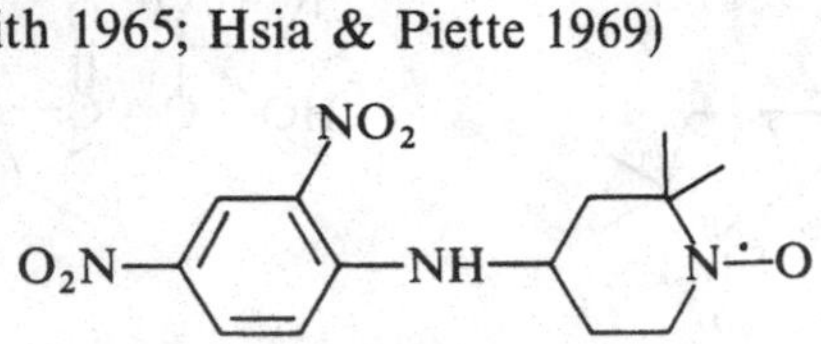

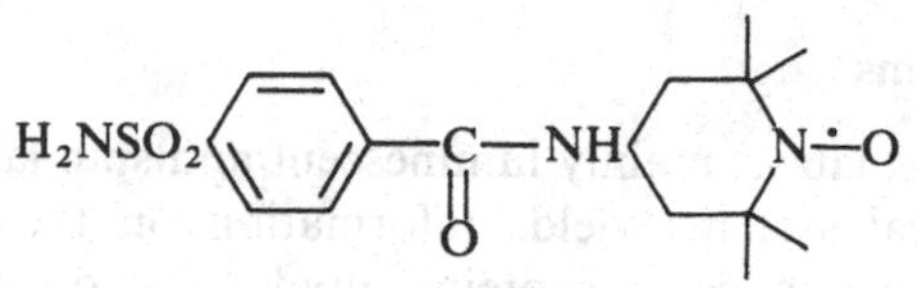

(2) morphine (Copelland *et al.* 1975)

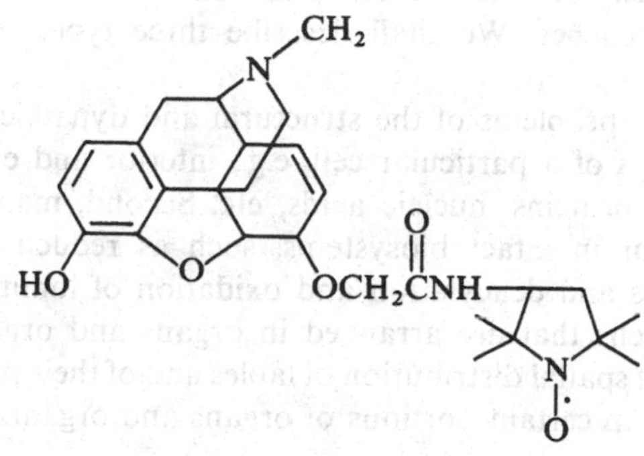

(3) local anesthetics (Zhdanov 1981)

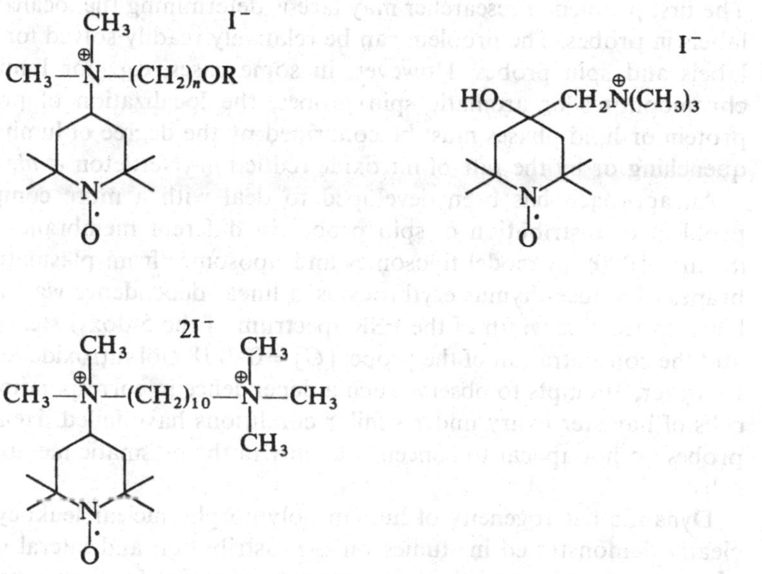

Advances in the synthetic chemistry of biologically active nitroxides have provided the basis for further developments in structural, dynamic, and biochemical investigations using spin-labeling methods.

8.4. Cells, tissues, organisms

Application on biophysical labels, mainly luminescent and spin labels, in investigations of biological systems yields information on the microstructure and molecular dynamics of proteins, lipids, and other components of cells, in analogy to the same information for individual components. In addition, a number of specific problems associated with the intrinsic inhomogeneity of cells and cell organization can be readily solved by labeling approaches. We shall describe three types of such problems.

The first type includes problems of the structural and dynamic states of individual components of a particular cell, e.g., interior and exterior surfaces of membranes, proteins, nucleic acids, etc. Second, many biochemical processes occur in intact biosystems, such as reduction and reoxidation of nitroxides and deacylation and oxidation of luminescent chromophores. Third, cells that are arranged in organs and organisms pose a special problem of spatial distribution of lables and of their physical and chemical properties in certain portions of organs and organisms.

8.4.1. *Distribution of labels: microcomponent localization of cells*

The first problem a researcher may face is determining the localization of labels in probes. The problem can be relatively readily solved for specific labels and spin probes. However, in some cases, e.g., for luminescent chromophores or aromatic spin probes, the localization of probes in protein or lipid phases must be confirmed by the degree of luminescence quenching or by the rate of nitroxide reduction (Nettleton *et al.* 1991).

An approach has been developed to deal with a more complicated problem of distribution of spin probes in different membranes of cells (Swartz 1990): In model liposomes and liposomes from plasmatic membranes of mouse-thymus erythrocytes, a linear dependence was observed between the line width of the ESR spectrum of the 5-doxyl stearic probe and the concentration of the probe: $[C] = 0$–0.18 mol nitroxide/mol lipid. However, attempts to observe such a dependence in corresponding intact cells of hamster ovary under similar conditions have failed. Hence, spin probes do not appear to concentrate only in the plasmatic membranes of cells.

Dynamic heterogeneity of human polymorphonuclear leukocytes was clearly demonstrated in studies on the distribution and lateral diffusion of various protein and lipid components modified by fluorescent labels (Petty *et al.* 1980). The FRAP method was used to monitor diffusion of the $F(ab')_2$ and Fab fragments of antibodies for the C3b receptor and the Ig of monoclonal antibodies for HLA. Diffusion of the lipid portions of the system under study was traced by fluorescent probes (3,3-diocta-decylindocyanide and 4-nitrobenzo-2-oxa-2,3-diazole-α-dimyristoylphos-phatidylamine). It was found that in the leukocyte molecules of the C3b receptor, immobilized clusters formed, the HLA antigens and Ig diffused, with $D_{tr} = 10^{-9} - 10^{-10}$ cm^2 s^{-1}, and spin probes were characterized by the diffusional coefficient $D_{tr} = 10^{-8}$ cm^2 s^{-1}.

In differentiated cells (ovary cells, spermatozoa) the FRAP method indicated the presence of protein domains and of different microcompart-ments in various parts of the cells (Devaus 1988). Each spot, about 1μm in size, different from the others in two parameters, namely, diffusion coefficient and mobile fraction (M). The latter was estimated by measuring the degree of recovery of the probe fluorescence, e.g., 9-(2-antrylnonamic)-acid, after photobleaching. Lateral diffusion in spermatozoa cells was found to vary in the head, central part, and tail of the cell. In erythrocytes, diffusion processes in the interior portions of the membrane proceeded 5–10-fold faster than in the exterior parts. In the plasmatic membrane in human fibroblasts, the values of M and D_{tr} decreased gradually from 90% to 27% when the radius of a photobleaching beam was increased from 0.35 to 5 μm. The M and D_{tr} parameters reportedly were different for various steps in the ovary cells' development and for myocytes of different ages (Yechiel & Edidin 1987).

The values of intracellular pH monitored by fluorescent probes (flu-oroescein acetate after enzymatic deacetylation) were found to be different in various compartments of the cells (Rotman 1973; Dixon & Wilson 1988; Roos & Slavik 1987). The local concentrations of Ca^{2+} in cells of lympho-cytes, thrombocytes, neurons, and muscle were determined with the aid of a highly sensitive fluorescent probe, fura II (Tsien & Zucker 1986).

Dextrans and proteins modified by fluorescent and spin labels were used by Balgavy (1988) and Jarnefelt *et al.* (1988) to study diffusion processes in cell preparations. Dextrans labeled by fluorescein isothio-cyanite diffused through a suspension of tightly packed cells with $D_{tr} = (1-4) \cdot 10^{-8}$ cm^2 s^{-1}, a value two orders of magnitude less than that in aqueous solution.

A promising potential for ESR tomography has been demonstrated in studies of the spatial locations of individual portions of complex biological structures using spin labels and probes (Smirnov *et al.* 1988; Golovina *et al.* 1991). In one example, single wheat grains were soaked with a solution of nitroxide (TEMPOL) and then put into an ESR tomograph. The distribution of a spin-labeled water solution within the grain endosperm was experimentally observed (Figure 8.3). The solution first of all pene-trated into the channel of the endosperm center from the germ side. Pene-tration of the solution into other portions of the grain took many days. Nitroxide radicals with very narrow signals, e.g., ^{2}H and ^{15}N derivatives,

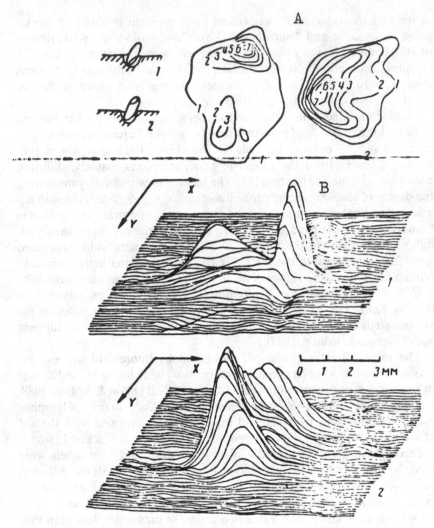

Figure 8.3. Space distribution of spin probe concentration in an individual wheat grain: (A) cross section. (B) space relief: 1, incubation at the germ side; 2, incubation at the opposite side. Incubation in $5 \cdot 10^{-2}$-M NR for 9 hours (Golovina *et al.* 1991).

appear to show promise for studying dioxygen distribution and corresponding oxidative processes in bioobjects by ESR tomography (Swartz 1990).

8.4.2. *Redox properties of cells*

The reactions of reduction and reoxidation of nitroxide radicals described in Chapter 1 are widely used for quantitative characterization of redox processes in biosystems. As long ago as 1968, reduction of nitroxide probes was observed in electron transport in mitochondria (Gendel *et al.* 1968; Koltover *et al.* 1968). Later such processes appeared to be typical of various biological systems, including bacteria, *Chlorella*, cells, spores of *Actinomyces*, and muscles and other fibers (Goldfeld & Tsapin 1974; Likhtenshtein 1976a; Quintanilha & Packer 1977).

Lipid spin probes with nitroxide fragments in different positions were found to be reduced in mitochondria in the presence of the correspinding biological substrates (Kerimov *et al.* 1978). Superficial probes were reduced at high rates; probes located in the lipid bilayers and in the aqueous phase reacted much more slowly. The reduction was inhibited by rotenone, rather than by antimycin and cyanide.

Coenzyme Q and vitamins K_2 and E proved to be good reducing agents. Nitroxide probes were used as indicators of redox processes in photosystems as well. Thus, in blue-green algae *Analaena variabilis* the probe γ-carboline (SLXIX) was reduced by photosystems I and II, and the corresponding hydroxylamine was oxidized by photosystem I (Goldfeld & Tsapin 1974).

The process of hydroxylamine oxidation to nitroxides has been thoroughly investigated in the aqueous phase and in membranes of mouse-thymus bone marrow and hamster ovary cells (Swartz 1987). Superficial probes in membranes, e.g.,

$$\text{(CH}_2)_2\text{CH}_3$$

O—N—O

were oxidized most efficiently. Probes that were deeply plunged into the lipid bilayers reacted noticeably more slowly. Water-soluble hydroxylamines were not oxidized at all. The oxidation was inhibited by cyanide, rather than by antimycin. The differences in the reduction rate of nitroxides were less definitely expressed. The latter process was suppressed by rotenone, but not by cyanide. All these facts suggested that the oxidation was being performed by cytochrome c oxidase, whereas the reduction was performed by semiquinones.

Other components of cells were involved in the reduction of nitroxides in cells; some of them appeared to be an SH-containing compound, glutathione, proteins, ascorbic acid, superoxide radicals, and Fe-containing proteins (Belkin *et al.* 1987). The ability of superoxide radicals to oxidize quantitatively hydroxylamines to nitroxides was successfully used to determine the rate of O_2 formation in various organelles rat liver. Microsomes were shown to be the main sources of radicals. Nevertheless, noticeable amounts of the radicals were initiated by electron transport particles in the presence of NADH, rotenone, and antimycin (Rasbha *et al.* 1986).

Hydrogen peroxide formed in metabolic processes is readily detected by fluorescent probes (Bass *et al.* 1983; Campbell *et al.* 1985; Fomenko & Deimbetova 1987). Thus, oxidation of the reduced form of dichlorofluorescein is accompanied by enhancement of the fluorescence intensity. Luminal may serve as a specific chemiluminescent probe for the detection of H_2O_2 in cells.

8.5. Biological assays

Nowadays a great variety of biological and diagnostic assays have been developed using fluorescent and spin probes. A detailed description of the results in this field is beyond the scope of this book. This section will be confined to a number of typical examples. First of all, the problems of biological activity and toxicity will be considered. Although the latter turns out to be a specific property of each compound, nevertheless some general tendencies may be outlined (Zhdanov 1981).

The quantitative characteristic of nitroxide radical toxicity is its LD_{50}, the concentration of the reagent in the organism under study (e.g., mg per kg of mouse) that will kill 50% of the animals. The value of LD_{50} was found to be determined not by the nitroxide fragment itself but rather by other functional groups of the label.

Thus, for the radical TEMPOL, $LD_{50} = 800\,mg\,kg^{-1}$, which corresponds to about 60 g of the radical per human body. For the radical

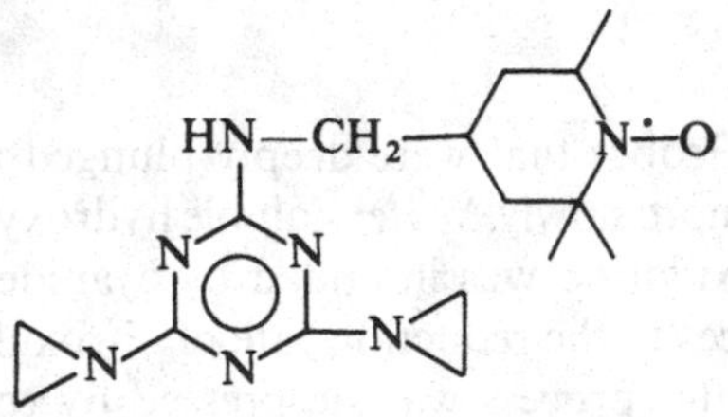

$$\text{HN—CH}_2 -$$

the value of $LD_{50} = 10\,mg\,kg^{-1}$. In general, piperidine nitroxides are less toxic than pyrrolidine nitroxides. As a rule, a nitroxide fragment in a biologically active compound decreases its toxicity, with the main biological activity of the spin-labeled reagent being retained.

The aforementioned properties of nitroxides allow for solutions to a number of medicobiological problems, including those of pharmacokinetics, the molecular mechanisms of pathological abnormalities, and diagnosis of diseases.

In a pioneer study of pharmacokinetics (Zbinden & Brandle 1975), a nitroxide analogue of an antitumor preparation of phosphamide was used to study the distribution of the preparation in organs and tissues of rats.

When the kinetics of a nitroxide analogue of the antitumor preparation rubomycin (ruboxyl) were studied in detail in various organs of mice, the main part of the preparation (up to 0.08 mg) was concentrated in the sperm and kidney (Emanuel & Konovalova 1986).

Nitroxide derivatives of diethyleneimine urethanic acids, phosphoorganic compounds, and nitrosourea were found to exert marked activity against tumor growth (Sosnovsky 1990). These compounds suppressed leukosis and solid tumors in animals. The antineoplastic activity of the nitrosourea derivatives *in vivo* against murine lymphocytic leukemia P388 increased with the preparation's hydrophobicity.

Nitroxide derivatives of morphine have revealed opiate activity (Copeland *et al.* 1975; Wu *et al.* 1977a). Introduction of $-(CH_2)_2NH_2$ and $-(CH_2)_2N(CH_3)_2$ moieties into nitroxide derivatives led to psychotropic anticonvulsive activity.

Hydrophobic nitroxides [di-*tert*-luthylnitroxide and 2,2,6,6-tetramethyl-piperidyl-1-oxyl derivatives of $(C_2H_5$-*o*-Ph-COCCH_2)N(CH_3)_2$ and diethyl-aminoethyl-4-alkoxybenzoates] appear to be local anesthetics. A number of nitroxide derivatives with tetraalkylamino cation groups and compounds of the

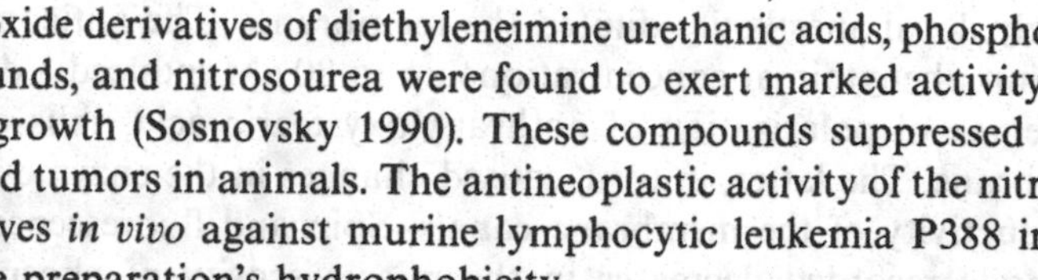

type, which act as β-adrenal blockers, have been proposed (Chignell 1979; Rozantsev 1990).

The nitroxides described earlier have been used to study the locations and mechanism of action of biologically active compounds. In another widespread approach, physical labels are used as indicators of structural changes in biological systems under the actions of physiologically active compounds.

The mechanism of action of the toxin melitin was studied by monitoring the fluorescence of egg lecithin liposomes loaded with a high concentration of a carboxyfluorescein probe (Ehen *et al.* 1988). If the lipsomes were pierced by toxin molecules (easily controlled by the increase in the fluorescence intensity), the probe flowed away. The position of a model peptide, Lys-Phe-Phe-Ile-Ile-Trp-OCH_3, in the lecithin membrane was located by means of a set of doxyl stearic acid probes and fluorescence depolarization techniques (Nagara 1988). Nagora placed a tryptophan group perpendicular to the peptide chain and parallel to the membrane plane.

Changes in membrane microviscosity under the actions of physiologically active compounds have been reported in numerous works; these changes were followed through spin and fluorescence labeling techniques. Cholesterol and steroids decreased the mobility of probes in the membranes. Conversely, local anesthetics diluted the membranes. The anticonvulsive preparation phenytoin at concentrations up to $80\,\mu$M induced a decrease in the degree of polarization of anthranyloxystearic acid, thus diluting the membrane. The latter, in turn, caused changes in the enzyme activity and conductivity of the membrane canals. Spin and fluorescence labeling methods recorded decreases in the microviscosity of phospholipid membranes under the actions of peptide transmitters, the thrombocyte activating factor, neuroaminidase, and N-acyldimethylamines (Kogtev *et al.* 1989; Ohyashiki *et al.* 1989). In the latter case, the mobility of doxyl stearic spin labels in synaptosome membranes from rat brain was found to be correlated with the biological activity of the amines.

A detailed investigation was performed to investigate the effects of pesticides and antibiotics on the kinetics and thermodynamics of hydrophobic probes introduced into erythrocyte membranes and the effects on the rate of probe reduction by ascorbate (Gendel & Krugliakova 1986). The pesticides linurone, phthalophos, chlorophos, ronnel, and dilor, at concentrations of 10^{-3}–10^{-2} M, affected a parameter of hydrophobicity of the probes. Nevertheless, the character of the influence depended on the structures of both the probe and the pesticide. Thus, the value of the solubilization parameter L increased upon introduction of phthalophos and ronnel and decreased in the case of chlorophos. The antioxidants ionol andd chlorohydrate 2-ethyl-6-methyl-3-oxypyridine, when embedded into the nuclear membranes of rat liver, caused an increase in the mobility of the spin probe in the hydrophobic portion of the membrane, but a decrease in the hydrophilic portion. Accumulation of the oxidation products was accompanied by enhancement of membrane rigidity. The latter result was confirmed with the help of the fluorescent probes pyrene and ANS (Dobretsov 1989).

The availability and high sensitivity of luminescence techniques led to wide use of fluorescent probes in studies of the molecular mechanisms of action of anesthetics, hormones, drugs, pesticides, etc. (Yeh *et al.* 1975; Waggoner 1986; Bloom & Webb 1983; Bergelson 1985; Bratton *et al.* 1988; Dobretsov 1989; Pringle & Hidalgo 1982).

Spin and fluorescence labeling methods have proved to be efficient tools for studying the molecular mechanisms of pathological processes.

The consequences of dyslipoproteinemia and ischemic heart disease (IHD) for the structure and microdynamics of blood lipoproteins of high density (HDLP), low density (LDLP), and very low density (VLDLP) have been studied (Ruuge & Gerasimova 1986). Lipoprotein lipolysis was shown to be accompanied by transfer of lipid fragments of VLDLP to HDLP particles. Changes in cholesterol concentrations and the occurrence of IHD led to a change in ability of HDLP to bind spin-labeled andorstane. Erythrocyte and thrombocyte membranes of patients with arteriosclerosis were characterized by high rigidity resulting from high concentrations of cholesterol. VLDLP preparations isolated from sick and healthy subjects affected the parameters of the ESR spectra of spin probes in the erythrocyte membrane differently. The parameters of fluorescent probes were found to be sensitive to the microstructure of HDLP as well (Vauhkonen & Somerharju 1989).

Parallel changes in microviscosity were observed for both hydrophilic and hydrophobic portions of mouse nuclear membranes monitored by spin-labeling techniques during the occurrence of Ehrlich carcinoma and hepatoma 27 (Burlakova & Goloshchapov 1986). The microviscosity of their membranes was markedly less sensitive to temperature and concentration of antibiotics than was that for membranes isolated from healthy animals. The degrees of fluorescence polarization of the probes were found to differ in membranes isolated from tumor cells and those from normal cells (Inbar *et al.* 1974; Dobretsov 1989).

In conclusion, it should be noted that the biophysical labeling method exhibits practically unlimited possibilities for investigations of pathological states of bioobjects and medical diagnostics.

8.6. Biological analyses

Some of the most promising applications of spin and fluorescence labeling appear to be for analysis of biologically active compounds, biopolymers, and enzymes, on the basis of the physical phenomena described in Chapters 1–4. Changes in rotational diffusion parameters and in the dipole–dipole electrostatic, spin–spin dipole, or exchange interactions under contact between physical probes and the compound being tested allow one to work out an appropriate analytical procedure in each particular case.

8.6.1. Biologically active ions and compounds

Nitroxides and fluorescent reagents sensitive to proton concentrations were described in Sections 1.6 and 3.5.1. A number of procedures for quantitative assays of Ca^{2+}, Ni^{2+}, Cu^{2+}, and Co^{2+} ions have been developed with the use of spin-labeled chelating reagents (Nagy *et al.* 1981, 1990).

The compound

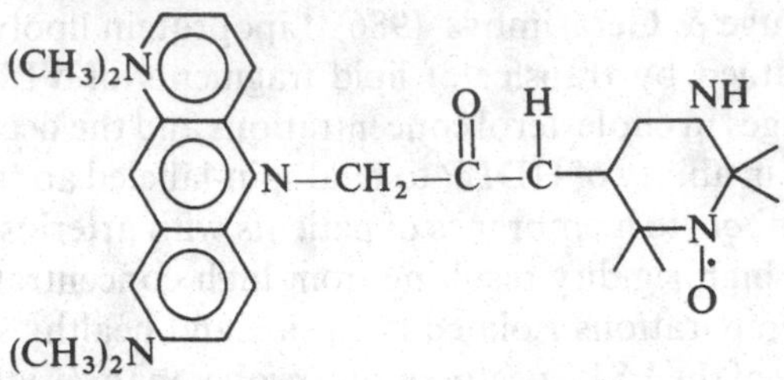

$$(CH_3)_2N \quad \overset{O \quad H}{\underset{||}{N-CH_2-C-C}} \overset{NH}{\cdots} \overset{|}{\underset{N}{\cdots}} \quad (CH_3)_2N$$

for example, has served as a chelating reagent for Zn^{2+}, Al^{3+}, Mo^{6+}, and Ga^{2+}. Upon chelation, the chromophore and nitroxide fragments move toward each other, thus drastically decreasing the intensity of fluorescence. This technique allows the detection of ion concentrations as low as 10^{-9} M.

The Na^+ ion concentration in cells was determined using the fluorescent probe rhodamine β-sulfonylphosphatidylethanolamine (Roe *et al.* 1989). Cl^- ions reacted quantitatively with a specific reagent, 6-methoxy-*N*-(3-sulfopropylquinolium) (Chao *et al.* 1989).

The ability of chromophores (propidium, Hoechst $+342$, acridine orange, etc.) to change the intensity of fluorescence upon binding with nucleic acids has been widely used in analytical biochemistry (Shapiro & Stephens 1986; Fogelsong 1989). Acridine orange bound with single-stranded DNA gave green fluorescence, whereas binding with RNA was characterized by red fluorescence. To estimate the concentration of guanine groups in defects of double-stranded DNA, it was proposed to measure the fluorescence intensity upon complexing these groups with Tb^{3+} ions. The proposed analytical technique was based on monitoring the fluorescence during enzymatic hydrolysis of DNA in the presence of topoisomerase, endonuclease, and the fluorescent probe (Fogelsong 1989). Qualitative analysis of oligoadenine fragments in RNA up to 1 mole was accomplished by tracing their interaction with spin-labeled polyuridine (Bobst 1979).

To estimate the dioxygen concentration in clumped cells, the spin-labeling techniques (Section 2.4.5), phosphorescence quenching and delayed fluorescence were used (Van der Kooi *et al.* 1990). In aqueous solution, the O_2 concentration ranged from 0.2 to 0.8 μM, and it was determined that the k_q [O_2] product equals 10^2–$10^4 s^{-1}$ in systems of unknown viscosity.

An analytical procedure has been developed based on substitution of the spin label

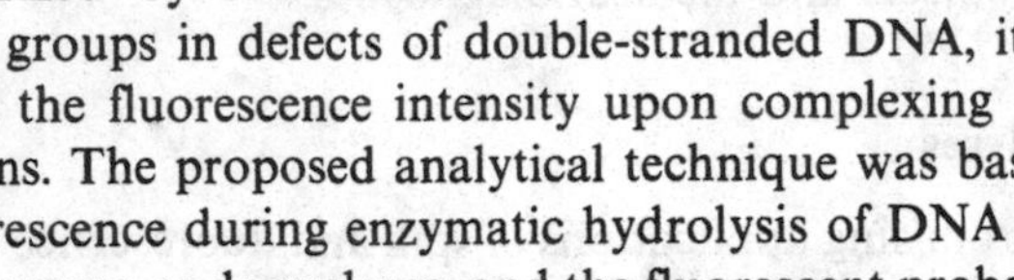

$$H_2N-SO_2 \overset{H}{\underset{\underset{O}{||}}{C-N}} \overset{\cdots}{N-O}$$

for sulfamides binding to the active center of anhydrase (Chignell 1979). The substitution was accompanied by narrowing and increased intensity in the spin probe ESR signal. To determine the concentration of bilirubin, this reagent was used to replace a nitroxide derivative of succinic acid from human serum albumin (Hsia *et al.* 1978).

A number of analytical procedures have been proposed for determination of enzymatic activity with the use of spin- and fluorescence-labeled substrates and inhibitors.

The amount of anhydrase in solution was measured by monitoring the typical binding for the nitroxide inhibitor (Chignell 1979).

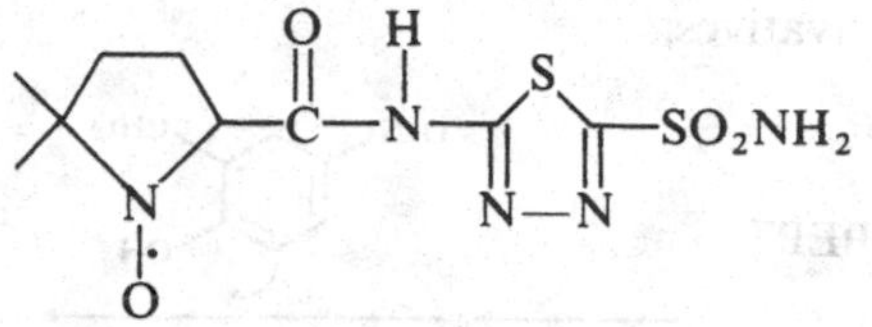

The mobilization of the nitroxide fragment lysozyme substrate (a spin-labeled derivative of peptide glycane) may result from the hydrolysis of 1,4-glucoside linkages and may be used in the procedure of enzyme analysis (Johnson & Neuhaus 1978). The compound 1-palmitoyl-2-(4-doxylpenta-noyl)glycerophospholine was split in micelles, followed by phospholipase assay. The decrease in the intensity of the ESR signal of octadecylnitro-phosphate due to the reaction of the nitroxide fragment with hydrogen peroxide and free radicals was used for determination of the lipoxidase of unsaturated fatty acids (Snipes *et al.* 1977).

Fluorescence techniques have been widely used in analytical assays of enzymes, mainly proteases. Most of the procedures are based on a drastic increase in the fluorescence intensity after hydrolysis of substrates modified by appropriate chromophores. Esterases, glycosidases, sulfatases, phosphatases, and phosphodiesterases can be analyzed by such procedures. Some of the reactions are exemplified as follows (Waggoner 1986):

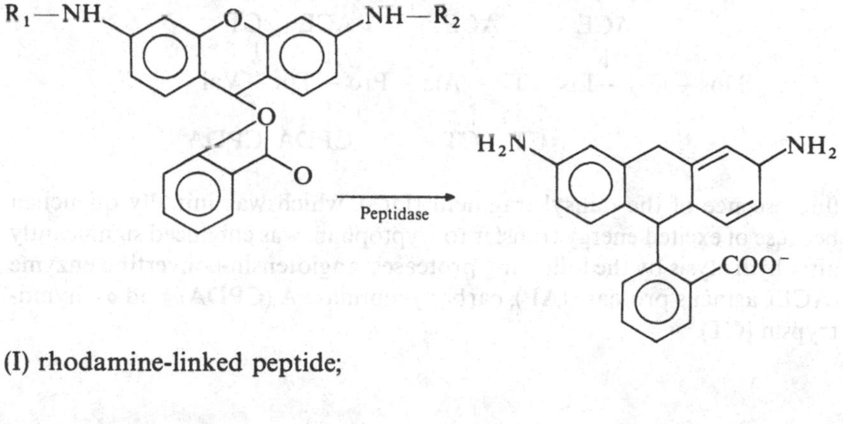

(I) rhodamine-linked peptide;

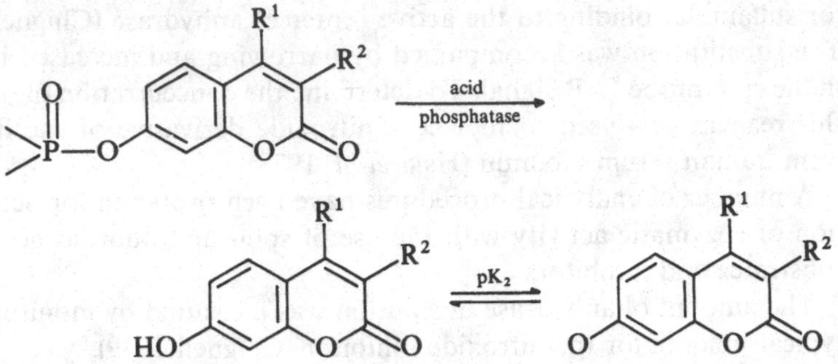

(II) coumarin phosphate derivatives;

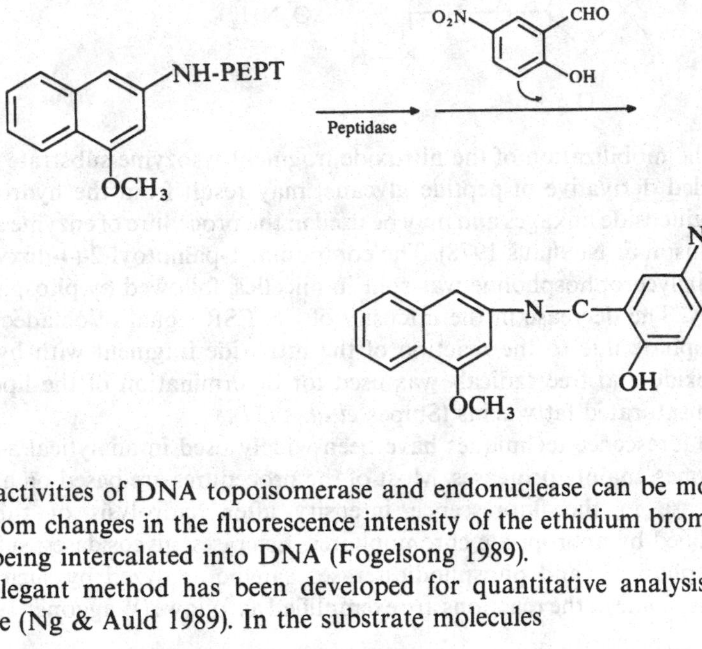

The activities of DNA topoisomerase and endonuclease can be monitored from changes in the fluorescence intensity of the ethidium bromide probe being intercalated into DNA (Fogelsong 1989).

An elegant method has been developed for quantitative analysis of protease (Ng & Auld 1989). In the substrate molecules

$$
\begin{array}{ccccc}
\text{ACE} & \text{ACE} & \text{ACE} & \text{CT} \\
\downarrow & \downarrow & \downarrow & \downarrow \\
\text{Dns} - \text{Gly} - \text{Lis} - \text{Tir} - \text{Ala} - \text{Pro} - \text{Tpr} - \text{Val} \\
\uparrow \quad \uparrow & & \uparrow \quad \uparrow \\
\text{CT} \quad \text{CT} & & \text{CPDA} \quad \text{CPDA}
\end{array}
$$

fluorescence of the dansyl fragment (Dns), which was initially quenched because of excited energy transfer to tryptophan, was enhanced significantly after hydrolysis by the following proteases: angiotensin-converting enzyme (ACE), astacus protease (AP), carboxypeptidase A (CPDA), and α-chymotrypsin (CT).

8.6.2 Immunological assays

Nowadays, immunological assays with application of fluorescent and spin labels and probes are widely used in medical, veterinary, and biological research (e.g., Wood 1984). Four main trends may be pointed out in this field: (1) covalent labeling of antibiotics and antigens with the aim of detecting these compounds after separation; (2) substitution of labeled haptens from the antibody active centers for natural haptens; (3) monitoring of the changes in the interactions between fluorescent or spin probes after destroying liposome sacks loaded by the probes upon complexation between haptens and antibodies; (4) recording the changes at complexation in the parameters of electrostatic or spin–spin interactions of labels on haptens and antibodies.

Next, some versions of immunofluorescence and immunospin assays based on the physical principles considered in Chapters 1–4 will be briefly reviewed. Data on applications of fluorescence techniques to visualization of labeled antibodies can be found elsewhere (Brigati *et al.* 1983; Waggoner 1986; Dobretsov 1989).

In such analysis, together with classical fluorescent chromophores, complexes of Eu^{3+} and Tb^{3+} ions with β-diketones have been used (Dakuba & Ekins 1985).

Complexation between a fluorescent substrate and antibodies can be followed using polarization techniques (Section 3.3). The phenomenon of inductive resonance energy transfer in donor–acceptor pairs (e.g., fluorescein–rhodamine, fluorescein–fikoerythrine) was used to monitor complexation of antigens and antibodies in solution without preliminary separation of the reagents (Ullman *et al.* 1976; Clarke *et al.* 1982; Kronick & Grossman 1983).

Free-radical assay techniques (FRAT) offer a number of advantages, namely, direct assays without separation of reagents, high speed of assay, relative simplicity, and absence of radioactivity. These advantages allow us to hope that the method will be widely introduced into the practice of clinical laboratories.

FRAT implies that a biologically active hapten competes with its spin-labeled analogue (SLA) for the active site of an antibody (Figure 8.4) (Leute *et al.* 1972; Montgomery *et al.* 1975; Wei & Almires 1975; Ashirov *et al.* 1980). At the initial stage of the procedure, an antibody-SLV complex with one immobilized nitroxide fragment is obtained. In the presence of a "true" hapten with a significantly higher affinity for the antibody, the substitution is accompanied by a drastic increase in the SLA ESR signal intensity due to the molilization of the label. After an appropriate calibration, the hapten concentration is determined by ESR techniques. This procedure was successfully used to assay morphine, cocaine, diphenyl-

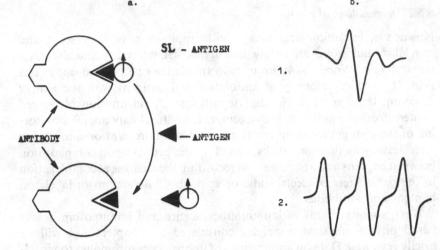

Figure 8.4. Schematic representation of spin immunoassays. (a) Scheme of the assay. (b) ESR spectra of spin label (SL) before (1) and after (2) addition of an antigen.

hydantoin, catecholamine, and hormones (progesterone, cortisone, prednisolone, etc.) in concentrations down to $3 \cdot 10^{-6}$ M. Formulas for some of the spin-labeled haptens are as follows:

The sensitivity of FRAT is limited to that of the ESR method. To extend the assay sensitivity, a new technique has been developed based on loading of the liposome "sacks" by highly concentrated hydrophobic nitroxide radical (e.g., spin-labeled acetylcholine) (Figure 8.5) (Chan *et al.* 1978). Because of the strong spin exchange interactions, the lines of the nitroxide ESR spectrum are broadened, and the resulting spectral intensity is very low. Then the surface of such a spin "sack" is modified with a lipid derivative of the hapten under analysis, followed by interaction with appropriate antibody molecules. The preparation thus obtained is extremely sensitive to the lytic effect of a complement or cardiotoxin. The process of lysis can be monitored by measuring the intensity enhancement of the ESR spectrum of the nitroxide that escapes from the destroyed spin

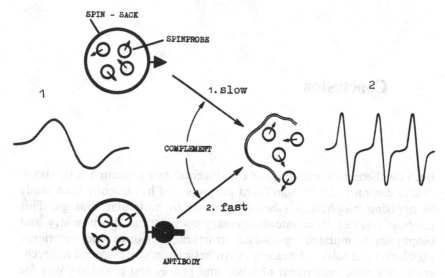

Figure 8.5. Schematic representation of "spin-sack" immunological assay; 1 and 2 are ESR spectra before and after the assay, respectively.

sacks. In the presence of outer haptens that capture antibodies, the spin sacks appear to be stable and the lysis is inhibited. After calibration, this procedure allows determination of concentrations of the compound under study down to 10^{-8} M.

At present, procedures have been developed to assay thyroxine, dinitrophenol, morphine, extol, and other compounds (Rosenquist & Vistnes 1977; Chan *et al.* 1978, Hsia & Piette 1969).

Such sacks are usually formed from mixtures of lipids (sphingomyelin, cholesterol, diacetylphosphate, etc.); ε-dinitrophenylaminocaprylethanolamine and the corresponding derivative of thyroxine are examples of the lipid hapten. In some cases, antibodies can induce liposome lysis. For instance, liposomes from cholesterol, lecithin, and cardiolipin are destroyed by antibodies against *Treponema pallidum* and therefore can be used in diagnosing syphilis.

In conclusion, the materials presented here show that the physical labeling method is an effective approach to the study of relatively simple, isolated biological objects, as well as very complicated structures, such as nucleic acids, glycoproteins, cells and their components, organs, and even organisms. The range of label applications to biological assays and medical diagnostics appears to be unlimited.

Conclusion

Both the theoretical material and experimental data presented in this book clearly demonstrate the significant progress that has recently been made in applying biophysical labeling methods to molecular biology. This progress resulted from interdisciplinary cooperation. Biochemistry and biophysics formulated up-to-date structural, dynamic, and functional problems to be solved. Advances in synthetic chemistry provided researchers with a wide assortment of labels and probes and paved the way for specific modifications of certain portions of biological objects.

Let us summarize the main advances, possibilities, and limitations of biophysical labeling methods.

Molecular dynamics of biosystems in the areas of spin, fluorescent, triplet, and Mössbauer labels can be investigated within a wide range of correlation times ($\tau_c = 10^2-10^{-10}$ s) and amplitudes. The values of the electron spin relaxation parameters, $1/T_{1e}$ and $1/T_{2e}$, of spin labels and the probability of recoilless absorption of γ-quanta (f') reflect the intensity of the high-frequency low-amplitude phonon processes. The parameters of Mössbauer spectra (the line width), ESR spectra (the line form, saturation transfer), and luminescence spectra (the relaxation shift, depolarization) are dependent on the processes of vibration, wobbling, and rotational diffusion of the label, as well as orientational relaxation of surrounding dipoles. The same parameters can be sensitive to such local properties as polarity, electric potential, ability to form the hydrogen bond and participate in redox reactions.

Analysis of spin–spin and exchange interactions between spin labels allows estimations of distances between the labels up to 6.0 nm and the depth of the radical immersion up to 4.0 nm. Measurements of energy transfer involving the excited triplet and singlet states of chromophores make it possible to estimate distances between the centers up to 6 nm and the depth of immersion up to 4.0 nm. Measurements of the parameters of energy transfer between the excited states of chromophores make it possible to determine their separation distances up to 1.6 nm in the case of the exchange mechanism and up to 6.0 nm for the inductive resonance mechanism. These ranges may be effectively widened by electron and

X-ray scattering by clusters of heavy metals (e.g., polymercarbide), with these effects being monitored by the methods of electron microscopy and low-angle scattering. Information on the surfaces of bioobjects can be obtained by a new method of total tritium labeling.

Translational diffusion of luminescent molecules with $D_{tr} = 10^{-6}$–10^{-10} cm^2 s^{-1} can be monitored by the FRAP technique, by fluorescence and phosphorescence quenching, and by determination of the luminescence parameters of excimers and exciplexes. The study of dynamic spin–spin interactions enables one to estimate the frequency of collision when $\tau_{ch} = 10^{-6}$–10^{-10} s. The latter range can be expanded up to $\tau_{ch} = 1$–10 s with the aid of the new approach of triplet probe–photochrome labeling.

The aforementioned methods have been successfully used for solving a number of problems in molecular biophysics involving proteins and enzymes. First of all, methods of specific modification for practically all functional protein groups have been developed. Nitroxide analogues of substrates, inhibitors, and coenzymes have been synthesized as well. More moderate progress has been made in the modification of bioobjects by luminescent, electron-scattering, and Mössbauer labels.

One of the first achievements of the biophysical labeling method was experimental monitoring of various types of conformational transitions: large-scale denaturation, transglobular and local effects. The depth of immersion and the relief of the enzymes' active centers were studied with the help of spin labels of various lengths and flexibility.

The method of labeling has contributed decisively to solving the problems of protein molecular dynamics. The parameters of rotational diffusion of whole protein globules in solution have been determined by luminescence and spin-labeling methods. The local segmental mobilities of protein globules within very wide ranges of temperature, humidity, and correlation time ($\tau_{ch} = 10$–10^{-10} s) have been measured by spin, fluorescence, triplet, and Mössbauer labeling. The results have been generalized as an "amorphous-crystalline" model of protein globule dynamics. The model suggests a hinge-like mobility of rigid blocks embedded in an amorphous medium that gradually softens as temperature and humidity increase.

Experiments have indicated a number of correlations between nanosecond protein dynamic parameters and biochemical reactions such as electron transfer, hydrolysis, and photoreactions. These experiments confirm the concept of dynamic adaptation of protein conformations to chemical processes in the active centers of enzymes and provide a theoretical basis for understanding the fine molecular mechanisms of protein complexation, recognition, and allosteric regulation.

The data on steric factors of heme proteins, local polarity, and electron conductivity of proteins obtained by spin and triplet labeling methods appear to be of significant interest for developing a quantitative theory of enzyme reactions such as electron transfer in proteins.

Experiments with fluorescence, triplet, and electron-scattering, and spin-labeling methods have succeeded in determining the mutual dispositions of catalytic and regulatory centers of enzymes. Thus, ensembles of ferredoxin-like clusters close to each other have been proved to exist in nitrogenase. NMR studies of spin-labeled enzymes enabled investigators to establish the locations of the substrate molecules with respect to functionally important groups of the enzymes' active centers.

Biological and model membranes are perhaps the main objects to which biophysical labeling methods have been applied. A whole collection of lipid derivatives of nitroxides, luminescent chromophores, and electron-scattering probes have been synthesized. The probes have been incorporated into various portions of biomembranes and monitored with suitable physical methods.

The microviscosities of various portions of biomembranes (a dynamic profile) have been estimated by measuring the rotational diffusion parameters of spin, fluorescent, triplet, and ^{2}D NMR probes. Under physiological conditions, typical values for the microviscosity of membranes' lipid layers were found to be several poise. The temperature dependences of the dynamic parameters of membranes over wide ranges of temperatures (77–330K) and correlation times ($\tau_c = 10$–10^{-10} s) have been measured, and correlations between certain values of τ_c and the kinetic parameters of a number of enzymatic processes (electron transfer, ATP hydrolysis, metal ion transport, hydroxylation, etc.) have been established. Numerous phase and prephase transitions in membranes caused by changes in temperature, humidity, composition, pH, salt content, and other factors have been recorded by fluorescence and spin-labeling methods.

One of the remarkable achievements of the spin-labeling method has been the discovery and detailed investigation of lateral diffusion and flip–flop transitions of spin-labeled lipids in biomembranes.

Qualitative estimations of the parameters of rotational diffusion of protein and lipid molecules modified by spin, fluorescent, and triplet probes have made it possible to study the processes of association of these compounds and to elucidate specific dynamic behaviors of lipids adjusted to the membrane proteins' surfaces. The method of electron-scattering labels has proved to be an effective approach in establishing the distribution of certain lipid and protein molecules in native and model membranes (microsomal, rhodopsin, erythrocyte membranes, vesicles, micelles).

Biological and model membranes modified by spin and fluorescent labels have served as adequate objects for studying the molecular mechanisms of physiological processes. With the help of these labels, the transmembrane potential, local pH, polarity, electrostatic charge density, and ability to participate in redox processes have been studied. The effects of physiologically active compounds (hormones, regulators of enzyme activity, cholesterol, drugs, etc.) on the physical properties of membranes have been found to be easily monitored.

Physical labeling methods have proved their effectiveness in studies of nucleic acids, polysaccharides, lipoproteins and glycoproteins, cells, and subcellular components, as well as intact organisms. In the latter cases, along with the usual information on specific components, new data can be obtained on the contributions of labeled preparations, including analogues of medicines, dioxygen, and water, to the cell organelles, and even to various parts of intact organisms. ESR tomography has been developed to study the latter systems.

A great variety of analytical assays of enzymes, proteins, antibiotics, nucleic acids, ions, etc., have been proposed on the basis of the labeling techniques.

Obviously the biophysical labeling methods have, along with their advantages, a number of limitations and drawbacks. An attentive and qualified reader will find in every particular case specific "reefs" that arise primarily from the danger of destroying the native form of a biological structure by introduction of the label. The information obtained with the aid of this method often is of an indirect nature, and the accuracy of determination of distances falls short of the accuracy provided X-ray analysis.

Nevertheless, the material cited here indicates that in some cases physical labels offer unique information on the structures and properties of biological systems, and in other cases the labeling technique is a useful and independent addition to the traditional biophysical and biochemical methods. It is our hope that biophysical labeling will continue to be an effective tool for solving exciting and complicated problems in molecular biology.

References

Abdallach, M. A., Andre, J. J., & Biellman, J. F. (1977). A new labelled analogue of nicotinamide adenine dinucleotide. *Bioorganic Chemistry*, 6, 157–163.

Abragam, A. (1961). *The Principles of Nuclear Magnetism*. London: Oxford University Press.

Abragam, A., & Bleaney, B. (1970) *Electron Paramagnetic Resonance of Transition Ions*. Oxford: Clarendon Press.

Afanasiev, A. M., Asch, L., & Kalvius, G. M. (1978). Frequency dependence of the spin–spin relaxation rate from Mössbauer spectroscopy. *Physical Review Letters*, 40, 816–819.

Ahl, P. L., & Cone, R. A. (1982). Photodichroism and the rotational motions of rhodopsin and bacteriorhodopsin. *Methods in Enzymology*, 88, 741–750.

Alakhverdiev, S., Kulikov, A. V., Bogatyrenko, V. R., Klimov, V. V., & Likhtenshtein, G. I. (1991). Study of the structure of the photosystem II reaction centre by continuous saturation technique. *Biofizika* (in Russian).

Alexander, S., Entin-Wohlman, O., & Orbach, R. (1986). Relaxation and nonradiative decay in disordered systems. II: Two fractions inelastic scattering. *Physical Review*, 33, 3935–3946.

Alfimova, E. Ya., & Likhtenshtein, G. L. (1979). Investigation of structure and conformational changes of proteins by the inductive resonance energy transfer method. *Advances in Molecular Relaxation and Interaction Processes*, 14, 47–89.

Alterman, M. A., Marakushev, S. A., Levchenko, I. A., Raevsky, A. V., Devichinsky, V. M., Bachmanova, G. I., Likhtenshtein, G. I., & Archakov, A. I. (1981). Study of the distribution of cytochrome P-450 in rat microsomes. *Biokhimiya*, 46, 222–229 (in Russian).

Anderson, K. W., Hochstrasser, R. M., Lutz, H., & Scott, C. W. (1974). Direct measurements of energy transfer between triplet states of molecules in liquids using picosecond pulses. *Journal of Chemical Physics*, 61, 2500–2506.

Anglister, J. (1989). Magnetic resonance study of the combining site structure of a monoclonal antispin-label antibody. In *Spin Labeling. Theory and Applications*, L. J. Berliner & J. Reuben (eds.), pp. 597–618. New York: Plenum Press.

Antsiferova, L. I. (1973). Theory of spin relaxation and EPR lineshape for inhomogeneous viscous media. *Zhurnal Fizicheskoi Khimii*, 157, 794–798 (in Russian).

Antsiferova, L. I., Belonogova, O. V., Kochetkov, B. B., & Likhtenshtein, G. I.

(1989a). Models for rotation spin labels and probes in proteins and membranes. *Izvestiya Akademii Nauk SSSR (Seria Biologicheskaya)*, No. 4, 494–501 (in Russian).

Antsiferova, L. I., Ivanova, A. N., & Lyubashevskaya, E. V. (1984). Calculation of 2-mm band EPR spectra for nitroxide radicals undergoing anisotropic rotations. *Khimicheskaya Fizika*, 3, 1476–1477 (in Russian).

Antsiferova, L. I., Kostina, N. V., Marupov, R. M., & Likhtenshtein, G. I. (1989b). Molecular mobility in spin-labeled cotton fibres. *Vysokomoleculyarnie soedyneniya (Highpolymers)*, 31, 539–545 (in Russian).

Antsiferova, L. I., & Lyubashevskaya, E. V. (1986). *Atlas of EPR Spectra of Nitroxide Radicals in the 2-mm Range*. Chernogolovka: Nauka (in Russian).

Antsiferova, L. I., & Lyubashevskaya, E. V. (1990). *Multifrequency EPR Spectroscopy of Nitroxide Radicals (Computer-based Experiments)*. Chernogolovka: Nauka.

Antsiferova, L. I., & Valova, E. V. (1992). Multifrequency 2ESR spectroscopy of nitroxide radicals. *Zhurnal Fizicheskoi Khimii* (in Russian).

Aplin, J. D., Bernstein, M. A., & Culling, C. F. A. (1976). Selective activation and chemical derivatization of the sialic acid and galactose components of glycoproteins. *Carbohydrate Researches*, 70, C9–C12.

Aragon, S. R., & Record, R. (1975). Fluorescence correlation spectroscopy and Brownian rotational diffusion. *Biopolymers*, 14, 119–138.

Aripov, T. F., Salakhutdinov, B. A., Salikhova, Z. T., & Ladykov, A. S. (1984). Structural changes of liposome phospholipid packing induced by cytotoxin of the Central Asian cobra venom. *General and Physiological Biophysics*, 3, 489–496.

Artukh, R. I., Katshalova, G. S., & Timofeev, V. P. (1990). Conformational transition in lamprey hemoglobin studied by spin-labeling methods. In *Magnetic Resonance in Biology and Medicine*. Abstracts of conference, Zvenigorod, USSR, May 1990, p. 116.

Arutyunyan, A. E., Timofeev, V. P., Dikenene, N. P., Akunite, O. U., & Lapuk, V. A. (1989). Segmental flexibility in F_c-region of spin-labeled Ig molecules of two subclasses from calves' blood serum and local conformational mobility of their carbohydrate components. *Biofizika*, 34, 946–948 (in Russian).

Ashirov, P. M., Likhtenshtein, G. I., & Smotrov, S. P. (1980). Assay of catecholamines in biological liquids. Patent SSSR N 789752.

Asakura, T., Leigh, J. S., Drott, H. R., Yonetani, T., & Chance, B. (1971). Structural measurement of hemoproteins: use of spin-labelled protoheme as a probe of heme environment. *Proceedings of the National Academy of Sciences, USA*, 68, 861–865.

Asmus, K. D., Nigam, S., & Willson, R. L. (1976). Kinetics of nitroxyl radical reactions. A pulse-radiolysis conductivity study. *International Journal of Radiation Biology & Related Studies in Physics, Chemistry, and Medicine*, 29, 211–219.

Atanasov, B. P., Postnikova, G. B., Sadykov, Yu. Ch., & Volkenstein, M. V. (1977). Study of electron transfer in haemoproteins. II: pH dependence of the reduction rate of ferricytochrome c by oxymyoglobin. *Molekulyarnaya Biologiya*, 11, 537–544 (in Russian).

Aurich, H. G. (1990). The chemistry of vinyl nitroxides. *Pure and Applied Chemistry*, 62, 183–188.

Azzi, A., Bragadin, M. A., Tamburro, A. M., & Santato, M. (1973). Site-directed spin labeling of the mitochondrial membrane. Synthesis and utilization of the adenosine triphosphatase inhibitor (*N*-(2,2,6,6-tetramethyl-piperidyl-1-oxyl)-*N'*-(cyclohexyl)-carbodiimide). *Journal of Biological Chemistry*, 248, 5520–5526.

Azizova, O. A., & Ritov, V. B. (1986). Study of structural-functional relationships of the sarcoplasmic reticulum Ca^{2+} ATPase. In *Methods of Spin Labels and Spin Probes*, M. H. Emanuel & R. I. Zhdanov (eds.), pp. 105–163. Moscow: Nauka (in Russian).

Azizova, O. A., Torkhovskaya, T. I., & Lopukhin, Yu. M. (1986). Structural-functional investigations of blood cell membranes at atherosclerosis. In *Methods of Spin Labels and Spin Probes*, N. M. Amanuel & R. I. Zhdanov (eds.), pp. 239–250. Moscow: Nauka (in Russian).

Backer, J. M., Budker, V. A., Eremenko, I., & Molin, Yu. N. (1977). Detection of the kinetics of biochemical reactions with oxygen using exchange broading in the ESR spectra of nitroxide radicals. *Biochimica et Biophysica Acta*, 460, 152–156.

Bakhshiev, N. A. (1972). *Spectroscopy of Intermolecular Interactions*. Leningrad: Nauka.

Bakhshiev, N. A. (ed.) (1989). *Solvatochromism. Problems and Methods*. Leningrad: Leningrad University Press.

Balcarova, Z., & Brabec, V. (1989). Reinterpretation of fluorescence of terbium ion–DNA complexes. *Biophysical Journal*, 33, 55–61.

Baldassare, J. J., Charach, S., Jones, R. T., & No, C. (1970). Study of conformational transitions in hemoglobin by the spin-labeling methods. *Biochemistry*, 9, 111–119.

Balgavy, P. (1988). Spin label study of the interaction of subfractions with model and biological membranes. In *Sixth CMEA Symposium on EDR Spectroscopy in Biochemistry, Molecular Biology, and Medicine*. Abstracts of conference, Smolenice Castle, Czechoslovakia, November 21–5, 1988, p. 7.

Bammel, B. P., Hamilton, D. D., Haygland, R. P., Hopkins, H. P., Schuette, J., Szalecki, W., & Smith, J. C. (1990). NMR, calorimetric, spin-label, and optical studies of a trifluoromethyl-substituted styryl molecular probe in DMPC vesicles and multilamellar suspensions; a model for location of optical probes. *Biochimica et Biophysica Acta*, 1024, 61–81.

Baroin, A., Bienvenue, A., & Devaux, P. F. (1979). Spin-label studies of protein–protein interaction in retinal rod outer segment membranes: saturation transfer ESR spectroscopy. *Biochemistry*, 18, 1151–1156.

Bass, D. A., Parce, J. W., Dechatelet, L. R., Szejda, P., Seeds, M. C., & Thomas, M. (1983). Flow cytometric studies of oxidative formation by neutrophils: a graded respose to membrane stimulation. *Journal of Immunology*, 130, 1910–1917.

Bauminger, E. R., Cohen, S. G., Nowik, I., Offer, S., & Yariv, J. (1983). Dynamics of heme iron in crystals of metmyoglobin and deoxymyoglobin. *Proceedings of the National Academy of Sciences, USA*, 80, 736–740.

Beddard, G. S. (1986). Picosecond laser absorption and emission techniques for fast rotation and lateral transport. *Biochemical Society Transactions*, 5, 827–832.

Beer, M., & Moudrianakis, E. N. (1962). Determination of base sequence in nucleic acid with the electron microscope: visibility of a marker. *Proceedings of the National Academy of Sciences, USA*, 48, 409–416.

Beer, M., Wiggins, J. W., Alexander, R., & Stoekort, C. J. (1978). Biological determination through atomic spectroscopy. *Chemica Scripta*, 14, 263–266.

Belkin, S., Mehlhorn, R. J., Hideg, K., Hankovsky, O., & Packer, L. (1987). Reduction and destruction rates of nitroxide spin probes. *Archives of Biochemistry and Biophysics*, 256, 232–243.

Belonogova, O. V., Frolov, E. N., Krasnopol'skaya, S. A., Atanasov, B. B., Gins, V. K., Mukhin, E. N., Levina, A. A., Andreeva, A. B., Likhtenshtein, G. L., & Goldansky, V. L. (1978). Study of the effect of the degree of hydration on the mobility of Mössbauer atoms at the active sites of metalloenzymes and metal carriers. *Doklady Akademii Nauk SSSR*, 241, 219–222 (in Russian).

Belonogova, O. V., Frolov, E. N., Illustrov, N. V., & Likhtenshtein, G. I. (1979). Intermolecular dynamics of proteins studied by spin-labeling method. *Molekulyarnaya Biologiya*, 13, 567–575 (in Russian).

Belonogova, O. V., & Likhtenshtein, G. I. (1992). Photoelectron transfer and dynamics in spin labeled α-chymotrypsin. *Biofizika* (in Russian), in press.

Belonogova, O. V., Likhtenshtein, G. I., Levashov, A. V., Khmel'nitskii, Yu. L., Klyachko, N. L., & Martinek, K. (1983). Use of the spin label method to study the state of the active site and microsurroundings of α-chymotrypsin, solubilized in octane, using surfactant aerosol T. *Biokhimiya*, 48, 379–386 (in Russian).

Bendersky, V. A., Blumenfeld, L. A., Stunzhas, P. A., & Sokolov, E. A. (1968). Double electron–electron resonance of triplet excitons in ion-radical salts. *Nature*, 220, 365–367.

Bennick, A., Campbell, J. D., Dwek, R. A., Price, N. C., Rudda, G. K., & Salmon, A. D. (1971). NMR investigation of spin-labeled phosphorylase b. *Nature (New Biology)*, 234, 140–147.

Berdnikov, V. M., & Bogdanchikov, G. A. (1979). On the calculation of the probability of the outersphere electron transfer in dilute aqueous solutions. *Zhurnal Fizicheskoi Khimii*, 53, 273–280 (in Russian).

Berdnikov, V. M., & Makarshin, L. L. (1982). Spin exchange between stable nitroxide radical and ions Fe^{3+}, Mn^{2+} and Gd^{3+} in aqueous solution. *Khimicheskaya Fizika*, 1, 1226–1232 (in Russian).

Berg, A. I., Kononenko, A. F., Noks, P. P., Khrymova, I. N., Frolov, E. N., Rubin, A. B., Likhtenshtein, G. I., Uspenskaya, N., & Khideg, K. (1979a). The conformational mobility and functional dependence of photosynthetic reaction centers from *Rhodospirillum rubrum*. *Molekulyarnaya Biologiya*, 13, 469–477 (in Russian).

Berg, A. I., Noks, P. P., Kononenko, A. A., Frolov, E. N., Khrymova, I. N., Rubin, A. B., Likhtenshtein, G. I., Goldansky, V. I., Parak, F., Bukl, M., & Mössbauer, R. L. (1979b). Conformational control of functional activity in photosynthetic membranes of purple bacteria. *Molekulyarnaya Biologiya*, 13, 81–89 (in Russian).

Bergelson, L. D. (1985). A novel approach to the mechanism of biological signal amplification. *Biologicheskie Membrany*, 2, 483–486 (in Russian).

Berliner, L. (ed.) (1976). *Spin Labeling. Theory and Applications*, Vol. 1. New York: Academic Press.

Berliner, L. (ed.) (1979). *Spin Labeling. Theory and Applications*. Vol. 2. New York: Academic Press.

Berliner, L. J. (1990). The use of spin labels in looking at subtle conformational changes in blood coagulation proteins. *Pure and Applied Chemistry*, 62, 247–254.

Berliner, L. J., & Fujii, H. (1985). Magnetic resonance imaging of biological specimens by electron paramagnetic resonance of nitroxide spin labels. *Science*, 227, 517–519.

Berliner, L. J., & McConnell, H. M. (1966). A spin-labeled substrate for α-chymotrypsin. *Proceedings of the National Academy of Sciences, USA*, 55, 708–712.

Bernardo, M., & Thomann, H. (1990). ELDOR-ENDOR spectrometry: pulsed electron nuclear and electron triple resonance. In *Thirty-second Rocky Mountain Conference*, Denver, Colorado. Abstract 141.

Bertini, I., & Luchinat, C. (1986). *Physical Bioinorganic Chemistry Series. Vol. 3: NMR of Paramagnetic Molecules in Biological Systems*. Menlo Park, Calif.: Benjamin/Cummings.

Best, L., John, E., & Jahnig, F. (1987). Order and fluidity of lipid membranes, as determined by fluorescence anisotropy decay. *European Biophysical Journal*, 15, 87–102.

Beth, A. H., & Robinson, B. H. (1989). Nitrogen-15 and deuterium substituted spin labels for studies of very slow rotational motion. In *Biological Magnetic Resonance. Vol. 8: Spin Labeling. Theory and Applications*, L. J. Berliner & J. Reuben (eds.), pp. 179–254. New York: Plenum Press.

Binder, H., & Dittes, K. (1987). Investigation on the pyrene fluorescence in phospholipid membranes. *Studia Biophysica*, 120, 59–71.

Blackwell, M. F., Gounaris, K., Zara, S. J., & Barber, J. (1987). A method for estimating lateral diffusion coefficients in membranes from steady-state fluorescence quenching studies. *Biophysical Journal*, 51, 735–744.

Blatt, E., & Sawyer, W. H. (1985). Depth-dependent fluorescent quenching in micelles and membranes. *Biochimica et Biophysica Acta*, 822, 43–62.

Bloom, J. A., & Webb, W. W. (1983). Lipid diffusibility in the intact erythrocyte membrane. *Biophysical Journal*, 42, 295–303.

Blume, M. (1967). Temperature-dependent spin–spin relaxation times: application to the Mössbauer spectra of ferric hemin. *Physical Review Letters*, 18, 305–308.

Bobodzhanov, P. Kh., & Likhtenshtein, G. I. (1974). Synthesis of spin labeled fibres of cotton, silk and wools by use of the nitroxide derivative of trichlortriazin. *Doklady Akademii Nauk Tadjiksroi SSR*, 17, 34–37 (in Russian).

Bobst, A. M. (1979). Application of spin labeling to nucleic acids. In *Spin Labeling. Theory and Applications*, Vol. 2, L. J. Berliner (ed.), pp. 291–345. New York: Academic Press.

Bobst, A. M., Pauly, G. T., Keyes, R. S., & Bobst, E. V. (1988). Enzymatic sequence-specific spin labeling of a DNA fragment containing the recognition sequence of EcoRI endonuclease. *FEBS Letters*, 228, 33–36.

Boldyrev, A. A. (1988). Causative factors of the nonlinearity of the Arrhenius plots for Na,K-ATPase. I: Role of the membrane lipids. In *Biomacromolecules in Methods of Spin Labels and Spin Probes*, G. I. Likhtenshtein & R. I. Zhdanov (eds.), pp. 80–92. Moscow: Nauka.

Bolin, J. T., Ronco, A. E., Mortenson, L. E., Morgan, T. V., Williamson, M., & Suong, Ng L. (1990). The structure of nitrogenase MoFe-protein: spatial distribution of intrinsic metal atoms determed by x-ray anomalous scattering. In *Nitrogen Fixation: Achievements and Objectives*, P. M. Gresshoff (ed.), p. L-10. New York: Chapman & Hall:

Bowman, M. K., & Michalski, T. J. (1990). Fourier-transform EPR and low-frequency EPR studies of nitroxides. *Pure and Applied Chemistry*, 62, 271–274.

Bradbury, E. M., & Nicolini, C. (eds.) (1986). *NMR in the Life Sciences*. New York: Plenum Press.

Bratton, D. L., Harris, R. A., Clay, K. E., & Henson, P. M. (1988). Effect of platelet activation factor on calcium–lipid interaction and lateral phase separations in phospholipid vesicles. *Biochemica et Biophysica Acta*, 161, 211–219.

Brigati, D. J., Myersow, D., Lenry, J. J., Spalholz, B., Travis, S. Z., Foug, C. K., Hsiung, C. D., & Wand, D. C. (1983). Detection of viral genomes in cultured cells and paraffin embedded tissue section using biotin hybridization probes. *Virology*, 126, 32–50.

Brown, G. H. (ed.) (1971). *Photochromisms*. New York: Wiley.

Buchachenko, A. L., & Wasserman, A. M. (1973). *Stable Radicals. Electronic Structure, Reactivity and Use*. Moscow: Khimia (in Russian).

Buchman, T., Kennedy, F. S., & Wood, J. M. (1969). Spin labeling of vitamin B_{12}. *Biochemistry*, 8, 4437–4442.

Bulger, J. E., Lloyd, R. V., & Struve, V. (1978). 9-TEMPOL aminoacrichine: a spin labeled fluorescent probe for acetylcholinesterase. *Analytical Letters*, 6, 393–398.

Burlakova, E. B., & Goloshchapov, A. N. (1986). Spin probes in study of normal and cancer cells. In *Methods of Spin Labels and Spin Probes*, N. M. Emanuel & R. I. Zhdanov (eds.), pp. 212–225. Moscow: Nauka (in Russian).

Burley, R. W., Seidel, J. C., & Gergely, J. (1972). Study of conformational changes in myosin by spin-labeling method. *Archives of Biochemistry and Biophysics*, 150, 792–800.

Burshtein, E. A. (1976). Luminescence of protein chromophores. In *Advances in Science and Technology. Biophysics*, Vol. 6, Yu. A. Vladimirov (ed.), Moscow: VINITI (in Russian).

Burstein, A. I. (1978). Molecular-kinetic aspect of chemical physics of the condense state. *Uspekhi Khimii*, 47, 212–234 (in Russian).

Bystryak, S. M., Likhtenshtein, G. I., & Kotelnikov, A. I. (1990). Estimation of parameters of the dielectric relaxation time distributions in liquids by luminescent techniques. *Zhurnal Prikladnoi Spektroskopii*, 52, 394–400 (in Russian).

Bystryak, S. M., Likhtenshtein, G. I., & Kotelnikov, A. V. (1992). Dynamic effects of media in the electron-transfer in bianthryl molecules in viscous liquids. *Khimicheskaya Fizika* (in Russian).

Bystryak, S. M., Likhtenshtein, G. I., Kotelnikov, A. I., Hankovskii, N. P., &

Hydeg, K. A. (1986). Influence of solvent molecular dynamics on photochemical reduction of nitroxide radicals. *Zhurnal Fizicheskoi Khimii*, 60, 2796–2802 (in Russian).

Campbell, A. K., Holt, M., & Patel, A. (1985). Chemiluminescence in medical biochemistry. *Recent Advances in Clinical Biochemistry*, 20, 1534–1538.

Chamorovsky, S. K., Kononenko, A. A., Petrov, E. G., Pottosin, I. I., & Rubin, A. B. (1986). Effects of dehydration and low temperature on the oxidation of high-potential cytochrome c by photosynthetic reaction center in *Ectothiorhodospira shaposhnikovii*. *Biochimica et Biophysica Acta*, 848, 402–410.

Champeil, P., Rigand, J. L., & Garybobo, C. M. (1980). Ca translocation mechanism in sarcoplasmic reticulum vesicles deduced from localization studies of protein bound spin labels. *Proceedings of the National Academy of Sciences, USA*, 77, 2405–2409.

Chan, S. W., Tan, C. T., & Hsia, J. C. (1978). Spin membrane immunoassay: simplicity and specificity. *Journal of Immunological Methods*, 21(1–2), 185–195.

Chao, A. C., Dix, J. S., Sellers, M. C., & Verkman, A. S. (1989). Fluorescence measure of chloride transport in monolayer cultured cells. Mechanism of chloride transport in fibroblasts. *Biophysical Journal*, 56, 1071–1081.

Chapman, D., & Benga, G. (1984). Concept of membrane fluidity. In *Biological Membranes*, D. Chapman (ed.), pp. 1–56. London: Academic Press.

Charkviani, G. G. (1988). Mutual disposition of some functional centres in myosin. In *Biomacromolecules in Methods of Spin Labels and Spin Probes*, G. I. Likhtenshtein & R. I. Zhdanov (eds.), pp. 570–580. Moscow: Nauka (in Russian).

Chen, K., & Swartz, H. M. (1988). Oxidation of hydroxylamines to nitroxide spin labels in living cells. *Biochimica et Biophysica Acta*, 161, 477–484.

Chen, X., & Blumenthal, R. (1990). On the use of self-quenching fluorophors in the study of membrane fusion kinetics. *Biophysical Chemistry*, 34, 283–292.

Cherepanova, E. S., Kulikov, A. V., & Likhtenshtein, G. I. (1990). EPR localization of paramagnetic centers in biological membranes relative to aqueous and lipid phases. *Biologicheskie Membrany*, 7, 51–56 (in Russian).

Cherry, R. J. (1979). Rotational and lateral diffusion of membrane proteins. *Biochimica et Biophysica Acta*, 559, 289–327.

Cherry, R. J. (1986). Effect of melittin and membrane potential on the mobility of band 3 proteins in human erythrocyte membranes. In *Dynamics in Biological Systems*, S. Damjanovich, T. Keleti, & L. Tron (eds.), pp. 487–499. Budapest.

Chignell, C. F. (1979). Spin labeling in pharmacology. In *Spin Labeling. Theory and Applications*, Vol. 2, L. Berliner (ed.), pp. 223–245. New York: Academic Press.

Chignell, C. F. (1990). Spin trapping studies of photochemical reactions. *Pure and Applied Chemistry*, 62, 301–306.

Chimelka, B. F., Mueller, K. T., Pines, A., Stebbins, J., Wu, Y., & Zwanziger, J. W. (1989). Oxygen-17 NMR in solids by dynamic-angle spinning and double rotation. *Nature*, 339, 42–43.

Chong, P. L. (1988). Effects of hydrostatic pressure on the locations of Prodan in lipid and cellular membranes. *Biochemistry*, 27, 399–405.

Churg, A. K., & Warshel, A. (1986). Control of redox potential of cytochrome c and microscopic dielectric effect in proteins. *Biochemistry*, 25, 1675–1681.

Clarke, J., Garcia-Borron, J. C., & Martinez-Carrcon, M. (1983). (1-pyren)sulfonila-zide: fluorescent probe for measuring the transmembrane topology of acetylcholine receptor subunits. *Archives of Biochemistry and Biophysics*, 256, 101–109.

Clarke, J., Halfman, F., & Wong, C. L. (1982). Quenching by electron transfer in immunoassays. *Clinical Chemistry*, 28, 1657–1666.

Cohen, M. D. (1975). Photochemie organischer Restkorper. *Angewandte Chemie*, 87, 439–447.

Colbert, D. L., Smith, D. S., Landon, J., & Sidki, A. M. (1984). Single-reagent polarization fluoro-immunoassay for barbiturates in urine. *Clinical Chemistry*, 30, 1765–1769.

Cone, R. A. (1972). Rotational diffusion of rhodopsin in the visual receptor membrane. *Nature (New Biology)*, 236, 39–43.

Coon, M. J., White, R. E., & Blake, R. C. (1981). Oxygen activation by cytochrome P-450. In *Oxidases and Related Systems*, Vol. 1, T. E. King (ed.), pp. 93–118. Baltimore: University Park Press.

Copeland, E. S., Boykin, M. B., & Kelley, J. A. (1975). An electron spin resonance study of synaptosome opiate receptors. The preparation and use of a spin labeled morphine. *Biophysical Journal*, 15, 1125–1139.

Corin, A. F., & Jovin, T. M. (1986). Proflavin binding to poly[d(A-T)] and poly[d(A-Br^{5}U)]: Triplet state and temperature-jump. *Biochemistry*, 25, 3995–4007.

Corker, G. A., Klein, M. P., & Calvin, M. (1966). Chemical trapping of a primary quantum conversion product in photosynthesis. *Proceedings of the National Academy of Sciences USA*, 56, 1365–1369.

Crewe, A. V., Wall, T., & Langmore, T. (1970). Visibility of single atoms. *Science*, 168, 1338–1340.

Criado, M., Vaz, W. C. L., Barrantes, F. J., & Jovin, T. M. (1982). Translational diffusion of acetylcholine receptor (of monomeric and dimeric forms) of *Torpedo marmoran* reconstituted into phospholipid bilayers studied by fluorescence recovery after photobleaching. *Biochemistry*, 21, 5750–5755.

Dakubu, S., & Ekins, R. P. (1985). The fluorometric determination of europium ion concentration as used in time-resolved fluoroimmunoassay. *Analytical Biochemistry*, 144, 20–26.

Dale, R. E., & Eisinger, J. (1976). Intramolecular energy transfer and molecular conformation. *Proceedings of the National Academy of Sciences, USA*, 73, 271–278.

Damerau, W. (1988). Characterization of active sites by spin labeling. In *Thiamine Pyrophosphate Biochemistry*, Vol. 2, R. L. Schowen & A. Schellenberger (eds.), pp. 137–146, New York: CRC Press.

Datema, K. P., Spujt, R. B., Co, J. A. M., & Hemminga, M. A. (1988). Deuterium nuclear magnetic resonance investigation of bacteriophage M13 coat protein in dimyristoylphosphatidylcholine liposomes using palmitic acid as a probe. *Biochimica et Biophysica Acta*, 944, 507–515.

Davidson, R. S., & Whelan, T. D. (1977). Geometrical requirement for fluorescence intramolecular excimer formation and fluorescent quenching. *Journal of the Chemical Society. Chemical Communications*, No. 11, 361–362.

Davoust, J., Seigneurtet, M., Herve, P., & Devaux, P. F. (1983). Collisions between nitrogen 14 and nitrogen 15 spin labels. 2: Investigation on the specificity of the lipid environment of rhodopsin. *Biochemistry*, 22, 3146–3151.

De Kruif, B., & Wirtz, K. W. A. (1977). Lipids transfer across membranes. *Biochimica et Biophysica Acta*, 467, 318–326.

Demas, J. N., Diemente, D., & Harris, E. W. (1973). Oxygen quenching of charge-transfer excited states of ruthenium (II) complex. *Journal of the American Chemical Society*, 95, 6864–6865.

Demchenko, A. P. (1986). Fluorescence analysis of protein dynamics. *Essays in Biochemistry*, 22, 120–157.

Devaux, P F. (1988). Phospholipid lipases. *FEBS Letters*, 234, 8–12.

Dexter, D. L. (1953). A theory of sensitized luminescence in solids. *Journal of Chemical Physics*, 21, 836–850.

Dixon, S. J., & Wilson, J. (1988). Plasma membrane proton transport in rat astroglia. *Journal of General Physiology*, 92, 26–27.

Dobretsov, G. E. (1989). *Fluorescent Probes in Investigation of Cells, Membranes and Lipoproteins*. Moscow: Nauka (in Russian).

Dobretsov, G. E., Spirin, M. M., Chekrygin, O. V., Karmansky, I. M., Dmitriev, V. M., & Vladimirov, Yu. A. (1982). A fluorescence study of apolipoprotein localization in relation to lipids in serum low density lipoproteins. *Biochemica et Biophysica Acta*, 710, 172–180.

Dobryakov, S. N., Gurovscaya, K. N., & Kayushin, L. P. (1988). Simulation of ESR spectra of spin-labeled biomacromolecules. In *Biomacromolecules in Methods of Spin Labels and Spin Probes*, G. I. Likhtenshtein & R. I. Zhdanov (eds.), pp. 251–260. Moscow: Nauka (in Russian).

Dodd, J. R., & Lewis, A. I. (1975). Spin label synthesis using *NN'*-thionyldiimidazole coupling reagent. *Journal of the Chemical Society. Chemical Communications*, No. 13, 520–521.

Dogonadze, R. R., & Zakharia, M. G. (1984). Theory of low temperature processes of electron transfer. *Biofizika*, 29, 548–552.

Dorokhov, K. E., Crigorian, E. L., & Zhdanov, R. I. (1988). Study of cholinesterase by spin-probe method. In *Biomacromolecules in Methods of Spin Labels and Spin Probes*, G. I. Likhtenshtein & R. I. Zhdanov (eds.), pp. 6–22. Moscow: Nauka.

Doster, W., Gusack, S., & Petry, W. (1989). Dynamic transition of mioglobin revealed by inelastic neutron scattering. *Nature*, 337, 754–756.

Druzhinin, S. Yu., Fogel, V. R., Syrtsova, L. A., Kotelnikov, A. I., & Likhtenshtein, G. I. (1988). Location study of cofactor-binding center in nitrogenase monitored by triplet probe method. *Biofizika*, 31, 16–20 (in Russian).

Dudich, I. V., Timofeev, V. P., Volkenstein, M. V., & Misharin, A. Yu. (1977). Maromolecule rotative correlation time measurement by ESR method for covalently bound spin label. *Molekulyarnaya Biologiya*, 11, 685–687 (in Russian).

Dwek, R. A. (1977). *NMR in Biology*. New York: Academic Press.

Dzuba, S. A. (1988). Hole-burning in pulsed EPR: spin diffusion in organic materials. In *Magnetic Resonance and Related Phenomena*; proceedings of XXIV Congress Ampere, Poland, p. C-68.

Dzuba, S. A., Maryasov, A. G., Salikhov, K. M., & Tsvetkov, Yu. D. (1984). Superslow rotations of nitroxide radicals studied by pulse EPR spectroscopy. *Journal of Magnetic Resonance*, 58, 95–117.

Dzuba, S. A., & Tsvetkov, Yu. D. (1987). Magnetization transfer in pulsed EPR of ^{15}N nitroxides: reorientational motion model of molecules in glassy liquids. *Zhurnal Strukturnoi Khimii*, 28, 15–38 (in Russian).

Eaton, G. R., & Eaton, S. S. (1987a). EPR imaging using T_1 selectivity. *Journal of Magnetic Resonance*, 71, 271–275.

Eaton, G. R., & Eaton, S. S. (1989). Resolved electron–electron spin–spin splittings in EPR spectra. *Biological Magnetic Resonance*, 8, 339–397.

Eaton, G. R., Eaton, S. S., & Maltempo, M. M. (1989). Three approaches to spectral-spatial EPR imaging. *Applied Radiation and Isotopes*, 40, 1227–1232.

Eaton, S. S., & Eaton, G. R. (1987b). EPR imaging. *Spectroscopy*, 1, 32–35.

Edidin, M. (1987). Rotational and lateral diffusion of membrane proteins and lipids: phenomena and function. *Current Topics in Membranes and Transport*, 29, 91–127.

Ehen, R. F., Scott, C. H., & Pisano, J. J. (1988). Fluorescence assay for membranolitic peptides. *Analytical Letters*, 21, 2017–2032.

Eisenback, C. D. (1979). Photochrome Prozesse zur Untersuchung der Kettenbeweglichkeit in Polymeren. *Die Makromolekulare Chemie*, 180, 565–571.

Eltsov, A. B. (ed.) (1982). *Organic Photochromes*. Leningrad: Khimiya.

Emanuel, N. M., & Konovalova, N. P. (1986). Nitroxides in chemiotherapy of tumours. In *Methods of Spin Labels and Spin Probes*, N. M. Emanuel & R. I. Zhdanov (eds.), pp. 5–20. Moscow: Nauka (in Russian).

Englander, S. W., & Kallenbach, N. R. (1983). Hydrogen exchange and structural dynamics of proteins and nucleic acid. *Quarterly Review of Biophysics*, 16, 521–552.

Ermolaev, V. L. (1963). Energy transfer in organic systems with participation of the triplet state. Solid solutions and crystals. *Uspekhi Fizicheskich Nauk*, 80, 3–41 (in Russian).

Ermolaev, V. L., Bodunov, E. N., Svechnikova, E. B., & Shakhverdov, T. A. (1977). *Radiationless Transfer of Electronic Excitation Energy*. Leningrad: Nauka (in Russian).

Eskelinen, S., & Peura, R. (1986). Preparation of lectin-gold complexes and localization of sugar residues on the erythrocyte surface with SEM-EDS system. *Journal of Electron Microscopy*, 31, 3463–2464.

Esmann, M., Hankovszky, H. O., Hideg, K., & Marsh, D. (1989). A novel spin-label for study of membrane protein rotational diffusion using STESR. Application to selectively labelled Class I and Class II-SH groups of the shark rectal gland Na$^+$/K$^+$-ATPase. *Biochimica et Biophysica Acta*, 978, 209–215.

Esmann, M., Marsh, D., Schwarzmann, G., & Sandhoff, K. (1988). Ganglioside–protein interactions: spin-label electron resonance studies with Na$^+$, K$^+$-ATP membranes. *Biochemistry*, 27, 2398–2403.

Esmann, M., Watts, A., & Marsh, D. (1985). Spin-labeled study of lipid–protein interactions in (Na$^+$, K$^+$)-ATPase membranes from rectal glands of *Squalus acanthias*. *Biochemistry*, 24, 1386–1393.

Ewert, U., & Freed, J. H. (1990). Fast ESR-imaging. In *Thirty-second Rocky Mountain Conference*, Denver, Colorado. Abstract 110.

Feigin, L. A., Gonchar, N. A., Likhtenshtein, G. I., L'vov, Yu. M., & Marakushev, S. A. (1978). Determination of the distance between heavy atom labels in the histidine decarboxylase molecule by means of small-angle X-ray scattering. *Soviet Physics, Crystallography*, 23, 420–424.

Ferrieres, X., Lopez, A., Atibelli, A., Dupon-Cezanne, L., Lagonanelle, J. L., & Tocanne, J. F. (1989). Continuous fluorescence microphotolysis of anthracene-labeled phospholipids in membranes. Theoretical approach of the simultaneous determination of their photodimerization and lateral diffusion rates. *Biophysical Journal*, 55, 1081–1091.

Filipescu, N., & Chang, D. S. C. (1972). Interaction between *p*-dimethoxy-benzene and norboxylene in a two-ring model compound. *Journal of the American Chemical Society*, 94, 5990–5996.

Fogel, V. R., Likhtenshtein, G. I., & Kotelnikov, A. I. (1985). An estimation of the electron transfer parameters in cytochrome c. *Biofizika*, 30, 394–399 (in Russian).

Fogelsong, P. D. (1989). Fluorometric methods employing low concentrations of ethidium bromide for Ser-DNA topoisomerase and endonuclease assays. *Analytical Biochemistry*, 182, 284–288.

Fomenko, B. S., & Deimbetova, G. K. (1987). Study on mechanisms of alternation of the pyren and diphenylhexatrien fluorescence in irradiated membranes. *Radiobiologiya*, 27, 85–87 (in Russian).

Forster, Th. (1948). Zwischenmolekulare Energiewanderung und Fluoreszence. *Annalen der Physik*, 2, 55–75.

Forster, Th., & Kasper, K. (1954). Ein Konzentrationrumichhschlag der Fluores-zent. *Zeitschrift für Physikalishe Chemie (BRD)*, 1, 275.

Fowler, V., & Branton, D. (1977). Lateral mobility of human erythrocyte integral membrane proteins. *Nature (London)*, 268, 23–26.

Frauenfelder, H. (1989). The Debye-Waller factor from villain to hero in protein crystallography. *International Journal of Quantum Chemistry*, 35, 711–715.

Freed, J. H. (1976). Theory of the ESR spectra of nitroxide. In *Spin labeling. Theory and Applications*, Vol. 1, L. Berliner (ed.), pp. 64–155. New York: Academic Press.

Frey, T., Anglister, J., & McConnell, H. M. (1988). Lineshape analysis of NMR difference spectra of an anti-spin-label antibody. *Biochemistry*, 27, 5161–5165.

Frolov, E. N., Belonogova, O. V., & Likhtenshtein, G. I. (1977). Investigation of the mobility of spin and Mössbauer labels bound to macromolecules. In *Equilibrium Dynamics of the Native Structure of Protein*, E. A. Burshtein (ed.), pp. 99–142. Pushchino: Izdatelstro Akademii Nauk SSSR (in Russian).

Frolov, E. N., Kharakhonicheva, N. V., & Likhtenshtein, G. I. (1974). Investigation of dynamic structure of human serum albumin by the method of spin-probe. *Molekulyarnaya Biologiya*, 8, 886–893 (in Russian).

Frolov, E. N., Mokrushin, A. N., Likhtenshtein, G. I., Trukhtanov, V. A., & Goldansky, V. L. (1973). Investigation of the dynamic structure of proteins using gamma-resonance labels. *Doklady Akademii Nauk SSSR*, 212, 165–168 (in Russian).

Gaffney, B. J. (1976). Chemistry of nitroxide. In *Spin Labeling. Theory and Applications*, Vol. 1, L. Berliner (ed.), pp. 207–263. New York: Academic Press.

Gailet, J. (1985). *Polymer Photophysics and Photochemistry: An Introduction to the Study of Photoprocesses in Macromolecules*. Cambridge University Press.

Gajewska, J., Bierzynski, A., & Bolewska, A. (1982). Fluorescence quenching and spin-label ESR studies of stacking self-association in aqueous solutions of 2-aminopurine riboside and its 5'-mono- and diphosphates. *Biophysical Chemistry*, 15, 191–204.

Galanin, M. D. (1951). Quenching of the fluorescence of solutions by absorbing substances. *Zhurnal Eksperimentalnoi i Theoreticheskoi Fiziki*, 21, 126–132 (in Russian).

Gantchev, T., Costadinova, Z., & Ganeve, E. (1988). Comparative spin-label study of the dynamic structure of human hemoglobins A and F and the influence of the antibiotic chloramphenicol. *Biochimica et Biophysica Acta*, 953, 218–225.

Gavish, B. (1980). Position-dependent viscosity effect on rate constants. *Physical Review Letters*, 44, 1160–1163.

Gavish, K., Arnold, K., & Balgavy, P. (1988). The influence of fusogenic polyethyleneglycols on model and biomembranes. In *Sixth CMEA Symposium on EPR Spectroscopy in Biochemistry, Molecular Biology, and Medicine*. Abstracts of conference, Smolenicc Castle, Czechoslovakia, November 21–5, 1988, p. 17.

Geertes, H. (1983). Experimental realization and optimalization of a fluorescence correlation spectroscopy apparatus. *Journal of Biochemical and Biophysical Methods*, 7, 255–261.

Gendel, L. Ya., Goldfeld, M. A., Koltover, V. K., Rosantsev, E. A., & Syskina, V. I. (1968). Study of conformational transitions in biomembranes by weakly bound paramagnetic probe. *Biofizika*, 113, 1114–1115 (in Russian).

Gendel, L. Ya., & Krugliakova, K. E. (1986). Structural-functional relationships at interaction of physiologically active compounds with biomembranes. In *Methods of Spin Labels and Spin Probes*, N. M. Emanuel & R. I. Zhdanov (eds.), pp. 163–193. Moscow: Nauka (in Russian).

Gendel, L. Ya., Panasenko, O. M., & Suskina, V. I. (19881). On the biomembrane's capacity to solubilize the benzo-γ-carboline spin probes. *Doklady Akademii Nauk SSSR*, 258, 220–224 (in Russian).

Gnewich, T., & Sosnovsky, G. (1986). Spin-labeling carbohydrates. *Chemical Review*, 86, 203–238.

Goldansky, V. I., Kononenko, A. A., Rubin, A. B., Frolov, E. N., & Chernavsky, D. S. (1981). On relationship between electron tunneling and conformational mobility and degree of freedom of biomolecules. *Doklady Akademii Nauk SSSR*, 257, 491–494 (in Russian).

Goldansky, V. I., & Krupyansky, Yu. F. (1988). Study of protein dynamics by Rayleigh scattering of Mössbauer radiation (RSMR). *Hyperfine Interactions*, 42, 223–232.

Goldansky, V. I., Krupyansky, Yu. F., & Flerov, B. N. (1983). Glass-like model of proteins. *Doklady Akademii Nauk SSSR*, 272, 978–981 (in Russian).

Goldberg, J. S., Rauchman, E. J., & Rosen, G. M. (1977). Bioreduction of nitroxides by *Staphylococcus aureus*. *Biochemical and Biophysical Research Communications*, 79, 199–202.

Goldfeld, M. G., & Tsapin, A. I. (1974). Photoreduction and photooxidation of nitroxides in chloroplasts. *Studia Biophysica*, 42, 195–202.

Golovina, E. A., Smirnov, A. I., Yakimchenko, O. E., & Akseonov, S. I. (1991). EPR-tomography of permeation pathways of water-dissolved nitroxide radical in the wheat caryopsis. *Fiziologia Rastenii*, 38, 119–124 (in Russian).

Görner, H., & Schulte-Frohlinde, D. (1979). Laser flash studies of thioindigo and indigo dyes. Evidence for a transconfiguration of the triplet state. *Chemical Physics Letters*, 66, 363–369.

Gorodisky, V. A. (1989). Solvation energetics and physicochemical properties of liquids and solutions. In *Solvatochromism. Problems and Methods*, N. A. Bakshiev (ed.), pp. 224–312. Leningrad: Leningrad University Press (in Russian).

Gray, H. B. (1987), Long-range electron transfer in proteins. *Aldrichimica Acta*, 23, 87–93.

Grdenic, D., Kamenar, B., Korpar-Coling, B., Sikirca, M., & Jovanovski, G. (1974). tetrakis(trifluoroacetoxymercuri)methane and tetrakis(acetoxymercuri)methane as the reaction products of Hofmann's base with the corresponding acid: X-ray crystallographic evidence. *Journal of the Chemical Society. Chemical Communications*, 16, 646–647.

Grebenshchikov, Yu. B., Charkviani, G. G., Gachechiladze, N. I., Kokhanov, Yu. V., & Likhtenshtein, G. I. (1972). Study of spin-labeled myosin. *Biofizika*, 17, 794–800.

Griffith, O. N., & Jost, P. C. (1976). In *Spin Labeling. Theory and Applications*, Vol. 1, L. Berliner (ed.), pp. 488–569. New York: Academic Press.

Grinberg, O. Ya., Dubinskii, A. A., & Lebedev, Ya. S. (1983). ESP of free radicals in the 2-mm range. *Uspekhi Khimii*, 52, 1490–1513 (in Russian).

Grinvald, A., Haas, E., & Steinberg, I. Z. (1972). Effect of distribution of distances on fluorescence quenching kinetics. *Proceedings of the National Academy of Sciences USA*, 69, 2237–2240.

Gupta, R. K., Koenig, S. H., & Redfield, A. G. (1972). On the electron transfer between cytochrome molecules as absorbed by NMR. *Journal of Magnetic Resonance*, 7, 66–73.

Haas, E., Wilcked, M., Katchalski-Katzir, E., & Steinberg, I. Z. (1974). Orientation factor in inductive resonance transfer. *Proceedings of the National Academy of Sciences USA*, 72, 1807–1812.

Halpern, H. J., & Peric, M. (1990). Towards a quantitatives EPR oxymetry in tissues of a living animal. In *Thirty-second Rocky Mountain Conference*, Denver, Colorado, Abstract 188.

Hara, H., Horiuchi, T., Saneyoshi, N., & Nishimura, S. (1970). 4-Tiouridine-specific spin-labeling of *E. coli* transfer RNA. *Biochemical and Biophysical Research Communications*, 38, 305–311.

Harris, W. E., & Stahl, W. L. (1988). Interaction of phenytoin with rat brain synaptosomes examined by fluorescent fatty acid probes. *Neurochemistry International*, 13, 369–377.

Haynes, D. H. (1974). 1-Anilino-8-naphthalen-sulfonate: a fluorescent indicator of ion binding electrostatic potential on the membrane surface. *Journal of Membrane Biology,* 17, 341–366.

Hemminga, M. A., & de Jager, P. A. (1989). Saturation transfer spectroscopy of spin label techniques and interpretation spectra. In *Biological Magnetic Resonance, Vol. 8: Spin Labeling. Theory and Applications,* L. J. Berliner & J. Reuben (eds.), pp. 131–178. New York: Plenum Press.

Hianik, T., Tichonov, A. N., & Tverdislov, V. A. (1986). A comparative study of microscopic and macroscopic parameters of lipid bilayers. *General and Physiological Biophysics,* 5, 205–209.

Hideg, K. (1990). Novel, potentially useful spin-label reagents. *Pure and Applied Chemistry,* 62, 207–212.

Hideg, K., & Hankowszky, O. H. (1989). Chemistry of spin-labeling amino acid and peptides. Some new mono and bifunctionalized nitroxide free radicals. In *Biological Magnetic Resonance. Vol. 8: Spin Labeling. Theory and Applications,* L. J. Berliner & J. Reuben (eds.), p. 428. New York: Plenum Press.

Highton, P. J., & Beer, M. (1964). Heavy metal markers in electron microscopy. In *Electron Microscopy. Proceedings of the Third European Conference B,* pp. 49–55, Prague.

Hofer, P. (1990). Development of pulsed ENDOR and application to organic radicals. In *Thirty-second Rocky Mountain Conference,* Denver, Colorado. Abstract 101.

Hoffman, J. F., & Laris, P. S. (1974). Potential of erythrocytes measured with the aid of dyes. *Journal of Physiology,* 237, 519–552.

Hoffman, W., & Restall, C. (1984). Rotational and lateral diffusion of membrane proteins as determined by laser technique. In *Biomembrane Structure and Transition,* D. Chapman (ed.), pp. 411–442. Weinheim: Verlag Chemie.

Holowka, D., Wensel, T., & Baird, B. (1990). A nanosecond fluorescence depolarization study on the segmental flexibility of receptor-bound immunoglobulin E. *Biochemistry,* 29, 4607–4612.

Holtzman, J. L. (ed.) (1984). *Spin Labeling in Pharmacology.* New York: Academic Press.

Hsia, J. S., Chen, W. L., Wong, L. T., Long, R. A., & Kalow, W. (1972). Synthesis of spin labeled 2,4-dinitrophenols and their activities in the uncoupling of oxidative phosphorylation in rat liver mitochondria. *Biochemical and Biophysical Research Communications,* 48, 1273–1279.

Hsia, J. S., & Piette, L. H. (1969). Spin-labeling as a general method in studying antibody active site. *Archives of Biochemistry and Biophysics,* 129, 296–307.

Hsia, J. S., Kwan, N. H., & Er, S. S. (1978). Development of a spin assay for reserve bilirubin loading capacity of human serum. *Proceedings of the National Academy of Sciences, USA,* 75, 1542–1545.

Hsia, J. S., Rosman, D. I., & Piette, L. H. (1969). Organophosphate spin-label studies of inhibited esterases, α-chymotrypsin and cholinesterase. *Biochemical and Biophysical Research Communications,* 36, 75–78.

Hubbell, W. L., & McConnell, H. M. (1971). Molecular motion in spin-labeled phospholipids and membranes. *Journal of the American Chemical Society,* 93, 314–322.

Hyde, J. S., Chien, J. C. W., & Freed, J. H. (1968). ELDOR of free radicals in solutions. *Journal of Chemical Physics*, 48, 4211–4226.

Hyde, J. S., & Dalton, L. R. (1979). Saturation-transfer spectroscopy. In *Spin Labeling. Theory and Applications*, Vol. 1, L. Berliner (ed.), p. 1070. New York: Academic Press.

Hyde, J. S., & Feix, J. B. (1989). Electron–electron double resonance. In *Biological Magnetic Resonance. Vol. 8: Spin Labeling. Theory and Applications*, L. J. Berliner & J. Reuben (eds.), pp. 305–339. New York: Plenum Press.

Hyde, J. S., Sczaniecki, P. B., & Froncisz, W. (1989). The Bruker lecture alternatives to field modulation ESR spectroscopy. *Journal of the Chemical Society. Faraday Transactions*, 85.

Hyde, J. S., & Subczinski, W. K. (1989). Spin-label oximetry. In *Biological Magnetic Resonance. Vol. 8: Spin Labeling. Theory and Applications*. L. J. Berliner & J. Reuben (eds.), pp. 399–426. New York: Plenum Press.

Hyde, J. S., Swartz, H. M., & Antholine, W. E. (1979). The spin probe–spin label methods. In *Spin Labeling. Theory and Applications*, Vol. 2, L. Berliner (ed.), pp. 72–113. New York: Academic Press.

Hyde, J. S., Yin, J. J., Feix, J. B., & Hubbell, W. L. (1990). Advances in spin label oximetry. *Pure and Applied Chemistry*, 62, 265–270.

Inbar, M., Shinitzky, M., & Sachs, L. (1974). Microviscosity in the surface membrane lipid layer of intact normal lymphocytes and leukemic cells. *FEBS Letters*, 38, 268–270.

Isaacson, R. A., & Feher, G. (1990). EPR/ENDOR 9-GHz spectrometer sensitivity enhancement for small samples using dielectric resonator. In *Thirty-second Rocky Mountain Conference*, Denver, Colorado. Abstract 120.

Ishida, S., Kumashiro, H., Tsuchihashi, N., Ogata, T., Ono, M., Kamade, H., & Yoshida, E. (1989). In vivo analysis of nitroxide radical injected into small animals by L-band ESR technique. *Physical Medical Biology*, 34, 1317–1323.

Islomov, S., Marupov, R., & Likhtenshtein, G. I. (1989). Temperature transitions in plasticised celluloses studied by spin-labeling. *Cellulose Chemistry and Technology*, 23, 13–21.

Ivanov, V. T., Miroshnikov, A. I., Snezhkova, L. G., Ovchinnikov, Yu. A., Kulikov, A. V., & Likhtenshtein, G. I. (1973). Spin labeled gramicidine. *Khimiya Prirodnych Soedinenii*, No. 1, 91–98 (in Russian).

Janick, F., Dell, V. A., Abrahamson, J. K., Watson, B. S., Miller, D. L., & Johnson, A. E. (1990). Fluorescence characterization of the interaction of various transfer-RNA species with elongation factor TuGTp: evidence for a new functional role for elongation factor Tu in protein biosynthesis. *Biochemistry*, 29, 4268–4277.

Janzen, E. G. (1971). Spin trapping. *Accounts of Chemical Researches*, 4, 31–45.

Janzen, E. G., Rotare, Y., & Oehler, U. M. (1987). Two-dimensional ENDOR imaging. In *Tenth International EPR Symposium, Rocky Mountain Conference*, Denver, August 1987. Abstract 141.

Jardetzky, O., & Roberts, G. C. (1981). *NMR in Molecular Biology*. New York: Academic Press.

Jarnefelt, J., Laurent, T., & Rigler, R. (1988). Diffusion of fluorescein labelled molecules in suspensions of erythrocyte ghosts. *FEBS Letters*, 242, 129–133.

Johnson, L. S., & Neuhaus, F. C. (1978). Spin-label assay for lysozyme. *Analytical Biochemistry*, 85, 56–62.

Johnson, M. E. (1979). Spin-label techniques for monitoring macromolecular rotational motion: empirical calibration under nonideal conditions. *Biochemistry*, 18, 378–385.

Josephson, B. D. (1960). Temperature dependent shift of γ-rays emitted by a solid. *Physical Review Letters*, 4, 341–342.

Jost, P. C., & Griffith, O. H. (1980). The lipid–protein interface in biological membranes. *Annals of the New York Academy of Sciences*, 348, 391–407.

Kaivarainen, A. I. (1985). *Solvent-dependent Flexibility of Proteins and Principles of Their Function*. Dordrecht: Reidel.

Kalamkarov, G. R., Raevsky, A. V., Salitra, I. S., Pivovarova, T. S., Ostrovsky, M. A., Levchenko, L. A., & Likhtenshtein, G. I. (1977). Study of rhodopsin distribution in photoreceptor membranes by electron scattering label technique. *Bioorganicheskaya Khimiya*, 3, 775–783 (in Russian).

Kamino, K., Komuro, H., Sakaj, T., & Hirota, A. (1989). Optical multiple site real time recording of neutral activity in the CNS using voltage sensitive dyes. *Neuroscience Research Supplement*, 9, 21–22.

Kamsolova, S. G., & Postnikova, G. B. (1981). Spin-labeled nucleic acids. *Quarterly Review of Biophysics*, 14, 223–288.

Kang, T. J., Kahlow, M. A., Giser, D., Srsallen, S., Nagarajan, A., Jurzeba, W., & Barbara, P. F. (1988). Dynamic solvent effects in electron-transfer kinetics of S_1 bianthryls. *Journal of Physical Chemistry*, 92, 6800–6807.

Kapinus, E. I. (1988). *Photonics of Molecular Complexes*. Kiev: Naukova Dumka (in Russian).

Kapinus, E. I., & Kucherova, I. Yu. (1984). Role of long distance electron transfer in the excited molecule's quenching. *Doklady Academii Nauk SSSR*, 279, 920–924 (in Russian).

Karplus, J. A., & McCammon, M. (1986). Dynamics of protein structure. *Scientific American*, 251, 4–12.

Kazuaki, P., & Ichiro, H. (1988). ESR studies on the ripple phase in multilamellar phospholipid bilayers. *Biochemica et Biophysica Acta*, 163, 77–78.

Keana, J. F. W. (1978). Newer aspects of the synthesis and chemistry of nitroxide spin labels. *Chemical Reviews*, 78, 37–64.

Keana, J. F. W. (1984). Synthesis and chemistry of nitroxide spin labels. In *Spin Labeling in Pharmacology*, J. L. Holtsman (ed.), pp. 2 86. New York: Academic Press.

Keana, J. F. W., Lex, L., Mann, J. S., May, J. M., Pank, J. H., Pou, S., Prubhu, V. S., Rosen, G. M., Sweetman, B. J., & Wu, Y. (1990). Novel nitroxides for spin-labelling-trapping and magnetic resonance imaging. *Pure and Applied Chemistry*, 62, 201–206.

Kevan, L., & Baglioni, P. (1990). Electron spin-echo modulation studies of ionic and nonionic micell structure via stearic acid nitroxide probes. *Pure and Applied Chemistry*, 62, 275–281.

Keintanar, A., Kunvar, A. O., & Oldfield, E. (1986). Deuterium nuclear magnetic resonance spectroscopic study of the fluorescent probe diphenyl hexatriene in model membrane systems. *Biochemistry*, 25, 6517–6524.

Kerimov, T. M., Mil'gram, Ya. M., Kozlov, I. A., & Ruuge, E. K. (1978). The study of the conformational changes in soluble mitochondrial ATPase by the spin probe method. *Biokhimiya,* 43, 1525–1531 (in Russian).

Khramtsov, V. V., Panteleev, M. V., & Weiner, L. M. (1989). ESR study of proton transport across phospholipid vesicle membranes. *Journal of Biochemical and Biophysical Methods,* 18, 237–246.

Khramtsov, V. V., & Weiner, L. M. (1988). Proton exchange in stable nitroxyl radical: pH sensitive spin-probe. In *Imidazoline Radicals,* Vol. 2, L. B. Volodarsky (ed.), pp. 37–80. Boca Raton: CRC Press.

Kiselev, P. A., Garda, G., Finch, S. A., Stier, A., Gurinovich, N. A., Khatyleva, S. Yu., & Akhrem, A. A. (1990). Rotation diffusion of cytochrome P-450 in the presence of NADPH–cytochrome reductase in lysosomes of different phospholipid composition. *Biologicheskie Membrany,* 7, 1026–1036 (in Russian).

Kivelson, D. (1960). Theory of ESP linewidths of free radicals. *Journal of Chemical Physics,* 33, 1094–1106.

Klug, A. (1978). Image analysis and reconstructions in the electron microscopy of biological macromolecules. *Chemica Scripta,* 14, 245–256.

Knox, P. P., Bystruak, I. M., Kotelnikov, A. I., Shaitan, K. V., Kononenko, A. A., Zakharova, V. I., Likhtenshtein, G. I., & Rubin, A. V. (1989). The influence of glycerol and sugars on electron phototransfer in a system of quinone acceptors of reaction centers of purple bacteria. *Izvestiya Akademii Nauk SSSR (Seria Biologicheskaya),* No. 5, 651–659 (in Russian).

Kochetkov, B. B., Likhtenshtein, G. I., Koltover, V. K., Knox, P. P., Kononenko, A. A., Grishanova, P. G., & Rubin, A. B. (1984). Intramolecular dynamics and electron transfer in photosynthetic reaction centres. Investigation by the method of spin-labels. *Izvestiya Akademii Nauk SSSR (Seria Biologicheskaya),* No. 4, 572–577 (in Russian).

Kogtev, L. S., Bogdanovich, Yu. V., Molotkovsky, J. G., Lazarovich, E. M., Utkin, Yu. N., Tsetlin, V. I., & Bergelson, L. D. (1989). Use of the fluorescence labeled lipid probes for analyzing substance P and its derivatives binding to the rat brain tachykinin receptors. *Biologicheskie Membrany,* No. 6, 34–41 (in Russian).

Kokorin, A. I., Zamaraev, K. J., Grigoryan, G. L., Ivanov, V. P., & Rozantsev, E. G. (1972). Distance estimation between nitroxyl radicals. *Biofizika,* 17, 34–41 (in Russian).

Koltover, V. K., Goldfeld, M. G., Gendel, L. Ya., & Rozantsev, E. G. (1968). Conformational transition in biomembrane studied by the spin label method. *Biophysical and Biochemical Research Communications,* 32, 421–425.

Koltover, V. K., Reichman, L. M., Jasaitis, A. A., & Blumenfeld, C. A. (1971). A study of spin-probe solubility in mitochondrial membranes correlated with ATP-dependent conformational changes. *Biochimica et Biophysica Acta,* 234, 306–310.

Koppel, D. E., & Sheetz, M. P. (1983). A localized pattern photobleaching method for the concurrent analysis of rapid and slow diffusion processes. *Biophysical Journal,* 43, 175–181.

Korstanje, L. J., Van Genkel, C., & Levine, Y. K. (1990). Effects of steroid molecules on the dynamical structure of dioleoylphosphatidylcholine and digalactosyl-diacylglycerol bilayers. *Biochimica et Biophysica Acta*, 1022, 155–162.

Koshland, D. E. (1959). Mechanisms of transfer of enzymes. In *The Enzymes*, Vol. 1, P. D. Boyer, M. Lardy, & K. Myrback (eds.), pp. 305–346. New York: Academic Press.

Kosman, D. J. (1972). ESR probing of macromolecules: a comparison of structure/function relationships in chymotrypsinogen, α-chymotrypsin and anhydro-tochymotrypsin. *Journal of Molecular Biology*, 67, 247–264.

Kosover, E. M. (1968). *Introduction to Physical Organic Chemistry*. New York: McGraw-Hill.

Kostina, N. V., Marupov, R., Antsiferova, L. I., & Likhtenshtein, G. I. (1987). Effect of UV-radiation on rotation diffusion of spin-labels in the cotton fibre. *Biofizika*, 32, 736–742 (in Russian).

Kotelnikov, A. I., Fogel, V. R., Bystyak, I. M., & Likhtenshtein, G. I. (1986). Factors determining electron transfer in proteins. In *Fundamental Research in Homogeneous Catalysis*, A. E. Shilov (ed.), pp. 773–784. New York: Gordon & Breach.

Kotelnikov, A. I., Fogel, V. R., Likhtenshtein, G. I., Postnikova, G. B., & Shlapnikova, E. A. (1981). The use of exchange deactivation of the triplet excited state for investigation of structure and electronic conductivity of proteins. *Molekulyarnaya Biologiya*, 15, 281–289 (in Russian).

Kotelnikov, A. I., Gvozdev, R. I., & Likhtenshtein, G. I. (1975). The use of the ESR signal saturation phenomenon for study of the macromolecular relief in the region of the paramagnetic centre. *Studia Biophysica*, 49, 215–221.

Kotelnikov, A. I., Kuznetsov, S. N., Fogel, V. R., & Likhtenshtein, G. I. (1979). Investigation of the microstructure of biological systems by the triplet label technique. *Molekulyarnaya Biologiya*, 13, 152–159 (in Russian).

Kotelnikov, A. I., Likhtenshtein, G. I., Fogel, V. R., Kochetkov, V. V., Noks, P. P., Kononenko, A. A., Grishanova, N. P., & Rubin, A. B. (1983). Molecular dynamics and electron transfer in photosynthetic reaction centers. An investigation by luminescence methods. *Zhurnal Prikladnoi Spectroskopii*, 17, 846–854 (in Russian).

Kotelnikova, R. A., Taryanenko, L. B., Mekler, V. N., & Kotelnikov, A. I. (1982). Study of Ca^{2+} ATPase of sarcoplasmic reticulum by triplet labeling method. *Molekulyarnaya Biologiya*, 16, 1188–1195 (in Russian).

Krainev, A. G., Weiner, L. M., Alferyev, I. S., & Slynko, N. M. (1985). Bifunctional compound study of the active-centre location of cytochrome P-450 in a microsomal membrane ("float" molecule method). *Biochimica et Biophysica Acta*, 818, 16–104.

Kratky, O. (1963). X-ray small angle scattering with substances of biological interest in diluted solutions. *Progress in Biophysics and Molecular Biology*, 13, 105–173.

Krinichny, V. I. (1990). Study of biological systems by high resolution ESR spectroscopy in the 2-mm range. *Zhurnal Prikladnoi Spektroskopii*, 52, 887–905 (in Russian).

Krinichny, V. I., Antsiferova, L. I., Lyubashevskaya, E. V., Belonogova, O. V., Grinberg, O. Ya., & Likhtenshtein, G. I. (1989). ESR-spectroscopy of revert micelles in the 2-mm range. *Zhurnal Fizicheskoi Khimii*, 63, 3015–3021 (in Russian).

Krinichny, V. I., Grinbert, O. Ya., Bogatyrenko, V. R., Likhtenshtein, G. I., & Lebedev, Ya. S. (1985). Study of microenvironment effects on magnetic resonance parameters of spin-labeled human serum albumin in a 2-mm range. *Biofizika*, 30, 216–219 (in Russian).

Krinichny, V. I., Grinberg, O. Ya., Dubinsky, A. A., Lifshitz, V. A., Bobrov, Yu. A., & Lebedev, Ya. S. (1987a). Study of anisotropic molecular rotation by 2-mm ESR spectroscopy. Saturation transfer technique. *Biofizika*, 32, 534–535 (in Russian).

Krinichny, V. I., Grinberg, O. Ya., Belonogova, O. V., Sudakova, E. I., Borin, M. L., Lebedev, Ya. S., & Likhtenshtein, G. I. (1987b). Study of molecular mobility in biological membranes by two millimeter band ESR spectroscopy. *Biofizika*, 1, 59–63 (in Russian).

Krinichny, V. I., Grinberg, O. Ya., Yudanova, E. I., Lyubashevskaya, E. B., Antsiferova, L. I., Likhtenshtein, G. I., & Lebedev, Ya. S. (1988). Study of spin-labeled lysozyme by 2-mm ESR spectroscopy. *Biofizika*, 32, 215–222 (in Russian).

Krinichny, V. I., Grinberg, O. Ya., Yusupov, I. Kh., Marupov, R. M., Bobodzhanov, P. Kh., Likhtenshtein, G. I., & Lebedev, Ya. S. (1986). Two-millimeter band ESR study of spin-labeled cotton fiber. *Biofizika*, 31, 482–485 (in Russian).

Kronick, M. N., & Grossman, P. D. (1983). Immunoassay techniques with fluorescent phycobiliprotein conjugates. *Clinical Chemistry*, 29, 1582–1586.

Kuhn, H. (1971). Interaction of chromophores in monolayer assemblies. *Pure and Applied Chemistry*, 27, 421–438.

Kulikov, A. V. (1976). Determination of the distance between label spins and paramagnetic centre spins in spin-labeled proteins from the saturation curve parameters of the EPR spectrum of the label at 77K. *Molekulyarnaya Biologiya*, 10, 132–141 (in Russian).

Kulikov, A. V., Cherepanova, E. S., & Bogatyrenko, V. R. (1981). Determination of the closest-approach distance between a radical and paramagnetic ion. *Teoreticheskaya i Eksperimentalnaya Khimiya*, 17, 788–797 (in Russian).

Kulikov, A. V., Cherepanova, E. S., Likhtenshtein, G. I., Uvarov, V. Yu., & Archakov, A. I. (1989). EPR study of localization of cytochrome P-450 heme group in microsomes relative to aqueous and lipid phases. *Biologicheskie Membrany*, 6, 1085–1094 (in Russian).

Kulikov, A. V., Kysel', M. A., Zyryanov, V. V., & Likhtenshtein, G. I. (1979). Study of surface of phospholipid bilayers by spin-labeling method. *Biofizika*, 24, 839–844 (in Russian).

Kulikov, A. V., & Likhtenshtein, G. I. (1977). The use of spin relaxation phenomena in the investigation of the structure of model and biological systems by the methods of spin labels. *Advances in Molecular Relaxation Interaction Processes*, 10, 47–78.

Kulikov, A. V., Likhtenshtein, G. I., Rozantsev, E. G., Suskina, V. I., & Shapiro, A. B. (1972). Distance estimation between nitroxide fragments in biradicals and spin labeled protein. *Biofizika*, 17, 42–48 (in Russian).

Kulikov, A. V., Yudanova, E. I., & Likhtenshtein, G. I. (1983). Investigation of the spin exchange of nitroxide radicals using the continuous EPR spectrum saturation technique. *Zhurnal Fizicheskoi Khimii*, 56, 2982–2987 (in Russian).

Kulikov, A. V., Yudanova, E. I., Likhtenshtein, G. I., Allachverdiev, S. I., & Klimov, V. V. (1988). Study of oxygen evolution in pea chloroplasts by spin labels. *Biofizika*, 6, 984–988 (in Russian).

Kusumi, A., & Hyde, J. (1982). Spin label saturation transfer electron spin resonance detection of transient association of rhodopsin in reconstituted membranes. *Biochemistry*, 21, 5978–5983.

Kuznetsov, A. N. (1974). *Spin Probe Technique*. Moscow: Nauka (in Russian).

Lakowicz, J. R. (1983). *Principles of Fluorescence Spectroscopy*. New York: Plenum Press.

Lakowicz, J. R., Gabor, R. L., Gryczynski, E., Szmacinski, H., & Wiczk, W. (1989). Frequency-domain fluorescence spectroscopy. Principles, biochemical applications and future development. *Berichte der Bunsengesellschaft Physicalische Chemie*, 93, 316–327.

Lamer, C. N., Holm, R., & Horrocks, W. D. (eds.) (1974). *NMR in Paramagnetic Molecules*. New York: Academic Press.

Lange, A., Marsh, D., Wassmer, K. H., Meier, P., & Kothe, G. (1985). ESR study of phospholipid membranes employing a comprehensive line-shape model. *Biochemistry*, 24, 4383–4392.

Lanzin, V. S., Asadchikov, V. E., Egorov, A. M., Marakushev, S. A., Popov, V. O., Likhtenshtein, G. I., & Berezin, J. V. (1985). Modification of bacterial formiatedehydrogenase by mercarbide electron-density labels. Small-angel X-ray scattering determination of distance between the enzyme active centers. *Doklady Akademii Nauk SSSR*, 281, 712–715 (in Russian).

Lapper, D. D., Paterson, S. J., & Smith, I. C. P. (1972). Spin probe of analog of steroids. *Canadian Journal of Biochemistry*, 50, 969–976.

Larsson, S. (1982). Some theoretical remarks concerning electron pathways. *Faraday Discussion. Chemical Society*, 74, 390–392.

Lassmann, G., Herrmann, A., Raikov, Z., & Muller, P. (1987). EPR spectroscopic analysis of the binding site of a cancerostatic agent on an erythrocyte membrane. *Cancer, Biochemistry and Biophysics*, 2, 169–178.

Law, P. W., & Asakura, T. (1976). Spin label studies on conformational change of apohemoglobin due to heme binding. *Journal of Biological Chemistry*, 251, 6838–6843.

Lazarev, G. G., & Lebedev, Ya. S. (1981). Study of rotational dynamics in solids by method of oriented spin probe. *Teoreticheskaya i Eksperimentalnaya Khimiya*, 17, 798–805 (in Russian).

Lebedev, Ya. S., & Muromtsev, V. I. (1972). *EPR and Relaxation of Stabilized Radicals*. Moscow: Khimiya (in Russian).

Lee, P. M., & Grant, C. W. H. (1980). Headgroup oligosaccharide dynamics of a transmembrane glycoprotein. *Canadian Journal of Biochemistry*, 58, 1197–1205.

LeMaster, D. M., & Richards, F. M. (1982). Preparative-scale isolation of isotopically labeled amino acids. *Analytical Biochemistry*, 122, 238–247.

Lemmetyinen, H., Ylipertula, M., Mikkova, J., Virtannen, J. A., & Kinnunen, P. K.

(1989). Kinetic study of monomer and excimer fluorescence of pyren-substituted phosphatidylcholine in phosphatidylcholine bilayer. *Journal of Chemical Physics*, 97, 7170–7175.

Leonhardt, H., & Weller, A. (1963). Electronübertragungsreaction des angeregten Perylens. *Berichte der Bunsengesellschaft Physicalische Chemie*, 61, 791–795.

Leute, R., Ellman, E. F., Goldstein, A., & Gerzenberg, L. A. (1972). Synthesis of spin-labeled morphine. *Nature*, 236, 93–94.

Levchenko, L. A., Likhtenshtein, G. I., Raevsky, A. V., & Salitra, I. S. (1979). Application of mercarbide labels is a new approach in electron microscopy of proteins and membranes. *Uspekhi Biokhimii*, 20, 257–274 (in Russian).

Levchenko, L. A., Raevsky, A. V., Sadkov, A. P., & Likhtenshtein, G. I. (1973). On the localization of an iron-containing active site on the nitrogenase macromolecule. *Doklady Akademii Nauk SSSR*, 211, 238–240 (in Russian).

Levental, V. I., Bekker, Zh. M., Molin, Yu. N., Kumarev, V. P., Grachev, M. A., & Knorre, D. G. (1972). Dioxygen assay with the acid of a spin label. *FEBS Letters*, 24, 149–153.

Levich, V. G., Dogonadze, R. R., German, E., Kuznetsov, A. N., & Kharakats, Yu. I. (1970). Theory of homogeneous reactions involving proton transfer. *Electrochemica Acta*, 15, 353–368.

Levi-Setti, R. (1988). Structure and microanalytical imaging of biological materials by scanning microscopy with heavy-ion probes. *Annual Review of Biophysical Chemistry*, 17, 325–347.

Libertini, L. J., & Small, E. W. (1990). Application of the method of moment analysis to fluorescence decay life-time distributions. *Biophysical Chemistry*, 34, 269–282.

Likhtenshtein, G. I. (1966). Regularities in the energy and entropy properties of enzymatic processes. *Biofizika*, 11, 24–32 (in Russian).

(1968). Determination of the topography of protein groups using specific paramagnetic labels. *Molecular Biology*, 2, 234–240 (in Russian).

(1976a). *Spin Labeling Methods in Molecular Biology*. New York: Wiley-Interscience.

(1976b). Water and protein dynamics. In *L'eau et les systemes Biologiques*, A. Alfsen & A. J. Bertrand (eds.), pp. 45–53. Paris: CNRS.

(1979a). Protein dynamics studied by spin-labeling, Mössbauer spectroscopy and NMR. In *Special Collogue Ampere on Dynamical Processes in Molecular Systems: Studies by Spectroscopy*, A. Losche (ed.), pp. 100–107. Leipzig: Karl Marx University.

(1979b). *Polynuclear Redox Metalloenzymes*. Moscow: Nauka (in Russian).

(1985). Amorphous-crystalline model of intramolecular dynamics of globular proteins. *Biofizika*, 30, 27–30 (in Russian).

(1986). Water and dynamics of proteins and membranes. *Studia Biophysica*, 111, 89–100.

(1988a). Structure and molecular dymanics of metalloenzymes studied by physical label methods. *Journal of Molecular Catalysis*, 48, 129–138.

(1988b). *Chemical Physics of Redox Metalloenzyme Catalysis*. Heidelberg: Springer-Verlag.

(1990a). The use of nitroxides in the solution of some problems of chemical biophysics. *Pure and Applied Chemistry*, 62, 281–288.

(1990b). Microstructure and dynamics of biomembranes studied by spin, triplet and photochrome labeling. In *Molecular Dynamics in Membranes*, pp. 17–20. Theses of FEBS satellite meeting, Lake Balaton, Hungary, August 15–9, 1990. Hungarian Academy of Sciences.

Likhtenshtein, G. I., Akhmedov, Yu. D., Ivanov, L. V., & Krinitskaya, L. M. (1974). Study of the lysozyme macromolecule by the spin-label technique. *Moleculyarnaya Biologiya*, 8, 45–58 (in Russian).

Likhtenshtein, G. I., & Bobodzhanov, P. Kh. (1968). Investigation of the structure and local conformational changes of proteins and enzymes using double paramagnetic labels. *Biofizika*, 13, 757–764 (in Russian).

Likhtenshtein, G. I., Bogatyrenko, V. R., & Kulikov, A. V. (1983). Study of the mobility of free radicals connected with serum albumins at 30–320K. *Biofizika*, 38, 585–589 (in Russian).

Likhtenshtein, G. I., Bystryak, S. M., & Kotelnikov, A. I. (1990). Role of molecular dynamics of media in processes of electron transfer in viscous liquids. *Khimicheskaya Fizika*, 9, 697–706 (in Russian).

Likhtenshtein, G. I., Grebenshchikov, Yu. B., Bobodzhanov, P. Kh., & Kokhanov, Yu. V. (1970). Investigation of the microstructure of proteins by the paramagnetic label–paramagnetic probe technique. *Molecular Biology*, 4, 682–691 (in Russian).

Likhtenshtein, G. I., & Kotelnikov, A. I. (1983). Study of the functional intramolecular mobility of proteins by the physical technique. *Moleculyarnaya Biologiya*, 17, 505–518 (in Russian).

Likhtenshtein, G. I., Kotelnikov, A. V., & Kulikov, A. V. (1982). New approaches to studies of water–protein and water–membrane interactions by physical labeling. *Studia Biophysica*, 91, 23–27 (in Russian).

Likhtenshtein, G. I., Kulikov, A. V., & Kotelnikov, A. I. (1986a). Use of relaxation processes with involvement of nitroxides in molecular biology. In *Methods of Spin Labels and Spin Probes*, N. M. Emanuel & R. I. Zhdanov (eds.), pp. 41–61. Moscow: Nauka (in Russian).

Likhtenshtein, G. I., Kulikov, A. V., Kotelnikov, A. I., & Levchenko, L. A. (1986b). Methods of physical labels – a combined approach to the study of microstructure and dynamics in biological systems. *Journal of Biochemical and Biophysical Methods*, 12, 1–28.

Likhtenshtein, G. I., Levchenko, L. A., Sadkov, A. P., Raevsky, A. V., Pivovarova, T. S., & Gvozdev, R. I. (1973). Study of the active site structure of nitrogenase by electron microscopy. *Doklady Akademii Nauk SSSR*, 213, 1442–1443 (in Russian).

Likhtenshtein, G. I., Pivovarov, A. P., & Smolina, N. S. (1968). Study of structural transitions in serum albumins by the luminescent active label technique. *Molekulyarnaya Biologiya*, 2, 291–302 (in Russian).

Linderström-Lang, K. H., & Schellmann, J. A. (1959). Protein structure and enzyme activity. In *The Enzymes*, Vol. 1, P. D. Boyer, H. Lardy, & K. Myrback (eds.), pp. 443–510. New York: Academic Press.

Lipka, J. J., Lippard, S. J., & Wall, J. S. (1979). Visualisation of polymercuri-methane-labeled bacteriophage in the scanning transmission electron microscope. *Science*, 206, 1419–1421.

Livshitz, V. A., Ivanova, P. O., & Kuznetsov, V. A. (1988a). Study of a reversible association of Ca^{2+}-ATPase in the sarcoplasmic reticulum by the spin labeling method. *Izvestiya Akademii Nauk SSSR (Seria Biologicheskaya)*, No. 6, 857–865 (in Russian).

Livshitz, V. A., & Kuznetsov, A. N. (1980). Study of superslow rotations of spin-labeled proteins using the effects of microwave saturation in RPR spectra. *Molekulyarnaya Biologiya*, 14, 182–189 (in Russian).

Livshitz, V. A., Pogazheva, I. D., Kuznetzov, V. A., & Ostrovsky, M. A. (1988b). Saturation transfer EPR-spectroscopy in study of rhodopsin molecular dynamics in the photoreceptor membrane. In *Biomacromolecules in Methods of Spin Labels and Spin Probes*, G. I. Likhtenshtein & R. I. Zhdanov (eds.), pp. 127–156. Moscow: Nauka (in Russian).

Lowe, D. J., & Smith, B. E. (1985). Nitrogenase. *Biochemical Society Transactions*, 13, 579–589.

Luckrust, E. R. (1976). Biradicals as spin probes. In *Spin Labeling. Theory and Applications*, Vol. 1, L. Berliner (ed.), pp. 133–181. New York: Academic Press.

Ludescher, R. D., Eads, T. M., & Thomas, D. D. (1988). Ionic effects on the rotational dynamics of cross-bridges in myosin filaments, measured by triplet absorption anisotropy. *Journal of Molecular Biology*, 200, 89–99.

Lumry, R., & Biltonen, R. (1969). Thermochemical and kinetical aspects of protein conformation. In *Structure and Stability of Biological Molecules*, Vol. 2, Biological Macromolecules Series, S. N. Timashev & C. D. Fosman (eds.), pp. 146–306. New York: Marcel Dekker.

Lumry, R., & Eyring, H. (1954). Conformation changes of proteins. *Journal of Physical Chemistry*, 58, 110–120.

Lumry, R., & Gregory, R. B. (1986). Free energy management in protein reactions: concepts, complications, and compensation. In *The Fluctuating Enzymes*, G. R. Welch (ed.), pp. 3–185. New York: Wiley.

Lumry, R., & Hershberger, M. (1978). Studies of indole photochemistry with special reference to biological applications. *Photochemistry and Photobiology*, 27, 819–840.

Luna, E. J., & McConnell, H. M. (1978). Multiple phase equilibrium in binary mixtures of phospholipids. *Biochemica et Biophysica Acta*, 509, 462–473.

Lurie, D. J., Foster, M. A., & Mallard, J. R. (1990). Imaging free radicals in-vivo in the rat using PERDI. In *Thirty-second Rocky Mountain Conference*, Denver, Colorado, Abstract 116.

Lyakhovich, V. V., Weiner, L. L., Khramtsov, V. V., Larionov, O. V., & Nagu, V. Yu. (1988). Study of cytochrome P-450 by spin-labeling methods. *Imidazoline Nitroxide*, Vol. 2, L. B. Volodarsky (ed.), pp. 216–278. New York: CRC Press.

McConnell, H. M. (1956). Effect of anisotropic hyperfine interactions on para-magnetic relaxation in liquids. *Journal of Chemical Physics*, 25, 709–711.

   (1978). Membrane spin-immune assay. *International Review of Biochemistry. Biochemistry of Cell Walls and Membranes*, 19, 91–136.

McConnell, H. M., & McFarland, B. G. (1970). Physics and chemistry of spin labels. *Quarterly Review of Biophysics*, 3, 91–136.

Magde, P., Elson, E. & Webb, W. (1974). Fluorescence correlation spectroscopy. II: An experimental realization. *Biopolymers*, 13, 29–61.

Malkin, S., & Fisher, E. (1964). Temperature dependence of photoisomerization. III: Direct and sensitized photoisomerization. *Journal of Physical Chemistry*, 68, 1153–1163.

Maniara, G., Van der Kooi, J. M., Bloomgarden, D. C., & Koleczek, H. (1988). Phosphorescence from 2-(*p*-toluidinyl)naphthalene-6-suffonate and *l*-anilinyl-naphthalene-8-sulfonate, commonly used fluorescence probes of biological structures. *Photochemistry and Photobiology*, 47, 207–208.

Maples, K. R. (1990). In vivo ESR spin-trapping investigations of xenobiotic metabolism by red blood cells. In *Thirty-second Rocky Mountain Conference*, Denver, Colorado, Abstract 147.

Marakushev, S. A., Levchenko, L. A., Likhtenshtein, G. I., Kaplun, A. P., Mekler, V. M., Raevskii, A. V., & Shvets, V. I. (1986). A new mercarbide electron dense label for the hydrocarbon moiety of lipid bilayers. *Biologicheskie Membrany*, 5, 423–427 (in Russian).

Marcus, R. A., & Sutin, N. (1985). Electron transfer in chemistry and biology. *Biochimica et Biophysica Acta*, 811, 625–632.

Marrot, G., Bureau, J. F., Roux, M., Maurin, L., Favre, E., & Devaux, P. F. (1987). Orientation and vertical fluctuations of spin-labeled analogues of cholesterol and androstanol in phospholipid bilayers. *Biochemica et Biophysica Acta*, 897, 341–345.

Marsh, D. (1989). Experimental methods in spin-label spectral analysis. In *Spin Labeling. Theory and Applications*, L. J. Berliner & J. Reuben (eds.), pp. 255–285. New York: Plenum Press.

(1990). Studies of membrane dynamics using nitroxide spin labels. *Pure and Applied Chemistry*, 62, 265–270.

Marsh, D., Watts, A., & Smith, I. C. P. (1983). Dynamic structure and phase behavior of dimyristoyl phosphatidylethanolamine bilayers studied by deuterium nuclear magnetic resonance. *Biochemistry*, 22, 3023–3026.

Marupov, R., Yusupov, I. Kh., Bobodzhanov, P. Kh., & Likhtenshtein, G. I. (1981). Study of molecular dynamics of cotton fibres of spin-labeling methods. *Doklady Akademii Nauk SSSR*, 256, 414–417 (in Russian).

Mason, R. P., Kennedy, C. H., & Marples, K. R. (1990). In vivo detection of free radical metabolites. *Pure and Applied Chemistry*, 62, 295–300.

Matesson, D. S., Castle, R. B., & Larson, G. L. (1970). Tetrakis(acetoxymercuri) methane. *Journal of the American Chemical Society*, 92, 231–232.

Matko, J., Szollosi, L., & Denjanovich, S. (1989). Luminescence spectroscopic approaches in studying cell surface dynamics. *Quarterly Review of Biophysics*, 21, 479–544.

Matsumoto, S., & Hammes, G. G. (1975). Fluorescence energy transfer between ligand binding sites on aspartate transcaramylase. *Biochemistry*, 14, 214–220.

Matsuzaki, K., Yokoyama, I., Komatsu, H., Handa, T., & Miyajima, K. (1989). A fluorescent probing study of microenvironments in bile site micelles and bile/

salt/phosphatidylcholine mixture in the absence and presence of cholesterol. *Biochimica et Biophysica Acta. Biomembranes*, 480, 371–378.

Mayo, K. H., Kucheyda, D., Parak, F., & Chien, J. C. (1983). Structural dynamics of human deoxyhemoglobin and hemochrome investigated by Mössbauer spectroscopy. *Proceedings of the National Academy of Sciences USA*, 80, 5294–5296.

Mazurenko, Yu. T. (1989). Spectroscopy of relaxation and statistical effects in solvent shells of molecules. In *Solvatochroism. Problems and Methods*, N. G. Bakhshiev (ed.), pp. 122–190. Leningrad: Leningrad University Press (in Russian).

Mazurenko, Yu. T., & Bakhshiev, N. G. (1970). The effects of orientational dipolar relaxation on the spectral, time and polarization characteristics of the luminescence of solutions. *Optika i Spektroskopiya*, 28, 905–913 (in Russian).

Meisel, D., & Czapski, G. (1975). One-electron transfer equilibria and redox potentials of radicals studied by pulse radiolysis. *Journal of Physical Chemistry*, 79, 1503–1509.

Mekler, V. M., Kotelnikov, A. I., & Likhtenshtein, G. I. (1983). Application of probes emitting annihilation delayed fluorescence for investigation of model and biological membranes. *Biofizika*, 28, 503–504 (in Russian).

Mekler, V. M., Kotelnikov, A. I., Likhtenshtein, G. I., & Berkovich, M. A. (1982). Application of luminescent probes to the investigation of model and biological membranes. *Biofizika*, 27, 641–645 (in Russian).

Mekler, V. M., Kotelnikov, A. I., Likhtenshtein, G. I., Kaplun, A. P., & Shvets, V. I. (1984a). Investigation of the quinone ring of ubiquinone Q-10 location in dipalmitoyl phosphatidylcholine liposomes by triplet probe method. *Biofizika*, 29, 779–782 (in Russian).

Mekler, V. M., Kotelnikov, A. I., Likhtenshtein, G. I., Kotelnikova, R. A., Tatyanenko, L. V., Shvets, V. I., & Kaplun, A. P. (1984b). Use of the annihilation delayed fluorescence phenomenon for studying the sarcoplasmic reticulum structural organization. *Biofizika*, 29, 809–813 (in Russian).

Mekler, V. M., & Likhtenshtein, G. I. (1986). Study of dynamic contacts of macromolecules according to the kinetics of isomerization of fluorescent labels sensibilized by the dye in a triplet state. *Biofizika*, 31, 568–571 (in Russian).

Mekler, V. M., & Umarova, F. T. (1988). Use of triplet-sensitized photochroism phenomena for investigation of dynamic interactions of proteins in membranes. *Biofizika*, 33, 720–722 (in Russian).

Mendel, D., & Cohen, J. (1975) Photochemie organischer Festkorper. *Angewandte Chemie*, 87, 439–447.

Michel, H., & Deisenhofer, J. (1986). X-ray diffraction studies on a crystalline bacterial photosynthetic reaction center. A progress report and conclusions on the structure of the photosystem II reaction center. In *Encyclopedia of Plant Physiology*, New Series, Vol. 19, L. A. Stachelin & C. J. Arntzen (eds.). pp. 371–381. Berlin: Springer-Verlag.

Mill, E. M., Kruglyakova, K. E., & Grigor'yan, G. L. (1986). Interaction of biologically active compounds with DNA. In *Methods of Spin Labels and Spin Probes*, M. N. Emanuel & R. L. Zhdanov (eds.), pp. 21–41. Moscow: Nauka (in Russian).

Mildvan, A. S., & Weiner, H. (1969). Interaction of a spin-labeled analogue of nicotinamide adenine dinucleotide with alcohol dehydrogenase. III: Thermodynamic, kinetic and structural properties of ternary complexes as determined by nuclear magnetic resonance. *Journal of Biological Chemistry*, 244, 2465–2475.

Mims, W. B., & Peisach, J. (1981). Electron spin echo spectroscopy and the study of metalloproteins. In *Biological Magnetic Resonance*, Vol. 8, L. Berliner & J. Reuben (eds.), pp. 213–263. New York: Plenum Press.

Mims, W. B., Peisach, J., & Davis, J. L. (1977). Nuclear modulation of the electron spin echo envelope in glassy materials. *Journal of Chemical Physics*, 66, 5536–5550.

Mitsova, I. Z., Skrypkin, A. Ju., Gvozdev, R. I., Likhtenshtein, G. I., & Shishkov, A. W. (1981). Study of the FeMo cofactor structure from the *Azotobacter* MoFe-protein. *Isvestiya Akademii Nauk SSSR (Seria Biologicheskaya)*, No. 2, 316–318 (in Russian).

Mock, D. M., Brono, G. V., Griffin, B. W., & Peterson, J. A. (1982). Low temperature EPR spectroscopic characterization of the interaction of cytochrome P-450 with a spin label analog of metyrapone. *Journal of Biological Chemistry*, 257, 5372–5380.

Montgomery, M. R., Holtzman, J. L., & Leute, R. K. (1975). Determination of diphenylhydantoin in juman serum by spin immunoassay. *Clinical Chemistry*, 21, 221–226.

More, J. K., More, K. M., Eaton, G. R., & Eaton, S. S. (1990). Metal–nitroxyl interactions. 55: Manganese(III)–nitroxyl electron–electron spin–spin interaction. *Pure and Applied Chemistry*, 62, 241–246.

Morgan, C. G., Sandhu, S. S., Yianni, Y. P., & Dodd, N. J. F. (1987). The phase behaviour of dispersions of bis-azo PC: photoregulation of bilayer dynamics via lipid photochromism. *Biochimica et Biophysica Acta*, 903, 495–503.

Morse, P. D., Petraszak, J. M., & Reminger, L. (1990). Metabolism of nitroxides by mouse lymphocytes. In *Thirty-second Rocky Mountain Conference*, Denver, Colorado. Abstract 164.

Moshkovsky, Yu. Sh. & Tat'yanenko, L. V. (1986). Testing of a chemical compound's biological activity by the spin labeling method. In *Methods of Spin Labels and Spin Probes*, N. M. Emanuel & R. I. Zhdanov (eds.), pp. 194–211. Moscow: Nauka (in Russian).

Motherwell, W. B., & Roberts, J. S. (1972). A convenient nitroxide radical synthesis. *Journal of the Chemical Society. Chemical Communication*, No. 6, 328–329.

Nagara, J. M. (1988). Interaction of a hydrophobic model peptide and its fatty acid derivative with lipid vesicles. *FEBS Letters*, 238, 411–414.

Nagy, V. Yu. (1988). Determination of metals in solution by chelate formation with spin-labeled reagents. In *Imidazoline Nitroxides*, Vol. 2, L. B. Volodarskii (ed.), pp. 115–135. Boca Raton: CRC Press.

Nagy, V. Yu., Bystryak, I. M., Kotelnikov, A. I., Likhtenshtein, G. I., Petrushin, O. M., Zolotov, Yu. A., & Volodarskii, L. B. (1990). Determination of metals in solution by chelate formation with intramolecular luminescence-quenching spin-labeled reagent. *Analyst*, 115, 839–841.

Nagy, V. Yu., Evstiferov, M. V., Petrushin, O. M., Volodarskii, L. B., & Zolotov,

Yu. A. (1981). Stable free-radical complexing reagents in applications of ESR to the determination of metals. *Analytica Chimica Acta*, 120, 85–99.

Narayanan, R., & Balaram, P. (1981). Fluorescence and NMR study of the binding of cholinergic fluorescence probes to horse serum cholinesterase. *International Journal of Peptide and Protein Researches*, 17, 170–175.

Navon, G., & Valensin, G. (1987). Nuclear relaxation times as a source of structural information. In *Metal Ions in Applications of NMR to Paramagnetic Species. Biological Systems*, Vol. 21, H. Sigel (ed.), pp. 1–45. New York: Marcel Dekker.

Neiman, M. B., Rosantsev, E. G., & Mamedova, Yu. C. (1962). Free radical reactions involving no unpaired electron. *Nature*, 196, 472–474.

Newton, M. D. (1982). Medium effect in electron transfer reaction. *Faraday Discussions of the Chemical Society*, 74, 110–111.

Nezlin, R. S. (1990). Internal movements in immunoglobulin molecules. *Advances in Immunology*, 48, 1–4.

Nezlin, R. S., & Sykulev, Y. K. (1982). Structural studies of immunoglobulins spin-labeled at the carbohydrate moiety. *Molecular Immunology*, 19, 357–366.

Nezlin, P. C., & Sykulev, Yu. K. (1988). Introduction of spin labels in carbohydrate components of glycoproteins: use for study of the immunoglobulins. In *Biomacromolecules in Methods of Spin Labels and Spin Probes*, G. I. Likhtenshtein & R. I. Zhdanov (eds.), pp. 181–219. Moscow: Nauka (in Russian).

Nezlin, R. S., Zagyansky, Y. A., & Tumerman, L. A. (1970). Strong evidence for the freedom of rotation of immunoglobulin C subunits. *Journal of Molecular Biology*, 50, 569–580.

Ng, M., & Auld, D. S. (1989). A fluorescent oligopeptide energy transfer assay with broad applications for neutral proteases. *Analytical Biochemical*, 183, 50–56.

Nishikawa, H., Fujii, H., & Berliner, L. J. (1985). Helices and surface coils for low-field in vitro ESR and EPR imaging applications. *Journal of Magnetic Resonance*, 62, 79–86.

Nowak, T., & Mildvan, A. S. (1972). Nuclear magnetic resonance studies of selectively hindered internal motion of substrate analogs at the active site of pyruvate kinase. *Biochemistry*, 11, 2813–2818.

Ogawa, S., & McConnell, H. M. (1967). Spin-label study of hemoglobin conformations in solution. *Proceedings of the National Academy of Sciences, USA*, 58, 19–26.

Ohno, K. (1985). ESR imaging and its applications. *Journal of Magnetic Resonance*, 64, 109–117.

Ohyashiki, T., Taka, M., & Mohru, T. (1989). Changes of the fluorescence characteristics of *N*(1-pyrene)maleimide bound to the intestinal brush-border membranes by neuraminidase treatment. *Chemical and Pharmacological Bulletin*, 37, 2165–2169.

Oldfield, E. (1988). Spectroscopy studies of lipids and biological membranes. *Biochemical Society Transactions*, 16, 1–10.

Ottensmeyer, F. P. (1982). Scattered electrons in microscopy and microanalysis. *Science*, 215, 461–466.

Owen, Y. (1962). Spin resonance of ion pairs in crystal lattices. *Journal of Applied Physics, Supplement*, 33, 355–357.

Owicki, J. C. & McConnell, H. M. (1980). Lateral diffusions in inhomogeneous membranes. Model membranes containing cholesterol. *Biophysical Journal*, 30, 383–398.

Parak, F., Knapp, E. W., & Kucheida, D. (1983). Protein dynamics in Mössbauer spectroscopy of deoxymyoglobin. *Journal of Molecular Biology*, 161, 177–194.

Park, J. H., & Trommer, W. E. (1989). Advantages of ^{15}N and deuterium spin probes for biomedical electron paramagnetic resonance investigations. *Biological Magnetic Resonance*, 8, 547–595.

Parker, C. A. (1968). *Photoluminescence of Solutions with Applications to Photochemistry and Analytical Chemistry*. Amsterdam: Elsevier.

Parmon, V. N., Kokorin, A. I., & Zhidomirov, G. M. (1980). *Stable Biradicals*. Moscow: Nauka (in Russian).

Patyal, B., Crepean, R. H., & Freed, J. H. (1990). Two-dimensional Fourier transform ESR in the slow motion regime. In *Thirty-second Rocky Mountain Conference*, Denver, Colorado, Abstract 104.

Pavlovich, V. S., Pershukevich, P. P., & Pikulik, L. G. (1979a). Decay kinetics and quenching of phtalimide acetyl-derivatives in polar solutions. *Optical i Spektroskopiya*, 46, 898–903 (in Russian).

(1979b). Relaxation shifts in phosphorescent and fluorescent spectra. *Zhurnal Prikladnoi Spektroskopii*, 39, 998–1005 (in Russian).

Permyakov, E. A., & Burshtein, E. A. (1977). Relaxation processes in forzen aqueous solutions. *Studia Biophysica*, 64, 163–172.

Perutz, M. F. (1989). Mechanisms of cooperativity and allosteric regulation in proteins. *Quarterly Review of Biophysics*, 22, 139–236.

Peters, R., Brunger, A., & Schulten, K. (1981). Continuous fluorescence microphotolysis: a sensitive method for study of diffusion processes in single cell. *Proceedings of the National Academy of Sciences, USA*, 78, 962–966.

Peterson-Kennedy, S. E., McCourty, J. L., & Hoffman, B. M. (1984). Temperature dependence of long-range electron transfer in [Zn, Fe^{3+}] hybrid hemoglobin. *Journal of the American Chemical Society*, 106, 5010–5012.

Petty, H. R., Smith, L. M., Feuron, D. T., & McConnell, H. M. (1980). Lateral distribution and diffusion of the C3b receptor of complement, HLA antigens and lipid probes in peripheral blood leukocytes. *Proceedings of the National Academy of Sciences, USA*, 77, 6587–6591.

Piette, L. H., & Hsia, J. C. (1979). Spin labeling in biomedicine. In *Spin Labeling. Theory and Applications*. Vol. 2, L. Berliner (ed.), pp. 291–345. New York: Academic Press.

Pink, D. A. (1989). Perturbing probes and the study of lipid protein bilayer membranes. *Chemistry and Physics of Lipids*, 30, 213–235.

Poo, M., & Cone, R. A. (1974). Lateral diffusion of rhodopsin in the photoreceptor membrane. *Natue (London)*, 247, 438–441.

Pratar, P. R., Novak, T. S., & Freedman, J. C. (1990). Two mechanisms by which fluorescent oxonols indicate membrane potential in human red blood cells. *Biophysical Journal*, 57, 835–849.

Pringle, M. J., & Hidalgo, C. (1982). Lipid–protein interactions in sarcoplasmic reticulum are not perturbed by ionophore A23187: and EPR and fluorescence study. *Biophysical Journal*, 37, 633–636.

Pullman, B., & Berthold, H. (1974). Quantum-mechanical studies on the conformation of phospholipids: the conformational properties of the polar head. *FEBS Letters*, 44, 266–269.

Quintanilha, A. T., & Packer, L. (1977). Surface localization of sites of reduction of nitroxide spin-labeled molecules in mitochondria. *Proceedings of the National of Academy of Sciences, USA*, 74, 570–574.

Rashba, Yu. E., Chernikov, V. A., Baider, L. M., & Vartan'yan, L. M. (1986). Use of hydroxylamines of piperidines and imidazolines for quantitative determination on the superoxide radicals. *Biologicheskie Membrany*, 3, 838–945 (in Russian).

Rassat, A. (1990). Magnetic properties of nitroxide multiradicals. *Pure and Applied Chemistry*, 62, 223–228.

Rassat, A., & Rey, P. (1971). Nitroxides: photochemical synthesis of trimethyl-isoquinuclidine *N*-oxylnuclidine. *Journal of the Chemical Society, Chemical Communications*, No. 19, 1161–1162.

Razi-Nagvi, K. (1974). Diffusion-controlled reactions in two dimensional fluids: discussion of measurement of lateral diffusion of lipids in biomembranes. *Chemical Physics Letters*, 28, 280–285.

Reichardt, C. (1969). *Losungsmittel-Effecte in der organischen Chemie*. Marburg: Verlag Chemie.

Robbey, F. A., Jamieson, G. A., & Hunt, J. B. (1979). Synthesis and use of a new spin-labeled analogue of ADP with platelet-aggregating activity. *Journal of Biological Chemistry*, 254, 1114–1118.

Robinson, D. (1988). Shedding light on lysosomes – Application of fluorescence techniques to cell biology and diagnoses of lysosomal disorders. *Biochemical Society Transactions*, 16, 11–16.

Rochev, V. Ya., Makarov, E. F., Kevdin, O. P., Kukushkina, L. B., & Kosova, G. N. (1985). Study of lipid multilayers by Mössbauer spectroscopy. *Biologicheskie Membrany*, 2, 95–103 (in Russian).

Roe, J. N., Szoka, F. C., & Verkman, A. S. (1989). Optical measurement of aqueous potassium concentration by a hydrophobic indicator in lipid vesicles. *Biophysical Chemistry*, 33, 295–302.

Roos, W., & Slavik, J. (1987). Measurement of pH with the aid of fluorescence dyes. *Biochimica et Biophysica Acta*, 899, 67–75.

Rosen, G. D., Kinkelstein, E., & Rauchman, E. J. (1982). Method for the detection of superoxide in biological systems. *Archives of Biochemistry and Biophysics*, 215, 367–378.

Rosenquist, E., & Vistnes, A. I. (1977). Immune lysis of spin label loaded liposomes incorporating cardiolipin: a new sensitive method for detecting anticardio-lipin antibodies in syphilis serology. *Journal of Immunological Methods*, 15, 147–155.

Roslyakov, B. Ya., & Churgin, Yu. I. (1972). Study of cyanamoyl-α-chymotrypsin hydrolysis in the solid state. *Biokhimiya*, 37, 493–447 (in Russian).

Rotman, B. (1973). *Fluorescence Techniques in Cell Biology*. Heidelberg: Springer-Verlag.

Rozantsev, E. G. (1970). *Free Nitroxyl Radicals*. New York: Plenum Press.

Rozantsev, E. G., & Zhdanov, R. I. (1986). Biologically active organic paramagnetics. In *Methods of Spin Labels and Spin Probes*, N. M. Emanuel & R. I. Zhadanov (eds.), pp. 250–264. Moscow: Nauka (in Russian).

Rozantsev, E. G. (1990). Some problems of nitroxyl chemistry. *Pure and Applied Chemistry*, 62, 311–316.

Rubinov, A. N., Tomin, V. I., & Bushuk, B. A. (1982). Nonhomogeneous broadening of fluorescent spectra in solution. *Journal of Luminescence*, 26, 377–384.

Rubtsova, E. T., Likhtenshtein, G. I., Fogel, R. I., & Kotelnikov, A. I. (1992). Photoelectron transfer and dynamics in model systems on a base of serum albumin. *Biofizika* (in Russian), in press.

Rudy, B., Dubois, H., Mink, R., & Trommer, W. E. (1989). Coenzyme binding by 3-hydroxybutyrate dehydrogenase, a lipid-requiring enzyme. *Biochemistry*, 28, 5354–5366.

Ruuge, E. K., & Gerasimova, E. N. (1986). Interaction of spin probes with the plasma lipoproteins. In *Methods of Spin Labels and Spin Probes*, N. M. Emanuel & R. I. Zhadanov (eds.), pp. 225–235. Moscow: Nauka (in Russian).

Sadkov, A. P., Kulikov, A. V., Likhtenshtein, G. I., Levchenko, L. A., & Pivovarova, T. S. (1990). An EPR study of the nature of protein ligands of the Mo- and Fe-containing component of nitrogenase. *Biokhimiya*, 55, 368–371.

Safronov, S. I., Mstislavsky, V. I., Safronova, Y. I., & Muromtsev, V. I. (1969). A method of determination of relaxation times from signal saturation curves under rapid passage conditions. *Zavodskaya Laboratoria*, 35, 1463–1466 (in Russian).

Sakaki, T., Tsuyi, T., Chang C. H., & Ohnishi, S. (1982). Rotational mobility of an erythrocyte membrane integral protein, band 3, in DMPC reconstituted vesicles and effect of binding cytoskeletal peripheral proteins. *Biochemistry*, 21, 2366–2371.

Salikhov, K. M., Doctorov, A. B., Molin, Yu. N., & Zamaraev, K. K. (1971). Exchange broadening of ESR lines for solutions of free radicals and transition metal complexes. *Journal of Magnetic Resonance*, 5, 189–205.

Salikhov, K. M., Semenov, A. G., & Tsvetkov, Yu. D. (1976). *Electron Spin-Echo and Its Applications*. Moscow: Nauka (in Russian).

Savitski, A. P. (1987). Fluorescent immunoassay. In *Nonisotope Methods of Immunoassay. Advances in Science and Biotechnology*, Vol. 3, pp. 117–165. Moscow: VINITI (in Russian).

Scaletfar, B. A., Selvin, P. R., Axelrod, D., Klein, M. P., & Hears, I. E. (1990). A polarized photobleaching study of DNA in agarose gels. *Biochemistry*, 29, 4790–4798.

Schmidt, P. G., & Kuntz, I. D. (1984). Distance measurements in spin-labeled lysozyme. *Biochemistry*, 23, 4261–4266.

Schwab, M. E., & Thoenen, H. (1978). Selective binding, uptake and retrograde transport of tetanus toxin by nerve terminals in the rat iris. An electron microscope study using colloidal gold as a tracer. *Journal of Cell Biology*, 77, 1–13.

Schwarz, D., & Pirrwitz, J. (1988). [^{15}N,^{2}H] spin label SP-EPR measurements of rotational dynamics of membrane-bound cytochrome P-450. In *Sixth CMEA Symposium on ESR Resonance Spectroscopy in Biochemistry, Molecular Biology and Medicine*. Abstracts of conference, Smolenice Castle, Czechoslovakia, November 21–5, 1988, p. 50.

Schwarz, D., Pirrwitz, J., & Ruckpaul, K. (1982). Rotational diffusion of cytochrome P-450 in the microsomal membrane – evidence for a cluster-like organization from saturation transfer EPR spectroscopy. *Archives of Biochemistry and Biophysics*, 216, 322–327.

Schweiger, A. (1990). Radiofrequencies in pulsed EPR. In *Thirty-second Rocky Mountain Conference*, Denver, Colorado. Abstract 99.

Seelig, J. (1977). Deuterium magnetic resonance. Theory and application to lipid membranes. *Quarterly Review of Biophysics*, 10, 353–418.

Seelig, J., & Seelig, A. (1980). Lipid conformation in model membranes and biological membranes. *Quarterly Review of Biophysics*, 13, 19–61.

Sergeev, P. V., Uliankina, T. I., Sejfulla, R. D., Grebenshchikov, Yu. B., & Likhtenshtein, G. I. (1974). Study of the interaction of steroids with human serum albumin by the spin label method. *Molekulyarnaya Biologiya*, 8, 206–217 (in Russian).

Serzhanov, V. F., Likhtenshtein, G. I., & Levchenko, L. A. (1992a). Determination of the form of electron-dense particles by electron microscopy. *Biofizika*.

Serzhanov, V. F., Likhtenshtein, G. I., Levchenko, L. A., Kovalenko, J., & Sadkov, A. P. (1992b). Statistical analysis of electron micrographs of electron-dense labeled bioobjects. *Biofizika*, in press.

Shaitan, K. V., & Rubin, A. B. (1980). Conformational mobility and the theory of the Mössbauer effect in biosystems. *Molekulyarnaya Biologiya*, 14, 1323–1334 (in Russian).

Shapiro, J. E. (1988). Paramagnetic hydrophilic probing of colloid systems by NMR spectroscopy. *Uspekhi Biokhimii*, 57, 1253–1272 (in Russian).

Shapiro, H. M., & Stephens, S. (1986). Flow cytometry of DNA content using oxazine 750 or related laser dyes with 633-nm laser excitation. *Cytometry*, 7, 107–110.

Shcherbitskaya, N. B., & Demchenko, A. P. (1989). Segmental mobility of phospholipids in the bilayer membrane in the region of binding of non-charged fluorescent probes. *Biofizika*, 34, 574–578 (in Russian).

Shimchick, E. J., & McConnell, H. M. (1972). Rotational correlation time of spin-labeled α-chymotrypsin. *Biochemical and Biophysical Research Communications*, 46, 321–327.

Shin, D. M., Schanze, K. S., Otruba, J. P., Brown, P. E., & Whitten, D. G. (1988). Charge transfer interactions in micelles and vesicles. Inter- and intramolecular probes of solubilization site polarity. *Israel Journal of Chemistry*, 28, 37–45.

Shin, Y. K., Moscicki, J. K., & Freed, J. H. (1990). Dynamics of phosphatidylcholine-cholesterol mixed model membranes in the liquid crystalline state. *Biophysical Journal*, 57, 445–459.

Shishkov, A. V., Filatov, E. S., Simonov, E. F., Unukovich, M. C., Goldansky, V. I., & Nesmeyanov, A. N. (1976). Synthesis of biological compounds labeled by tritium. *Doklady Akademii Nauk SSSR*, 228, 1237–1239 (in Russian).

Shofield, P., Hoffman, B. M., & Rich, A. (1970). Spin-labeling studies of aminoacyl transfer ribonucleic acid. *Biochemistry*, 9, 2525–2533.

Sider, P., Cave, R. J., & Marcus, R. A. (1984). A model for orientation effects in electron transfer reactions. *Journal of Chemical Physics*, 81, 5613–5625.

Siegel, S., & Judeikis, H. S. (1964). Triplet–triplet energy transfer in rigid glasses: lack of a solvent effect. *Journal of Chemical Physics*, 4, 648–652.

Sinha, B. K., & Chignell, C. F. (1975). Acridine spin labels as probes for nucleic acids. *Life Sciences*, 17, 1829–1836.

Sixl, F., Brophy, P. J., & Watts, A. (1984). Selective protein–lipid interactions at membrane surfaces: a deuterium and phosphorus NMR study of the associa-tion of myelin basic protein with bilayer head groups of dimyristoyl-phosphatidylcholine and dimyristoylphospatidyl glycerol. *Biochemistry*, 23, 2032–2039.

Sixl, F., & Watts, A. (1982). Interactions between phospholipid head groups at membrane interfaces: a deuterium and phosphorus nuclear magnetic resonance and spin-label electron spin resonance study. *Biochemistry*, 24, 6446–6452.

Skrypkin, A. Yu., Blazhchuk, I. S., Gvozdev, R I., Likhtenshtein, G. I., & Shishkov, A. V. (1985). Determination of amino acids localized on the surface of nitrogenase by total tritium labeling. *Doklady Akademii Nauk SSSR*, 281, 595–599.

Smirnov, A. I., Yakimchenko, O. E., Aksenov, S. I., Golovina, E. A., Likhtenshtein, G. I., & Lebedev, Ya. S. (1988). EPR-tomography study of the destination of water-dissolved probes in wheat grain during imbibition. *Fiziologia Rastenii*, 3, 663–668 (in Russian).

Smith, J., Kuczera, K., Tidor, B., Daster, W., Cusack, S., & Kurplus, M. (1989). Internal dynamics of globular proteins: comparison of neutron scattering measurements and theoretical models. *Physica*, B156, 437–443.

Smith, L. M., & McConnell, H. M. (1981). Pattern photobleaching of fluorescent lipid vesicles using polarized laser light. *Biophysical Journal*, 33, 139–146.

Snipes, W., Kenny, K., & Keith, A. (1977). A spin-reduction assay for lipoxidase. *Analytical Biochemistry*, 81, 425–431.

Soloman, J., & Blombergen, N. (1956). Nuclear magnetic interaction in HF molecule. *Journal of Chemical Physics*, 25, 261–266.

Solozhenkin, P. M. (1986). *Electron Paramagnetic Resonance in Analysis.* Dushanbe: Donish (in Russian).

Sombyo, A. P. (1986). *Recent Advances in Electron and Light Optical Imaging in Biology and Medicine.* New York Academy of Sciences.

Sosfenov, N. I., Andrianov, B. I., Vagin, A. A., Strokopytov, B. V., Vainstein, B. K., Shilov, A. E., Gvozdev, R. I., Likhtenshtein, G. I., Mizova, I. Z., & Blazhchuk, I. S. (1986). X-ray study of the nitrogenase MoFe protein of the nitrogenase from *Azotobacter vivelandii. Doklady Akademii Nauk SSSR*, 291, 1123–1127.

Sosnovsky, G. (1990). The guess for a predictive design of anticancer drugs. *Pure and Applied Chemistry*, 62, 289–294.

Sosnovsky, G., & Konieczny, M. (1970). Synthesis of phosphoorganic derivatives of nitroxide radicals. *Naturforschung*, 33b, 792–804.

Spartenstein, A., Robinson, B. H., & Hopkins, P. B. (1989). Sequence- and structure-dependent DNA base dynamics: synthesis, structure and dynamics of site and sequence specificity of spin-labeled DNA. *Biochemistry*, 28, 9484–9495.

Squier, T. C., & Thomas, D. D. (1989). Selective detection of the rotational dynamics of the protein-associated lipid hydrocarbon chains in sarcoplasmic reticulum membranes. *Biophysical Journal*, 56, 735–748.

Stone, T. J., Buchman, T., Nordio, P. L., & McConnell, H. M. (1965). Spin-labeled biomolecules. *Proceedings of the National Academy of Sciences, USA*, 54, 1010–1017.

Strothkamp, K. G., Lehmann, J., & Lippard, S. J. (1978). Tetrakis(acetoxymercuri)-methane: a polymetallic reagent for labeling sulfur in nucleic acids. *Proceedings of the National Academy of Sciences, USA*, 75, 1181–1184.

Strothkamp, K. G., & Lippard, S. J. (1976). Platinum binds selectively to phosphorothiate groups in mono- and polynucleotides: a general method for heavy metal staining of specific nucleotides. *Proceedings of the National Academy of Sciences, USA*, 73, 2536–2540.

Stryer, L. (1968). Fluorescence spectroscopy of proteins. *Science*, 162, 526–533.

(1978). Fluorescence energy transfer as a spectroscopic ruler. *Annual Review of Biochemistry*, 47, 819–846.

Stryer, L., & Griffith, O. H. (1965). A spin-labeled hapten. *Proceedings of the National Academy of Sciences, USA*, 54, 1785–1791.

Subczynski, W. K., & Hyde, J. S. (1981). The diffusion-concentration product of oxygen in lipid bilayers using the spin-label method. *Biochimica et Biophysica Acta*, 643, 283–291.

Subkhankulova, T. N., Lyakhovich, V. V., Resnikov, B. A., Elinova, B. I., & Vainer, L. M. (1986). Study of the interaction of lipophilic spin-labeled analog of the substrate of cytochrome P-450 with liposomes and microsomes. *Biologicheskie Membrany*, 3, 720–729 (in Russian).

Sudha, B. P., Dixit, N., Moy, V. P., & Vanderkooi, J. M. (1984). Reaction of excited-states cytochrome c derivatives. Delayed fluorescence and phosphorescence of zinc, tin and metal-free cytochrome c at room temperature. *Biochemistry*, 24, 2103–2107.

Sueki, M., Eaton, G. R., & Eaton, S. S. (1990). Multidimensional EPR imaging of nitroxide. *Pure and Applied Chemistry*, 62, 229–234.

Sukhorukov, B. I., & Petrov, A. I. (1988). Study of model polynucleotides by spin-labels. Dynamics and thermodynamics of helixes. In *Biomacromolecules in Methods of Spin Labels and Spin Probes*, G. I. Likhtenshtein & R. I. Zhadanov (eds.), pp. 220–251. Moscow: Nauka (in Russian).

Sukhorukov, B. I., Vasserman, A. M., Kozlova, L. I., & Buchachenko, A. L. (1967). Use of the spin-labeling method for study of the nucleic acid–water system. *Doklady Akademii Nauk SSSR*, 177, 454–457 (in Russian).

Suzdalev, I. P. (1979). *Dynamic Effects in Gamma-Resonance Spectroscopy*. Moscow: Atomisdat (in Russian).

(1988). *Gamma-Resonance Spectroscopy of Proteins and Model Compounds*. Moscow: Nauka (in Russian).

Swartz, H. M. (1987). Measurement of pertinent oxygen concentrations in biological systems. *Acta Biochimica et Biophysica of Hungary*, 22, 277–293.

(1990). The use of nitroxides in viable biological systems: an opportunity and challenge for chemists and biochemists. *Pure and Applied Chemistry*, 62, 235–240.

Swartz, H. M., Glockne, F. J., Gast, P., & Glarksan, R. (1990). Progress in EPR oximetry of viable biological systems. In *Thirty-second Rocky Mountain Conference*, Denver, Colorado. Abstract 187.

Tatikolov, A. S., & Kuzmin, V. A. (1975). Quenching of the triplet states of aromatic hydrocarbons by stable nitric acid radicals. *Doklady Akademii Nauk SSSR*, 223, 403–406 (in Russian).

Taylor, J. C., Leigh, J. S., & Cohn, M. (1969). Magnetic resonance studies of spin-labeled creatine kinase system and interaction of two paramagnetic probes. *Proceedings of the National Academy of Sciences, USA*, 64, 219–226.

Taylor, J. S., Mushak, P., & Coleman, J. N. (1970). Anhydrase: transition metal ions and spin-labeled sulfonamides. *Proceedings of the National Academy of Sciences, USA*, 67, 1414–1416.

Temkin, S. I., & Jacobson, B. J. (1984). Diffusion controlled reaction of chemically anisotropic molecules. *Journal of Chemical Physics*, 88, 2679–2685.

Thomas, D. D., Dalton, L. R., & Hyde, J. S. (1976). Rotational diffusion studied by passage saturation transfer EPR. *Journal of Chemical Physics*, 65, 3006–3024.

Thosar, B. V., & Iyengar, A. K. (eds.) (1983). *Studies in Physical and Theoretical Chemistry. Vol. 25: Advances in Mössbauer Spectroscopy. Applications to Physics, Chemistry and Biology*. Amsterdam: Elsevier.

Timofeev, V. P., Polyanovskii, O. L., Volkenstein, G. I., & Likhtenshtein, G. I. (1970). Study of conformational transition of aspartateglutamate transaminase by spin labeling method. *Biochimica et Biophysica Acta*, 220, 357–364.

Toublanc, D. B., Fessenda, R. W., & Hitachi, A. (1989). Rapid polarity reversal in the charge-transfer excited state of 9,9'-bianthryl. *Journal of Physical Chemistry*, 93, 2893–2896.

Träuble, H., & Sackman, E. (1972). Studies of the crystalline phase transition of lipid model membranes. III: Structure of steroid-lecitin system below and above the lipid phase transition. *Journal of the American Chemical Society* 94, 4499–4506.

Tsetlin, V. I., & Ovchinnikov, Yu. A. (1986). Spin labeled neurotoxin and its interaction with the acetylcholine receptor. In *Methods of Spin Labels and Spin Probes*, N. M. Emanuel & R. I. Zhdanov (eds.), pp. 105–124. Moscow: Nauka (in Russian).

Tsien, R. Y., & Zucker, R. S. (1986). Control of cytoplasmic calcium with photolabile tetracarboxylate 2-nitrobenzhydrol chelators. *Biophysical Journal*, 50, 843–853.

Tsvetkov, Yu. D., & Dikanov, S. A. (1987). Electron spin echo: applications to biological systems. In *Metal Ions in Biological Systems*, H. Sigel (ed.), pp. 207–263. New York: Marcel Dekker.

Tsymbal, L. B., & Moiseev, V. A. (1985). Study of thermoinduced structural changes of the erythrocyte membrane by spin probe methods. *Biologicheskie Membrany*, 2, 190–194 (in Russian).

Ulanov, B. P., Malysheva, L. P., & Moshkovsky, Yu. Sh. (1967). On the determination of the base sequence in nucleic acids. *Biofizika*, 12, 326–330 (in Russian).

Ulanov, B. P., Matorina, T. I., & Emanuel, N. M. (1976). Double modification of cytidine residues in DNA. *Molekulyarnaya Biologiya*, 10, 1211–1219 (in Russian).

Ulanov, B. P., Matorina, T. I., & Gonikberg, E. M. (1979). Electron microscopic study of DNA of bacteriophage CD modified by O-β-diethylaminoethylhydroxylamine. *Doklady Akademii Nauk SSSR*, 245, 1009–1013 (in Russian).

Ullman, E. F., Schwarzberg, M., & Rubinstein, K. E. (1976). Fluorescent excitation transfer immunoassay. A general method for determination of antigens. *Journal of Biological Chemistry*, 251, 4172–4178.

Vainshtein, B. K., L'vov, Yu. M., Feigen, L. A., Gvozdev, R. I., Karakushev, S. A., & Likhtenshtein, G. I. (1980). Determination of the distance between heavy atom labels in hemoglobin by means of small-angle X-ray scattering. *FEBS Letters*, 16, 107–110.

Vainshtein, B. K., Sosphenov, N. I., & Feigin, L. A. (1970). X-ray method for determination of the spacing between heavy atoms in macromolecules in solution, and its application for the investigation of gramicidin C derivatives. *Doklady Akademii Nauk SSSR*, 190, 574–577 (in Russian).

Van der Kooi, J. M., Wright, W. W., & Ericinska, M. (1990). Oxygen gradients in mitochondria examined with delayed luminescence from excited-state triplet probes. *Biochemistry*, 29, 5328–5332.

Van Ginkel, G. (1989). Diphenyl hexatrienes as fluorescent membrane probes. *Applied Fluorescence Technology*, 1, 1–8.

Vasserman, A. M., Kuznetsov, A. N., Kovarskii, A. L., & Buchachenko, A. L. (1971). Anisotropic rotation of nitroxide radicals in solution. *Zhurnal Strukturnoi Khimii*, 12, 609–616 (in Russian).

Vauhkonen, M., & Somerharju, P. (1989). Parinaroyl and Pyienyl phospholipids as probes for the lipid surface layer of human low-density proteins. *Biomembranes*, 9, 81–87.

Vavilov, S. I. (1945). On the photoluminescence of solutions. *Izvestiya Akademii Nauk SSSR (Seria Fizicheskaya)*, 9, 283–304 (in Russian).

Vaz, W. L. C., Criado, M., Madeira, V. M. C., Schoelimann, G. & Iovin, T. M. (1982). Size dependence of the translational diffusion of large integral membrane proteins in liquid-crystalline phase lipid bilayers. A study using fluorescence recovery after photobleaching. *Biochemistry*, 21, 5608–5612.

Vladimirov, Yu. A., & Dobretsov, G. E. (1980). *Fluorescent Probes for the Investigation of Biological Membranes*. Moscow: Nauka (in Russian).

Vogel, S. V., Slepneva, I. A., & Backer, J. M. (1975). Influence of Mn^{2+} ion coordination of τ RNA 1 macrostructure. *Biopolymers*, 14, 2445–2456.

Vogel, V. R., Rubtsova, E. T., Kotelnikov, A. I., & Likhtenshtein, G. I. (1986). Investigation of protein diffusion collisions by triplet labels. *Biofizika*, 31, 152–153 (in Russian).

Volodarsky, L. B. (ed.) (1988). *Imidazoline Radicals*. Boca Raton: CRC Press.

Volynskaya, A. V., Shishkov, A. V., Skrypkin, A. Yu., Dzhafarov, E. S., Rumyanzev, Yu. M., & Goldansky, V. I. (1985). Determination of the accessible surface of lysozyme and serum albumin by tritium labeling. *Molekulyarnaya Biologiya*, 19, 1294–1300 (in Russian).

Waggoner, A. S. (1986). Fluorescent probes for analysis of cell structure, function, and health by flow and imaging cytometry. In *Applications of Fluorescence in the Biomedical Sciences*, pp. 3–28. New York: Alan R. Liss.

Ware, W. R. (1975). Photophysics of exciplexes: some kinetic aspects. *Pure and Applied Chemistry*, 41, 635–660.

Watts, A. (1985). Spin-labels in biochemistry. *Biochemical Society Transactions*, 13, 588–592.

(1990). Mechanisms and dynamics of protein–protein associations in membranes. In *Molecular Dynamics in Membranes*, pp. 43–4. *Theses of FEBS satellite meeting*, Lave Balaton, Hungary, Hungarian Academy of Science, August 25–9, 1990.

Weber, C., & Teal, F. G. (1959). Electronic energy transfer in heme proteins. *Discussions of Faraday Society*, 27, 134–141.

Weber, G. (1989). Perrin revisited: parametric theory of the motional depolarization of fluorescence. *Journal of Physical Chemistry*, 93, 6069–6073.

Wei, R., & Almires, R. (1975). Spin immunoassay of progesterone. *Biochemical and Biophysical Research Communications*, 52, 510–516.

Weidman, S. N., Drysdale, D. R., & Wildvan, A. S. (1973). Interaction of a spin-labeled analog of acetyl coenzyme a with citrate synthase. Paramagnetic resonance and proton relaxation rate studies of binary and ternary complexes. *Biochemistry*, 12, 1874–1883.

Weiner, L. M. (1986). Magnetic resonance study of the structure and functions of cytochrome P-450. *CRC Critical Reviews in Biochemistry*, 20, 139–200.

Weinkam, R. J., & Jorgensen, E. C. (1971). Free radical analogs of histidine. *Journal of the American Chemical Society*, 93, 7028–7038.

Wenzel, H. R., & Trommer, W. E. (1977). 8-spin-label nicotinamide adenine dinucleotide synthesis and properties of a new spin-labelling coenzyme. *FEBS Letters*, 78, 184–188.

Wien, R. W., Morrisett, J. D., & McConnell, H. M. (1972). Spin-label induced nuclear relaxation distance between bound saccharides, histidine-15 and tryptophan-123 on lysozyme in solution. *Biochemistry*, 11, 3707–3716.

Wolfs, C. J., Horvath, L. I., Marsh, D., Watts, A., & Hemminga, M. (1989). Spin-label ESR of bacteriophage M13 coat protein in mixed lipid bilayers. Characterization of molecular selectivity of charged phospholipids for the bacteriophage M13 coat protein in lipid bilayers. *Biochemistry*, 28, 9995–10001.

Wood, W. G. (1984). Luminescence immunoassays: problems and possibilities. *Journal of Clinical Chemistry and Clinical Biochemistry*, 22, 905–918.

Woolley, G. A., Kapral, M. K., & Deber, Ch. M. (1987). Potential-sensitive membrane association of a fluorescent dye. *FEBS Letters*, 224, 337–342.

Wu, W. V., Abood, L. C., Gales, M., & Kreilick, R. W. (1977a). Spin-labeled narcotics. *Molecular Pharmacology*, 13, 766–773.

Wu, E. S., Jacobson, K., & Papahadjopoulos, D. (1977b). Lateral diffusion in phospholipid multibilayers measured by fluorescent recovery after photo-bleaching. *Biochemistry*, 16, 3936–3941.

Wütrich, K. (1986). *NMR of Proteins and Nucleic Acids*. New York: Wiley-Interscience.

Yamazaki, I., Winnik, F. M., Winnik, M. A., & Tazuve, S. (1987). Picosecond fluorescence studies. *Journal of Physical Chemistry*, 91, 4213–4216.

Yeagle, P. L., Albert, A. D., Boesze-Battaglia, K., Yong, J., & Frye, J. (1990). Cholesterol dynamics in membranes. *Biophysical Journal*, 57, 413–424.

Yechiel, E., & Edidin, M. (1987). Micrometer-scale domains in fibroblast plasma membranes. *Journal of Cellular Biology*, 105, 755–760.

Yguerabide, J., & Foster, M. C. (1981). Theory of fluorescent recovery after photobleaching. In *Membrane Spectroscopy*, E. Grell (ed.), pp. 199–269. Heidelberg: Springer-Verlag.

Yoh, J. Z., Takeno, K., Rosen, G. M., & Narahashi, T. (1975). Ionic mechanism of action of a spin-labeled local anesthetic on squid axon membranes. *Journal of Membrane Biology*, 25, 237–247.

Yudanova, E. I., & Kulikov, A. V. (1984). Determination of spin-exchange frequency of nitroxide radicals and oxygen by the method of continuous saturation of ESR spectra of radical bound to proteins. *Biofizika*, 29, 925–929 (in Russian).

Yudanova, E. I., Meckler, V. L., Fogel, V. R., Kulikov, A. V., Kotelnikov, A. I., Likhtenshtein, G. I., Berkovich, M. A., Karyakin, A. V., Archakov, A. I., Kaplun, A. P., & Schvets, V. I. (1986). Haem location in haemoproteins by spin and triplet tools. *European Journal of Biochemistry*, 156, 541–544.

Yusupov, I. Kh., Fogel, V. R., Kotelnikov, A. I., & Likhtenshtein, G. I. (1988). Molecular dynamics of cotton fibre studied by the method of luminescent labels. *Biofizika*, 33, 508–511 (in Russian).

Zamaraev, K. I., Molin, Yu. N., & Salikhov, K. M. (1981). *Spin Exchange. Theory and Physicochemical Applications*. Heidelberg: Springer-Verlag.

Zbinden, G., & Brandle, E. (1975). Toxicological screening of daunorubicin, adriamycin and their derivatives in rats. *Cancer Chemotherapy Reports*, 59, 707–716.

Zhdanov, R. I. (1981). *Paramagnetic Models of Biologically Active Compounds*. Moscow: Nauka (in Russian).

Zubarev, V. E. (1984). *Spin-Trapping Methods: Applications in Chemistry, Biology and Medicine*. Moscow: Moscow University (in Russian).

Zymbal, L. B. Gavrilova, I. I., & Moiseev, V. A. (1975). Study of the polyalcohol effect on the dynamics of biomembranes by the method of spin probe. *Kriobiologiya*, 3, 20–24.

Index

absorption spectra, 81, 101
acetylcholine receptor, 228
acetylcholinesterase, 161
acidity, 38, 99
alkoloids, 244
aminoacids, 243
assays, 252
  H^+ concentration, 44, 99
  immunological, 259
  "spin-sack", 261
ATPase
  Ca^{2+}-ATPase, 223
  K^+, Na^+-ATPase, 231
  in the nitrogenase, 165
  in redox processes, 165
  spin-labeled, 166

biradicals, 53, 54
broadening, inhomogeneous, 36

carbonhydrase, 171
cells, 248
cellulose, 241
chemical shift
  in NGR spectra, 137
  in NMR spectra, 75
cholesterol, 218
clusters
  iron, 165
  iron-molybdenum, 165
coefficient
  diffusion
    translational, 127
    rotational, 95
  transmission, 196, 198
coenzymes, 245, 246
collision
  between chromophores, 111, 126, 135
  between macromolecules, 69, 135
  between paramagnetics, 46, 62
constant, dielectric, 6, 38, 99
correlation time, 2, 5, 21, 34, 93, 214
cotton fiber, 241

cytochrome c, 198
cytochrome P-450, 168

dehydrogenase, 166
denaturation, 175
density
  electrostatic charges, 99
  spin, 9, 137, 196
depolarization
  luminescence, 93
  phosphorescence, 116
depth of immersion
  active center, 160, 165, 168, 170
  chromophor, 110
  radical immersion, 70
diffusion
  rotational, 2, 5, 21, 33, 116
  translational, 2, 6, 62, 126, 135
dioxygen
  concentration, 71
  diffusion, 71
distances
  between chromophores, 105, 124
  between spins, 6, 57, 70
DNA, 233, 252
double-label techniques, 46
double resonance
  electron nuclear (ENDOR), 16
  electron spin–spin (ELDOR), 16, 19, 213
drugs, spin-labeled, 246
dynamic adaptation, 198
dynamics
  intramolecular, 180, 186
  of membranes, 201, 209, 230
  of proteins, 180

effects
  electrostatic, 67, 100, 197
  orbital overlap, 57
  sterical, 67, 197
  transglobular, 177
electron paramagnetic resonance, 6, 16
  Fourier, 19

high resolution, 14, 15, 29, 185, 211
imaging, 41, 250
nitroxide radicals, 10
suturation transfer, 25
electron scattering labels
in electron microscopy, 146, 148
in X-ray scattering, 146
electron spin echo, 17
electron transfer
estimation of parameters, 198
long-distance, 195
multi-electron, 165
energy activation, 196
energy migration
singlet–singlet, 104
singlet–triplet, 117
triplet–singlet, 117
triplet–triplet, 117
erythrocytes, 221
ESE MT, 19, 25
exchange interaction
dynamic, 62, 126
static, 58, 124

factors
electrostatic, 67
g-factor, 6, 15
nuclear statistical, 64
steric, 67, 197
fluorescent label, 80, 85
formulas
Forster, 106
Marcus, 196
rate constant, 196
free energy
activation, 196, 198
reorganization, 196, 198
frequency
collision, 62, 113, 126, 135
oscillation, 139

Hamiltonian
exchange interaction, 50
general, 9
hyperfine interaction, 9, 54
spin–spin interaction, 50
haptens, 246
heme, 67
hemoglobin, 26, 176
hyperfine structure of ESR spectra
anisotropic, 9, 23
isotropic, 9, 23

immunoglobulins, 239
integral
exchange, 47, 48, 58, 62, 120, 126
overlap, 195

resonance, 195
interaction
anisotropic, 9, 23
dipole–dipole, 9, 13, 46, 48
electron–nuclear, 2
electrostatic, 67, 81, 100
exchange, 47, 48, 120
hyperfine, 9
isotropic, 9

labels
electron-scattering, 144, 154
fluorescent, 80
Moessbauer, 136
NMR, 141
phosphorescent, 116
spin, 3, 5
triplet, 116
tritium, 143
line shape, 11
Gaussian, 13
Lorentz, 13
lipids label, 88, 92, 142, 204
liposomes, 205, 217
luminescence, 82
lysozyme, 177
macrosomal oxidation system, 226
magnetic hyperfine structure
ESR spectra, 9, 23
NGR spectra, 137
magnetic momentum, 9
membranes, 73, 129, 201
chromatophores, 229
erythrocytes, 221
microsomal, 226
mixed, 217
model, 203
rhodopsin, 225
sarcoplasmic reticulum, 223
spin-labeled, 73
microsomes, 226
mitochondrial respiratory chain, 231
Moessbauer labels, 136
Moessbauer spectroscopy, 136
MW radiation, 10
myoglobin, 179, 189
myosin, 170

nitrogenase, 161
nuclear magnetic resonance, 74, 141
contact shift, 75
deuterium, 141
probe, 141
pseudocontact shift, 75
nucleic acid
assay, 252
spin-labeled, 233

nucleotides, 245

orbital overlap, 57
oxymetry, 71

paramagnetics
  complexes, 53, 56, 58, 67
  ions, 255
  radicals, 3, 5
parameters
  fluorescent, 83
  Moessbauer, 137
  order, 34, 210
  phosphorescent, 83, 116
  spin labels, 6
peptides, 243
phonone dynamics, 36
phosphoorganic compounds, 224
phosphorescence, 82, 117
phosphorescent labels, 116, 216
photobleaching, 114
photochrome labels, 130
polarity, 6, 38, 99
  dynamical, 102
proteases, 158, 258
  acetylcholinesterase, 161
  α-chymotrypsin, 158
  leucinaminopeptidase, 160
  thrombin 160
pulse techniques
  ESR, 17
  luminescence, 83

quadrupole splitting in NGR, 137
quenching, 111

radical
  nitroxide, 1, 3, 5
  polyradicals, 54
radical pair, 37, 183
reaction center, 190, 229
recovery
  after photobleaching, 114
  saturation ESR, 18
redox probe, 38, 250
relaxation
  orientation, 83
  rotation, 21, 93
  see also spin relaxation
relaxation time
  electric dipoles, 83, 96

longitudinal, 7, 12
  spin–lattice, 7, 12, 76
  spin–spin, 7
  transverse, 7, 76

sarcoplasmic reticulum, 223
saturation curve, 11, 12, 74
serum albumins, 180, 182
  bovine, 176, 182
  human, 176, 182
shift
  chemical, 137
  fluorescent spectra, 103
  phosphorescent spectra, 103
spin, 6
spin labels, 1, 3, 5
spin relaxation
  electron lattice, 7, 61
  electron transverse, 7, 61
  nuclear, 76
spin–spin interaction
  dipole–dipole, 46, 48, 53
  exchange, 47, 49, 54
spin-traps, 43, 196

terpens, 244
tissues, 248
transfer
  electron, 117, 195
  energy, 104, 124
transition
  allosteric, 175
  conformational, 175
  local, 159
  Seeman allowed, 6
  Seeman nonallowed, 60
  transglobular, 177
triplet label, 116, 216
triplet-photochrome label, 133

vesicles, 203
viscosity
  membrane, 209, 217
  protein, 182

wave function, 120, 195

X-ray
  analysis, 163, 167, 173, 178
  scattering, 146